DITCHES ACROSS THE DESERT

DITCHES ACROSS THE DESERT

Irrigation in the Lower Pecos Valley

STEPHEN BOGENER

Texas Tech University Press

This book is typeset in Walbaum. The paper used in this book meets the minimum
requirements of ANSI/NISO Z39.48-1992 (R1997). ∞

Printed in the United States of America

Library of Congress Cataloging-in-Publication Data
Bogener, Steve.
Ditches across the desert : irrigation in the lower Pecos Valley /
Stephen Bogener.
p.   cm.
Includes bibliographical references and index.
ISBN 0-89672-509-X (cloth : alk. paper)
1. Irrigation—Pecos River Valley (N.M. and Tex.)—History. I. Title.

TC824.N6B64 2003
333.91'3'097649—dc21
2003008313

03 04 05 06 07 08 09 10 11 / 9 8 7 6 5 4 3 2 1

Texas Tech University Press
Box 41037
Lubbock, Texas 79409-1037 USA
800.832.4042
ttup@ttu.edu
www.ttup.ttu.edu

for Sharon

CONTENTS

ILLUSTRATIONS

PHOTOGRAPHS

MAPS

ACKNOWLEDGMENTS

I could not have completed this project without the helpful support and assistance of many individuals and organizations. I would like to especially thank William Tydeman, director, and his staff at the Southwest Collection/Special Collections Library, and Catherine Miller, Department of History, Texas Tech University, for their support. The following fine people also lent assistance: the Southeastern New Mexico Historical Society and the Eddy County Clerk's Office in Carlsbad, New Mexico; the Special Collections and Archives, Tutt Library, Colorado College, Colorado Springs; the Carnegie Special Collections, Penrose Public Library, Colorado Springs; the James W. Starsmore Center for Local History, Pioneers Museum, Colorado Springs, Colorado; the Western History Collection, Denver Public Library; the Salida Regional Library's Local History Archive, Salida, Colorado; Patrick Rand, president and curator, Sacramento Mountains Historical Society, Cloudcroft, New Mexico; Bill Boehm and Dennis Daily of the Archives and Special Collections Department, New Mexico State University Library, Las Cruces, New Mexico; the New Mexico State Engineer's Office and the State Archives and Records Center in Santa Fe, New Mexico; the National Archives and Records Administration, Rocky Mountain Region, Lakewood, Colorado; John Mulvihill; and Noel Parsons, Texas Tech University Press.

DITCHES ACROSS THE DESERT

INTRODUCTION

S ETTLEMENT of the American West came slowly, based on advances in technology and the harnessing of nature, especially water. In the arid environment of New Mexico's Pecos Valley, the exploitation of water, combined with corporate accumulation of land through manipulation of federal law, led to speculation and development in the frontier region of southeastern New Mexico. Beginning in the late 1870s, and through 1925, a succession of people tried to transform the river and the desert embracing it.

First used as open range for cattle grazing, the Pecos Valley became the scene of ever more ambitious plans to establish an agricultural mecca based on irrigation and funded by wealthy investors from Chicago, New York, Colorado Springs, and Europe. Following a natural disaster and financial downturns in 1893, settlers and investors fled the valley, making its future uncertain. A series of financial reorganizations to attract much-needed capital brought a major railroad to the valley, but the heyday of corporate irrigation was over.

Instead, the moribund irrigation company turned to the federal government for help. Under pressure from President Theodore Roosevelt, the U.S. Reclamation Service, although reluctant to rehabilitate the valley's irrigation system, agreed to take on the project. Reclamation began a long, sometimes contentious relationship with water users in the valley, caught between its own policies promoting small farmers and the agendas of corporate and absentee landowners. At the heart of contentiousness in the Pecos Valley lay the competition for precious water and its usage in an area fraught with natural limitations.

The people, events, and transformation of the Pecos Valley from rustic cattle territory to towns and irrigated farmland form the framework for a rich story of the American West. The economic, political, and social connections that reached well beyond the confines of the Pecos watershed transform what might otherwise be a pedestrian story of building ditches across New Mexico. Herculean efforts to create an oasis in the middle of the desert ran counter to the natural environment of southeastern New Mexico, and in the end nature subdued technology despite local and national political maneuverings and convoluted investment strategies to secure water in the Chihuahuan Desert.

Historically a crossroads for Native Americans, Spanish explorers, American military operations, and Anglo migration and transportation networks, the arid Pecos country was settled slowly. Like much of the Great American Desert, it was a place to cross on the way to somewhere else. A few adventurous Mexican Americans from the Manzano Mountains near Albuquerque ventured farther east of the Rio Grande in the mid-1800s. In the valleys, they established self-sufficient communities where they found water and practiced a mixed planting-grazing economy. During the late nineteenth century, ranching, not farming, dominated the Pecos economy. The presence of nomadic Indians delayed Anglo incursions into the region, but in 1867 Charles Goodnight and Oliver Loving trailed a herd of cattle up the Pecos to sell to miners, the military, and Indian reservations.

Across the West, cattlemen and land speculators took advantage of the various laws created by Congress for disposing of the public lands. New Mexico was no exception. Since little arable land was available in New Mexico, pasturelands, along with surface waters, which gave the owner control of thousands of acres of the public domain, were the primary objective. By the 1870s, competition for grazing lands along the Pecos River in part sparked the Lincoln County War.[1]

Optimistic promoters of western lands in the early 1880s became concerned by the end of the decade about plans to survey the West and withdraw from entry lands that might be irrigated. In 1878 John Wesley Powell's report on arid lands was released. Powell recommended classifying western land based on its irrigation potential.[2] His concept met with stiff resistance from land speculators and western congressmen, who denounced Powell's plan to survey and remove from sale irrigable lands across the West. Investors stood to lose millions of dollars in the event that Powell's proposed survey withdrew lands where irrigation works and land developments were already in progress. While Congress debated the merits of Powell's survey, James J. Hagerman of Colorado and others invested close to a million dollars in irrigation works along the Pecos River.

As Powell's survey idea threatened to disrupt the image of turning the desert into a corporate garden, many Americans—not the least of whom were western congressmen—considered irrigation companies as America at its best. Farmers laboring side by side to transform a wasteland into something productive harked back to Jeffersonian notions of yeoman democracy.

Much of the historiographical debate over water in the West has centered on the degree to which democracy actually existed within the frameworks established for controlling water. Well before the Reclamation Service arrived, the Pecos Valley depended heavily on eastern and foreign capital for its survival. Investors from England, Switzerland, New Bedford (Massachusetts), New York City, Chicago, and Colorado Springs, a mountain playground for wealthy easterners, bankrolled irrigation in the valley. When the new Reclamation corps of engineers descended on the valley in 1904, they perpetuated a dependency on outside capital already in existence for fifteen years. Because of the availability of capital, southeastern New Mexico initially jumped ahead of what were seen as more backward-looking parts of the Territory. To support a viable agricultural colony in a region that received twelve inches of rain a year meant recruitment of, and permanent ties to, eastern financial and political capital.

The men who invested in the valley and their interrelationships with one another are essential to the story of Pecos Valley irrigation. A visionary sheriff made famous for shooting Billy the Kid joined forces with an indefatigable New York promoter named Charles Eddy whose older brother was a cattle pioneer in the mountains west of Colorado Springs. With the help of professional promoter and newspaperman Charles W. Greene, they attracted investors whose money would irrigate the desert. Tubercular millionaires, engineers, reclamationists, government bureaucrats, cigar manufacturers, steel magnates, and mining and railroad capitalists invested vision, time, and money in the valley. Primary investor James J. Hagerman, a close friend of politico Marcus Hanna, had made millions from steel, mining, and railroad building before he ever set eyes on the Pecos Valley.

Other investors included William Bonbright, co-owner of Hood, Bonbright, and Company, the precursor of Wanamaker's Department Stores; New Yorker Amos Bissell; Frederic Stevens, a member of the board of directors of Chemical Bank of New York, and his son, Joseph Sampson Stevens, a descendant of Albert Gallatin; and Richard Bolles, who made a fortune from investments in the famous Mollie Gibson Mine. Charles Otis of Otis Steelworks; Arthur and Eli Mermod, heirs to the internationally known Mermod-Jaccard jewelry business; and Robert Weems Tansill, who manufactured Punch cigars, were all tied to the Pecos Valley irrigation projects. Some players in the Valley had political or social ties to Theodore Roosevelt's family and his Rough Rider cavalry unit. Others like Jay Gould,

George Pullman, Benjamin Cheney, Harriman and Company, Charles Head, and William McMillan, among other industrial and economic leaders, played a role in the Valley's irrigation or railroad financing.

These prominent individuals and their efforts paralleled those of cattle thieves, land swindlers, Swiss and Italian immigrants, settlers looking for a better life, Mexican laborers, saloon keepers, prostitutes, lawyers, and lawmen who changed the face of the Pecos Valley. The characters and their political and economic influences and connections set this work apart from other studies of water in the West.

Many have studied the history of the American West and its relationship to water. A number of historians have pointed to the aridity of the American West as a determining factor in the region's development.[3] Walter Prescott Webb maintained that settlers failed to subdue the frontier until the last decades of the nineteenth century, when technology shifted the balance.[4] In the case of Carlsbad, one of Reclamation's earliest projects, certain natural, private, political, and corporate elements intervened to mitigate the Reclamation Service's authority. Reluctant to jeopardize its uncertain status in Washington and the West, Reclamation approached the early projects gingerly, establishing and changing engineering and administrative policies as it went along.

More recently, much of the historiographical discussion over water in the West has centered around who controlled water and how. Norris Hundley addresses the issue in his survey *The Great Thirst: Californians and Water, 1770s–1990s*. Hundley shows how politics, legal wrangling, and changing constituencies carved out the California waterscape.[5] In the Pecos Valley, control was vested in those who either had capital of their own or could raise it.

This capital, combined with manipulation of federal land laws, placed an appreciable amount of property with corporate investment companies. The Pecos Valley became the site of one of the most ambitious irrigation projects in the United States. However, following a natural disaster and financial downturns in 1893, settlers and investors fled the valley, making its future uncertain. A series of financial reorganizations to attract much needed capital brought a major railroad to the valley, but the heyday of corporate irrigation was over.

Instead, the moribund irrigation company turned to the federal government for help. William Smythe, foremost promoter of the irrigation movement at the turn of the century, advocated irrigation as a form of cooperative capitalism utilizing the natural laws of God to compel the arid lands to produce. Smythe expected material well-being and moral progress to follow this conquest of nature. By 1902, political pressure led to passage of the Newlands Act, under which the federal government developed water projects and by extension established a federal presence across the West.[6] Under pressure from President Theodore Roosevelt, the U.S. Reclamation Service reluctantly agreed to rehabilitate the valley's irrigation system.

The twentieth-century story of Carlsbad is enmeshed in the transition from private sector sponsorship to that of the federal government. Karen Smith, in describing Arizona's Salt River Project, discusses how that project set the pattern for subsequent projects across the West.[7] A distinct difference between Salt River and Carlsbad is that while Reclamation looked with favor on Salt River because of the valley's small-acreage landholders, such was not the case in southeastern New Mexico. What is consistent in the two projects is how Reclamation implemented policy in the two valleys and how locals adapted, rebelled, and compromised with engineering bureaucrats over issues like repayment policy and construction costs. At Carlsbad these issues were compounded by the constant bullying of engineers by nineteenth-century remnant landholders trying to secure more storage facilities to water their properties. Control was splintered among engineers, small local water users, and the holdovers of corporate irrigation enterprises in the valley.

Other historians have examined the contentiousness between various water constituencies in the twentieth century. Daniel Tyler's "history from the ground up" examination of the Colorado–Big Thompson Project and the Northern Colorado Water Conservancy District is a thorough treatment of that subject and covers a huge span of time, revealing many struggles between water proponents and environmentalists in the twentieth century.[8] Still others see collusion between the government and the more powerful groups of water users across the West. In *Rivers of Empire*, Donald Worster discusses how nineteenth-century inroads by the federal government laid a framework for more ambitious goals in the twentieth century. After World War II, according to Worster, capitalism and technological expertise produced federally centralized, authoritarian control over local water resources. Worster argues that the federal government worked in concert with powerful economic elites to create a "hydraulic empire" in the West.[9]

In the case of Carlsbad, the focal point of irrigation in the Pecos Valley and one of Reclamation's earliest projects, there does not appear to have been a concerted effort to work with economic elites to centralize power. In fact, Reclamation only grudgingly made improvements to the project, which favored holdover nineteenth-century speculative interests. The relationship between the two was rocky at best. It is important to note that when the federal government entered the Pecos Valley as reclamationist, it did so reluctantly. Once in, however, it slowly plodded ahead, its actions based on the prevalent belief in bureaucratic expertise and efficiency, and rehabilitated and improved the valley anyway. This is not to say that speculative interests did not have an effect on Reclamation decision making. They did. The ongoing struggle for more water in the twentieth century at Carlsbad has much to do with this relationship.

Unlike Worster, Donald J. Pisani sees the emergence of water associations and other quasi-governmental structures as a natural consequence of trial and error. To Pisani, the West established whatever worked best, based on the collective experience of a locality or region, without the assistance or persuasion of some cabal bent on centralizing power.[10] At least in the Pecos Valley, the Reclamation Service, whose resources were increasingly in demand, established its presence while methodically addressing the needs of local water users. Reclamation effectively created a system of capital and engineering resources that responded to local and federal political pressure, project exigencies, and Reclamation's own agenda. Reclamation established a long-term identity in almost every valley across the western United States capable of capturing a trickle of water. Drinking at the same trough, local interests and water users fought over the resources of the federal government, which nineteenth-century private interests could no longer provide.

Other writers have examined this shift from private to government-sponsored reclamation, and beyond. In *Salt Dreams: Land and Water in Low-Down California*, William deBuys, after examining the implications of pre-reclamation natural and human history, moves on to discuss how the great Colorado River floods in the first decade of the twentieth century ended private reclamation efforts in the area encompassing the Salton Sea. DeBuys sees the floods as pivotal moments in the region's history, when a succession of schemers, agriculturalists, and others tried to take advantage of the government's noblesse oblige.[11] Similarly, the Pecos River floods of 1893 and 1904 practically ended all hope of private capital-supported irrigation in southeastern New Mexico. In the Pecos Valley, destruction of dams and irrigation works by floodwaters was compounded by the serious nationwide economic downturn of 1893.

Like deBuys, Dan Flores and John Miller Morris have written from a cultural and bioregional perspective about the vast Llano Estacado, the 50,000-square-mile island in the sky that dominates West Texas and eastern New Mexico just east of the Pecos Valley. Although Flores acknowledges the importance in bioregionalism of emphasizing watersheds and John Wesley Powell's unrealized plan for divvying up such waters across the country, his focus does not include the Pecos Valley.[12]

Likewise, John Miller Morris's geographical journey is focused by and large on El Llano Estacado.[13]

James Sherow, examining the conflict over the waters of the Arkansas River, stressed the significance of market values in turning river water into a commodity. Sherow noted that engineers became an instrument of expansion, providing for ever increasing control, development, and use of water in the West.[14] In the Pecos Valley, using the natural environment to maximize the land's productive value eventually required federal help. In the twentieth century, differing opinions between federal experts and water users on how best to use the Pecos led to an often uneasy relationship that lasted for years. Government improvement of existing irrigation systems, like those in the Pecos Valley, came with new costs for many. Local farmers and land promoters lost much of their autonomy. As on the Arkansas River, at Carlsbad the proliferation of water districts and other jurisdictions along the same water source led to both competition and cooperation for precious water.

Although initially subscribing to a plan more in keeping with the social philosophies of Progressive reformers Elwood Mead, William E. Smythe, and George Maxwell, Reclamation engineers often became dam builders instead. The choice of the types of dams built across the West was not entirely based on engineering practicality, as Donald Jackson has shown in the experiences of John Eastwood.[15] Rather, power struggles often determined what kind of dam was built. But beyond dam building, Reclamation often had no clear sense of where it was headed. This is abundantly clear at Carlsbad. As Pisani points out, after passage of the Reclamation Act as a means of relieving urban pressures in the East, Reclamationists Frederick Newell, A. P. Davis, and others did not show a great interest in social reform.[16] At Carlsbad, the Pecos Valley's speculative ties to the nineteenth century tempered Reclamation's decision making. Lingering corporate interests firmly grafted themselves onto the administrative and political framework established by Reclamation vis-à-vis the local water users association. Reclamation was often caught between providing water for small resident landowners and for speculative and absentee holdovers from the previous century.

Although the Pecos Valley never lived up to the expectations of promoters or reclamationists of the nineteenth or twentieth centuries, its failure in many ways was the result of the very thing that promised to make the desert productive, the Pecos River itself. Early promoters did not know or did not acknowledge the river's drawbacks in an era when unbridled optimism and the investment of millions of dollars carried the day.

The once formidable Pecos River of New Mexico and Texas is today a mere shadow of its former self. Dammed in many places for irrigation, its springs pumped dry in others, the Pecos River of today leads a precarious existence. Careful observers recognized the fate of the river prior to the 1920s. Irrigators along the middle Pecos River began to ensure that their priority rights were intact, forming an ad hoc coalition to protect their water. Leaders in Roswell and Carlsbad worked together to encourage federal development of storage facilities—first close to existing reservoirs near Carlsbad, and later in concert with irrigators near Fort Sumner. What forged such cooperation among these New Mexico communities were requests by the state of Texas to apportion part of the Pecos River based on irrigators' priority claims in Texas. In 1925, a committee of water experts from Texas and New Mexico arrived at a compact to apportion the river's water supply. The plan lingered until a new one was adopted in 1948. The Pecos River Compact between New Mexico and Texas did not end the contest over water—either within New Mexico or between the two states.

1

VISIONS OF AN IRRIGATION EMPIRE

WORKING from the nineteenth-century American mindset of subduing nature and making money at the same time, cattleman Charles Eddy and lawman Pat Garrett set in motion the concept of transforming a cattle country into an agricultural mecca based on irrigation. Building dams and canals to water the desert would cost money, so Eddy recruited capitalists from Chicago and Colorado Springs, the latter a retreat for some of the country's industrial and political elite. Investors with strong ties to business and political circles joined Eddy in an ever more ambitious plan to reclaim the desert and profit through its development.

The Pecos River of the nineteenth century, unlike its faint twentieth-century shadow, was a formidable watercourse. The river stretches for some 755 miles, from the Sangre de Cristo Mountains just northeast of Santa Fe to its eventual merger with the Rio Grande near present-day Lake Amistad in Texas. The stream flows in an east-southeast direction from its source 13,000 feet above sea level, where numerous streams and heavy snows contribute to its volume. For forty miles the Pecos rushes clear and cold through mountainous terrain frequented for years by the Pecos and Picuris Indians, and later by the religious Penitentes who painfully trudged to the high country bearing crosses and suffering other hardships during Holy Week each year.[1]

Following a mostly southerly route after reaching Fort Sumner, the Pecos flows past the New Mexico communities of Roswell, Dexter, Greenfield, Hagerman, Lake Arthur, Artesia, Atoka, Lakewood, and finally Carlsbad before veering southeast toward the Texas state line. The land on either side of the Pecos changes dramatically from alpine forests and meadows along its upper reaches to scrubland from Fort Sumner to near Carlsbad. Below Fort Sumner the topography of the valley changes.

Canyon-like walls disappear and turn into low rolling hills and then to prairie. South of Roswell is a country of limestone and gypsum, marking what is now considered premium agricultural land. The topography changes again as the river nears Texas to reflect the flora and fauna of the Chihuahuan Desert.[2]

In the nineteenth century, as today, several sources of water fed New Mexico's Pecos River: the Hondo, itself fed by the Rio Bonito and the Rio Ruidoso, the Felix, the Peñasco, and the Seven Rivers, the Black and Delaware Rivers near Texas, the North and South Spring Rivers, and the Berrendo. A nineteenth-century traveler moving from Roswell south to Texas along the Pecos River would cross each of these tributaries. The Spring Rivers were located close to Roswell and, as their name indicates, acquired their water from rich springs in the area. Both the Berrendo and the North Spring Rivers emptied into the Hondo before reaching the Pecos, but the South Spring flowed directly into the river. The Hondo, largest tributary on the Pecos, was formed at the confluence of numerous brooks rising out of the White Mountains to the west, continuing through the foothills, and finally crossing the prairie just west of Roswell.[3]

Continuing south of Roswell, the Felix River was the first stream of importance. It rose among the southeastern foothills of the White Mountains and after a few miles sank and did not rise again until some four miles from its mouth where it appeared again in a series of springs.[4]

Moving south, the Peñasco River rose in the Sacramento Mountains, flowed for forty miles before entering marshland ten to twelve miles long, and then disappeared only to join the Pecos some ten to twelve miles later. Farther south, the Seven Rivers, formed as a series of small springs on the western side of the Pecos, turned into a small stream and flowed for a short distance before disappearing into the ground. Some thirty-five miles south of Seven Rivers, the Black River drained part of the eastern slope of the Guadalupe Mountains. Larger than the Berrendo, it carried an unfailing water supply. The Black River ran thirty-five miles, its volume considerably increased by springs located in its banks. A smaller stream, the Blue River, flowed into the Black a few miles from its mouth. The very last stream to enter the Pecos in New Mexico was the Delaware, which flowed in from Texas about seven miles from its mouth. This river was larger than the Berrendo at Roswell. Dark Canyon and other draws west of the Pecos held only floodwaters. Normally they were dry. Noticeably, no rivers drained from the eastern side of the Pecos on the Llano Estacado.[5]

The abundance of surface water attracted indigenous peoples, especially those of the Pueblo culture, who settled along the upper reaches of the Rio Grande and Pecos Rivers as early as C.E. 1200 when drought forced them to leave their homes in the Four Corners region. At roughly the same time, Apache and Navajo peoples from western Canada arrived. The Navajo stayed in the Four Corners region. The Apache drifted east along the foothills of the Sangre de Cristo Mountains and onto the plains of what is today eastern New Mexico and West Texas. The Apache presence was strongly mitigated in the late eighteenth century by an alliance between the Spaniards and the Comanche. The Comanche obtained weapons and horses from the Europeans and effectively forced the Mescalero Apache into strongholds in the Sangre de Cristo Mountains.[6] Ranging from Kansas to the southern edge of the High Plains, the Comanche often disrupted Mexican and later Anglo settlements east of the high country.[7] Chasing bison, the Comanche joined Kiowa and Ute crossing the Llano Estacado, riding south into Mexico.[8]

Before the Mexican War in 1846, the Mescalero saw only small pockets of Spanish and Mexican settlers and few Americans in the region stretching from the Sangre de Cristo, Guadalupe, Delaware, and Organ ranges east to the Pecos River. There, the Mescalero retained their raiding lifestyle, preying on scattered Spanish and, later, Mexican and Anglo settlers in the nineteenth century. After the Mexican War, the United States sent explorers to the American Southwest to study the land gained from Mexico through the Treaty of Guadalupe-Hidalgo (1848).

In 1849, First Lieutenant J. H. Simpson, in the company of California-bound emigrants and a military escort conducted by Captain R. B. Marcy, traveled from Fort Smith, Arkansas, to Santa Fe. The federal government wanted to establish a wagon road from Fort Smith to Santa Fe, and Simpson was to explore and survey the route. On reaching New Mexico, Simpson learned that he was to abandon plans for proceeding to California and instead trek south along the Rio Grande and then eastward across the flat expanse of eastern New Mexico and western Texas.[9]

The report submitted by the lieutenant one year later included both Simpson's largely topographical information and Marcy's journal entries describing the flora, fauna, and water of southeastern New Mexico.[10] Moving eastward from the Organ Mountains near El Paso to the Sacramento range thirty miles farther, and finally encountering the Guadalupe Mountains just west of the Pecos Valley, the party reached the spring-fed waters of the Delaware River near the present-day southern New Mexico–Texas border.[11]

Foreshadowing the notions of promoters four decades later, Marcy asked, "Is it not within the scope of probabilities that these springs may be found to possess valuable medicinal properties, and that this place may yet . . . become a place of fashionable resort for the 'upper-ten-thousand' of

New Mexico?"[12] In another prophetic entry, Marcy noted on September 15 that the soil was extremely poor, covered with "decomposed gypsum," but that the land was covered with luxuriant grama grass, offering excellent forage for animals.[13] Marcy's descriptions offered an early glimpse into the potential successes and failures of the Pecos Valley.

The first non-Indian settlers in the area were Mexican Americans who worked, in part, to provide food for the military as farmers and sheep-herders. By the mid-1850s, Mexican Americans had settled into the Hondo Valley from the Manzano Mountains near Albuquerque. The first settlement in the region began near Fort Stanton on the Rio Bonito. Called La Placita, the settlement attracted Anglos as well. The town, now called Lincoln, became the seat of Lincoln County in 1869. By then, some thirty to forty Mexican families had moved to the Hondo Valley, fifteen miles southwest of Roswell. Since most of the men had been employed as freighters from Albuquerque to St. Joseph, Missouri, they began calling their settlement La Plaza de Missouri, or San Jose or Missouri Bottoms. The name Missouri Plaza stuck.[14]

Mexican settlers brought with them their Spanish-based community and agricultural, land, and water-distribution systems. Spanish custom and law provided that people could use water for domestic purposes through ditches (*acequias*) even if the ditches crossed another's property. As the former freighters irrigated 400- to 500-acre farms near Missouri Plaza, New Mexico's territorial legislature sanctioned these centuries-old customs in 1851. Nonetheless, Anglos moving into the Hondo Valley did not accept Spanish community traditions as a means of agricultural production. And, along the lower stretches of New Mexico's Pecos Valley tributaries, no Spanish land grants interrupted Anglo settlement. Anglo settlers began appropriating the water, and in the 1870s the citizens of Missouri Plaza moved away because of Indian problems, racial conflict, and upstream irrigation.[15]

New Mexico attorney and historian William Keleher characterized the Pecos Valley during the last half of the nineteenth century as wide open for adventure, danger, and opportunity. Since the American occupation of New Mexico during the Mexican War, New Mexico had suffered from cultural conflicts among Anglos, Mexican Americans, Comanches, Apaches, Utes, Cheyennes, Arapahos, Kiowas, and Navajos.[16] Keleher attributed the "color and adventure" of Lincoln County in the 1870s and 1880s to cowboys, camp followers, saloon keepers, gamblers, gunfighters, and horse thieves—most of whom came from Texas.[17] A number of such characters filtered into the region. The first major Anglo settlement in what would become Eddy County was Seven Rivers, a village of "dozens of hard-core drifters, drovers, gunmen and fugitives. . . . [G]unfire was frequent [and] it is said that the first four persons buried in the cemetery died of gunshot wounds." Until the 1890s, Lincoln County encompassed most of eastern New Mexico, and settlements relied on the federal government for law enforcement, which was stretched over an area larger than Massachusetts.[18]

The largest influx of Anglos into the area followed on the heels of the military presence at Bosque Redondo following the 1863 campaign against the Mescalero Apache. The U.S. military established a Pecos River post called Fort Sumner near the present-day town of the same name.[19] The irrigation system built here by Indians represented the first major effort at irrigation along the Pecos River itself. The fort was laid out next to Bosque Redondo, a round grove of cottonwoods on the river and a favorite camping site of the Apaches. The Mescalero had the choice of moving to a reserve established near the fort, or being further hounded by U.S. soldiers. The Apache, who were organized into small semi-autonomous bands, trickled into the region one group at a time. By March 1863 the military, having established contracts with local cattlemen and Texans like Charles Goodnight, was feeding over four hundred men, women, and children at the fort. Although some one hundred Mescalero fled west to join their Gila cousins, by the summer of 1863 the Apaches at Bosque Redondo settled into laying out fields, planting crops, and setting up shelters.[20]

At this point, the American military, using questionable logic, embarked on a series of campaigns to place other Indians of New Mexico and the Southwest on reservations. The military began settling some nine thousand Navajos on the lands at Bosque Redondo—lands that by then the Mescalero considered their own. The march of these Navajos from the Four Corners region—known as the "long walk"—was carried out by Kit Carson, who had previously joined in rounding up the Mescalero. Carson thought the Indian policy foolish, but followed orders.[21]

The Mescaleros' growing resentment of the Navajos did not initially stymie the work already in progress. The Mescalero built a slaughterhouse, dammed the river, and irrigated crops and trees, which they planted nearby. Unfortunately for the Indians and military alike, the irrigation venture was a dismal failure. Worms, blight, hail, floods, drought, and complaints about the effects of Pecos water on the human digestive system all combined to undermine irrigation efforts in Fort Sumner.[22]

Despite the army's failure in settling the Apache and Navajo at Bosque Redondo and the danger inherent in a frontier setting, Texas cattlemen saw the grazing potential of lands in the Pecos Valley of New Mexico.[23] Charles Goodnight and Oliver Loving, working to fill government contracts for beef at Bosque Redondo and to stock ranges in Colorado, trailed cattle along the Pecos River beginning in the 1860s. Goodnight, who owed much of his success to feeding Indians, shared the Indians' dislike for the waters of the Pecos. One of the most dangerous and celebrated junctions of the river was at Horsehead Crossing in Texas. The crossing gained its name and reputation because horses drowned in the rushing water or drank the brine and died. Indians found targets for attack in unwary travelers at Horsehead and many other crossings up and down the Pecos. Goodnight's partner Oliver Loving found himself holed up just south of present-day Carlsbad in a bend of the river, which still bears his name. Indians shot Loving in the arm, an injury that later cost him his life.[24]

The military removal of the Indians from eastern New Mexico opened the land to Anglo settlement. Goodnight and others, including John Chisum, who settled near Roswell, New Mexico, weathered the uncertainties of the Pecos frontier and prospered. In the late nineteenth century, Chisum probably controlled the largest piece of ranching property in the country.[25] In addition to Chisum's cattle range, his Southspring Ranch headquarters included 1,500 acres planted in fruit trees and 150 acres devoted to alfalfa. Chisum did not own outright all of the rangeland that he controlled, but his control extended over an area equal to three or four New England states.[26] Like most Anglos, he regarded the land as public domain and open to use by those who got there and controlled it first.[27] Control over the public domain of southeastern New Mexico came from controlling the scant water sources of the region, including the Pecos, its tributaries, and springs. The Lincoln County War in the late 1870s developed partially out of competition over grazing lands and water sources in the region.[28] Whoever controlled the water controlled the land.

Chisum's success and ultimately his departure in 1875 attracted New Yorkers John and Charles Eddy, who in 1881 began running cattle on a 20,000-acre ranch at the southern end of Chisum's former kingdom, near Seven Rivers.[29] The Eddy brothers learned about ranching in Colorado, where in 1879 they established a cattle operation along the high mountains of South Park, west of Colorado Springs.[30] In partnership with George and Amos Bissell, the Eddys established their ranch on lush pastureland. During the winter of 1879 and 1880, the Eddys, through Amos Bissell, had access to ready supplies of cash. According to the Eddy brothers' journal, besides $7,000 in seed money, they could get "as much more as we need, subject to our own draft."[31]

While the Eddys had sufficient capital to develop their Colorado operations, in the early cattle days the brothers did not live in opulence. Both men, Charles in his twenties and John in his thirties, worked hard on the ranch, one serving as cook and the other riding after livestock.[32] An 1880 New Year's Day journal entry recorded their inventory as eleven horses, a year's grub, twenty-seven calves, two mavericks, and about 2,500 pounds of grain. During the year just ended, the brothers sold seven steers for $30 apiece. After borrowing $100, they had $150 on hand by the end of their first year of operation. The Eddys made a number of improvements that year, fencing pastures and building corrals. They built a small cabin, which became a cook shack for their hands, and later a log ranch house similar to homes in New York villages such as Milford, their hometown.[33]

Descendants of the Eddys surmise that their intent in establishing a ranch as high as 10,000 feet was to create a "natural feedlot—at the center of perhaps sixty square miles of lush mountain grass in the open range— on which to fatten longhorns brought in by cattle drives."[34] Getting cattle to their ranch necessitated herding the animals from West Texas or New Mexico.[35] Early in June 1880, the Eddys noted that they were "leaving for Dodge [City] in the morning,"[36] indicating that they were going there to secure stock to improve their herd or to sell cattle. On Charles Eddy's first route bringing cattle north to Colorado, he followed the Pecos River along the Goodnight-Loving Trail, to near Santa Fe, then to the lower San Luis Valley, over Poncha Pass, and up Ute Trail to his property at South Park. Thus, Eddy learned about the Pecos Valley.[37]

The Eddys knew Charles Goodnight, who settled on a ranch near Pueblo, Colorado, in the 1880s, and the men may have trailed cattle together. At one point during a cattle quarantine near Las Animas, Goodnight and other cattlemen, including the Eddy brothers, had to select a more westerly direction, crossing through Raton Pass or up through the San Luis Valley to avoid being stopped.[38] Local sources mention the Eddy brothers associated with various cattle drives in the 1880s. Apparently, many of the early pioneers in the area of Colorado near South Park first arrived driving cattle for the Eddys from the Pecos Valley or western Texas.[39]

Beyond the initial support of investors like Amos Bissell, the Eddys first made money in the 1880s selling cheap Texas cattle to hungry miners. In those years, a second mining bonanza occurred in parts of Colorado. Mining camps and new townsites sprang up along the South Platte, upper Arkansas, and upper Colorado river basins to the west and southwest of Denver. The Eddys, located on a ranch in the midst of such mining activity, were in a position to supply hungry miners with meat on the hoof, since most of the big game had been hunted out during the heyday of placer mining years before. The Eddys charged their customers thirty dollars a head straight from the cattle drive, and given a good summer of fattening

The remains of buildings erected by the Eddy brothers on their VVN Ranch at the foot of Black Mountain in Park County, Colorado, in the 1870s. The Eddys, like others of their day, controlled vast acreages beyond their 160-acre homestead by using Black Mountain as a natural barrier to wandering cattle. The property is today controlled by the Bureau of Land Management.
(Photograph by the author, 1999.)

on lush mountain grasses, the animal might realize twice that amount in the mining camps.[40]

The Eddy brothers selected a ranch that provided excellent grass for their cattle in the summer and early fall, but the property, which they called the "High and Lonesome," was located on the upwind side of Black Mountain some 1,500 feet above the roads and later railroads that would cross the area.[41] John Eddy, returning from the plains one summer, found that summers did not last as long in the Rocky Mountains as at lower elevations. Running a thousand cattle in the early to mid-1880s, Eddy remarked that "the grass [was] very short [with] . . . hard winter weather since October 17. If we are to have any cattle, I am glad they are [not here now]."[42] The Eddys put the ranch in South Park to best use, making it an upper ranch for summer feeding. But the property was too high and isolated to allow man or beast a fighting chance of survival in winter. Consequently, the Eddys acquired the use of the IM Ranch at the foot of the mountain.[43] Here they leased land and paid others to feed their cattle hay during the cold winter months.[44] John Eddy, who became the manager of the brothers' Colorado operations, began in the late 1880s to close the high ranch property in winter and move his family to town—either Salida, where he maintained an office, or Denver or El Paso—while neighbors such as the Mulocks, who owned the IM Ranch, fed the cattle. The Eddys also placed cattle on the Poncha, Monarch, Marshall, and Tincup Passes, all high mountain passes accessible to Salida.[45]

As they later did in the Pecos Valley of New Mexico, the brothers took advantage of liberal land laws in Colorado. Observers estimated that some 10 to 50 percent of the patented land in Colorado had been obtained through fraud. In 1884, some 300,000 acres were illegally fenced in South Park, Colorado, the area where the Eddys had their cattle ranch. An 1884 House of Representatives committee investigating illegal fencing in Colorado discovered that two-thirds of the 193 cases brought to the attention of the General Land Office occurred in only twelve counties, including Park County where the Eddys established their ranch.[46] At the beginning, the Eddy brothers owned only 160 acres of deeded land, but they fenced a 1,000-acre pasture on government land next to their deeded property, using logs to hold their cattle.[47] In the 1870s and 1880s, inventive ranchers amassed large chunks of land by fencing three sides of the public property. As long as they did not totally enclose the property, ranchers acted within the law. In order to ostensibly obey the law and at the same time effectively enclose their animals, the Eddys, aided by associates who took out eighty-acre preemption claims, fenced everything except Black Mountain itself, which served as an effective natural barrier to wandering cattle.[48] Black Mountain lay just south of the Eddys' VVN Ranch in Park County. As two of the pioneer ranchers in the area, the Eddys let their cattle drift down into largely uncontrolled and unoccupied lands.[49] When the weather got cold, the cattle naturally drifted down into the Arkansas Valley, where cowboys rounded them up in the spring or butchered those that had become too wild.[50] Some of their cattle drifted as far as the San Luis Valley in the winter.[51]

In the early 1880s, buyers from the midwestern Corn Belt came in the spring and contracted for fall delivery, by rail, of steers and yearlings.[52] By 1890 the Eddys shipped cattle by the trainload to markets connected by new rail lines, one of which was James Hagerman's Colorado Midland Railway.[53] Prior to Hagerman's rail line in 1887, other rails existed near South Park. In October 1879, just as the Eddys were acquainting themselves with the area, the Denver South Park and Pacific Railway (DSP&P) completed a line that intersected an old wagon road through the park. The DSP&P continued building southwest to connect with the Denver and Rio Grande line, which extended from Buena Vista to Leadville, west of South

Park. In the 1880s, the DSP&P, completed in June 1880, stretched the existing line from Pueblo along the Arkansas River to Salida, Colorado, where the Eddys set up winter headquarters for their cowboys and built an addition to the town, providing close access to regional markets.[54] The Colorado Midland provided even better access for cattle ranchers at South Park. The Midland became known as a "stockman's railroad" because of fast, careful handling of stock.[55]

When, in 1881, Charles Eddy established a herd of cattle near Seven Rivers, New Mexico, he envisioned using the New Mexico ranch as a winter headquarters to augment summer operations in Colorado. For much of the decade of the 1880s, the Eddys operated dual headquarters. John Eddy spent much of the rest of his life as a Colorado cattleman and townsman. He was elected mayor of Salida in 1889. Charles involved himself not only in cattle speculation, but also in irrigation, railroads, town building, and mining,[56] primarily in New Mexico.[57]

A side note to all the pioneering cattle efforts of the Eddy brothers is the business conducted by their neighbors. In the 1880s a great deal of money was generated through fattening cattle on free grass. Many people jumped into the cattle business with little knowledge or experience, hiring cowboys to do the work for twenty-five to thirty dollars a month.[58] Among these were Arthur Jr. and Eli Mermod, sons of the owner and president of the Mermod-Jaccard Jewelry Company of St. Louis. In 1887 the company was one of the largest and best-established jewelry firms in the country. Arthur Mermod Sr. was quite wealthy; his sons developed reputations in St. Louis for their hard gambling, drinking, and womanizing. The elder Mermod, determined to break his sons of their habits, purchased a large cattle ranch northwest of Fremont County, Colorado, for $78,000. Called the Stirrup Ranch, it sat close to the Eddy brothers' ranch. The Eddys knew the Mermods and, in light of the Mermods' subsequent activities in Colorado, may have lost a few cowboys to them. The Mermod brothers missed their former St. Louis lifestyle; they began to convert their ranch headquarters into a gambling hall, complete with roulette wheel, whiskey, and local ranch girls, who began affecting the jewelry and trappings of women from large cities like St. Louis. Local cowboys bet their monthly paychecks at the Mermods' roulette wheel, and consequently many of them worked for free.[59]

The Mermods operated the Stirrup Ranch for seven or eight years and sold a number of steers, but they consistently failed to show a profit. The elder Mermod spent $155,000 purchasing and improving the ranch, but sold it for $16,000.[60] Arthur A. Mermod Jr., while an incompetent rancher in Colorado, would later join the Eddy brothers in New Mexico, speculating in land and real estate. Mermod made heavy investments in the Pecos Valley, purchasing a number of lots within the town of Eddy and elsewhere. While the Mermod brothers dabbled in Colorado ranching, the Eddys expanded into New Mexico.

The Eddy brothers established their New Mexico cattle operation opposite a large spring system near Seven Rivers, major tributaries to the Pecos.[61] The Eddys bought some land from Mexican settlers[62] and in 1884 bought the herd of Laramore and Van Wyck, who had run cattle in the area for several years. L. Wallace Holt, representing the Holt Cattle Company, bought out cattleman Thomas Gardner and ran cattle near a big spring at the head of North Seven Rivers. Until the ranch was deserted because of later artesian pumping that dried up the spring, the Holt Ranch was a showplace for the entire valley.[63] Holt and the Eddys agreed to obtain title to opposite banks of the Pecos River near where the future McMillan and Avalon Reservoirs would be built, north of present-day Carlsbad. While Holt developed his water rights on North Seven Rivers, the Eddys moved their headquarters to the east bank of the Pecos, just north of the future town of Eddy (Carlsbad). The area later became known as La Huerta, or "the orchard."[64]

By 1884, the valley was heavily overstocked. The Eddys themselves branded thousands of calves each year, the roundup commencing the first part of March. Cowboys cut out the young steers, collecting about two thousand and sending them north to Colorado. Sometimes the Eddy brothers sent two-year-old heifers up the trail and spayed them when they arrived so they would fatten quickly for market.[65] The drive to Black Mountain took from two to three months and employed several men, including a trail boss, six herders, two wranglers, and a cook. The Eddys' trail boss, Jeff Chisum, gained a reputation for "always choos[ing] the poorest men and horses to send to Colorado." The brothers had no option but to send the cattle north; no outlet for cattle existed east of the Pecos due to the lack of surface water.[66]

During the "Big Die" of 1886, over a third of the cattle in the Seven Rivers area perished when drought dried up the springs and winter cold killed the weakened cattle. The disaster, which bankrupted many small ranchers, encouraged Eddy to look for a perpetual water supply.[67] In 1886 he began expanding the Halagüeño Ditch, a diversion from the east side of the Pecos, building south to irrigate his ranch holdings in La Huerta.[68] Charles Eddy was not the first to irrigate in the area. A group of Mormons attempted to irrigate from the Pecos in 1880, but the attempt failed after only one year.[69] Local ranchers built small irrigation ditches extending from the main stream of Seven Rivers and its branches, and from the Black

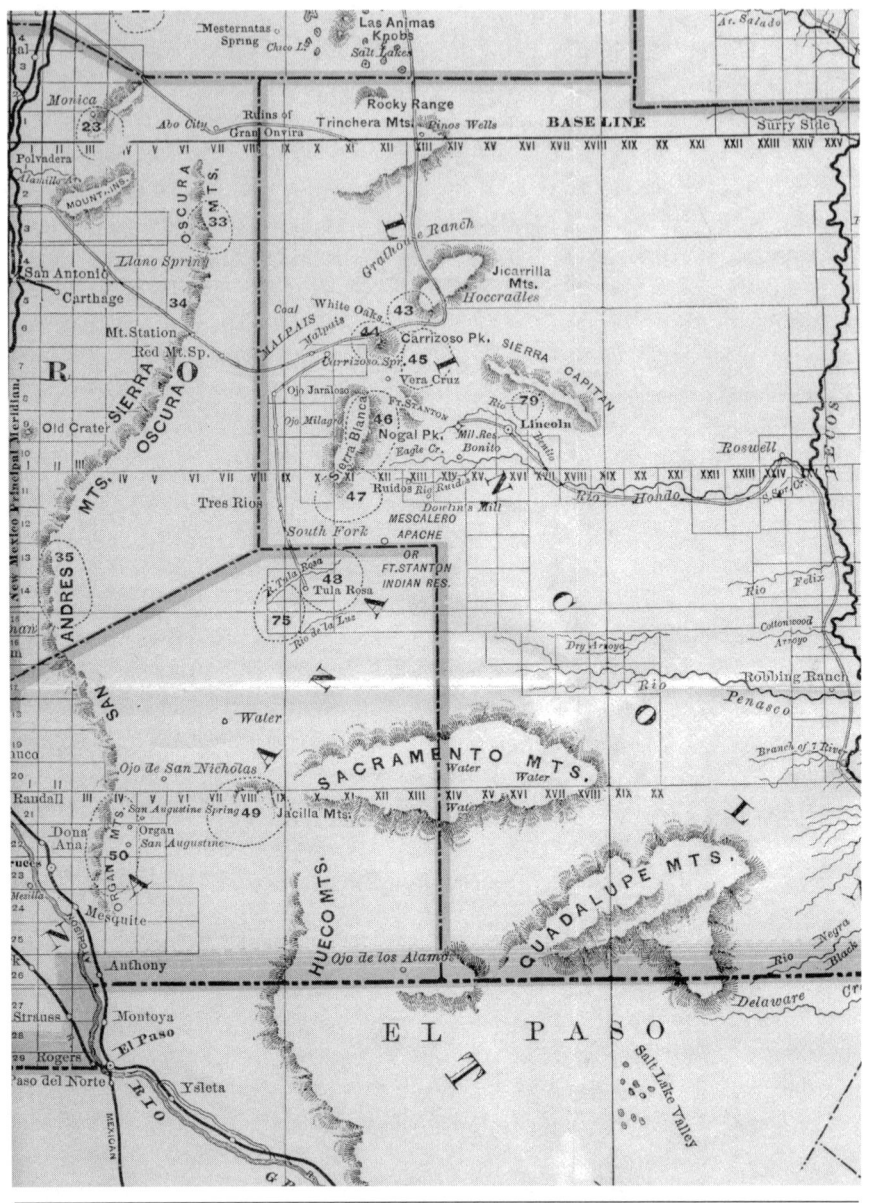

Tributaries of the Pecos River in southeastern New Mexico.
(From W. G. Ritch, Illustrated New Mexico, 5th ed., 1885; courtesy Southwest Collection/Special Collections Library, Texas Tech University.)

New Mexico and Colorado.
(From United States of America: Physical and Political *[Chicago: Rand McNally, 1923].)*

River, as early as 1873. William Brady, sheriff of Lincoln County, attempted a one-man operation, which he abandoned shortly before being shot to death by William Bonney, a.k.a. Billy the Kid.[70] Although Dan Harroun took out the first permanent diversion ditch in 1885, little is known about it.[71] Large-scale irrigation for the valley still lay in the future. Despite the disaster of 1886 and the limited access to surface water, by the end of the decade almost all inhabitants of Lincoln County depended on ranching to make a living.[72]

Although many hardships faced early ranchers, the population of southeastern New Mexico continued to grow. Many of the settlers up and down the Pecos came from Texas. Land in New Mexico attracted Texas cowboys into the region as early as the 1870s, but an equally important push factor caused many Texans to leave their native state. Texas cowboys running from the law poured into Lincoln County, leaving an indelible imprint on southeastern New Mexico from 1879 to 1890. In 1878, the state of Texas issued an official 226-page book to peace officers across the West, listing and describing fugitive outlaws from various counties. Texas offered generous bounties totaling almost $100,000 for the return of such fugitives to justice. Out of over four thousand criminals, peace officers wanted at least three hundred for murder. Texas effectively rid the state of desperadoes, but they reappeared in New Mexico and other western territories.[73]

One of the Texans who moved to the Pecos Valley in the 1870s was Patrick Floyd Jarvis Garrett, sometime cowboy, buffalo hunter, gunfighter, gambler, horse breeder, lawman, customs agent, and irrigator. Not only would Garrett become a legend for killing Billy the Kid, but he proved instrumental in first promoting irrigation in the valley. Born in Alabama, Garrett grew up in post–Civil War Louisiana until, at the age of nineteen, he moved to Texas. There Garrett cultivated his skills as a cowboy and buffalo hunter, drifting west to Tascosa, Texas, and then to Fort Sumner, New Mexico, by 1878. In 1880, Lincoln County stockmen elected Garrett sheriff to deal with cattle rustlers up and down the Pecos Valley, including William Bonney. Following the shooting of Bonney in July 1881, the same year Eddy established his ranch at Seven Rivers, Garrett lost his reelection bid for Lincoln County sheriff. Garrett turned to raising his own cattle while actively pursuing New Mexican cattle rustlers as a Texas Ranger. In the mid-1880s, he managed the vast New Mexican holdings of Captain Brandon Kirby, an Englishman who took advantage of cheap land to acquire cattle and rangeland near the Pecos. In 1886, after Texas Fever,[74] drought, and frigid temperatures hit New Mexico, Garrett had a falling-out with the Kirby outfit and purchased a small ranch near Roswell.[75]

Garrett's ranch consisted of 1,800 acres of bluegrass, roses, oak and pecan trees, and an adobe home, three miles east of Roswell on the Hondo River. Using a crude irrigation ditch, Garrett grew alfalfa and grapes and planted hundreds of apple and peach trees, gradually widening his attention to irrigating the entire Pecos Valley.[76] Unlike cattleman Charles Eddy and other ranchers in the Seven Rivers area seventy-five miles south, Garrett envisioned an agricultural region of farmers. His interest in farming was not a novel one. Small-acreage farmers along the Hondo had initiated early ditch efforts long before Garrett's arrival.[77] On January 15, 1887, Garrett purchased one-third interest in Texas Irrigation Ditch from William H. Holloman. On August 15, Garrett and Holloman formed the Holloman and Garrett Ditch Company, with capital stock of $5,000 and fifty shares at

Pat Garrett, known as the lawman who gunned down William Bonney in the wake of the Lincoln County War, played an early role in the irrigation ventures along the Pecos Valley of New Mexico. At various times throughout his life he served as sheriff, horse trader, farmer, rancher, U.S. commissioner, and livery and hack operator. *(Courtesy Southeastern New Mexico Historical Society, Carlsbad, New Mexico.)*

$100 each. When the corporation dissolved, Garrett bought interest in the Pioneer Ditch Company.[78]

Garrett wanted to build a dam across the Hondo River, the chief tributary of the Pecos. By building a flume, Garrett reckoned he could send water across the desert to areas previously thought worthless for farming.[79] Garrett needed investors to pull off the Hondo scheme, and he recruited Charles Greene, a newspaperman from St. Louis and Santa Fe who had made Garrett famous as the slayer of Billy the Kid. When Greene was sent by his newspaper, the *Santa Fe New Mexican*, to examine the resources of the Pecos Valley, he found cattle ranchers and the potential to make a fortune. Garrett and Greene undoubtedly thought that Garrett's reputation might enhance land and water sales.[80]

Greene was born in Providence, Rhode Island, on March 14, 1839. He taught school in Pennsylvania, served in the Twelfth Connecticut Infantry during the Civil War, then moved to the Midwest, where he became a publisher of several newspapers. In the 1880s, Greene moved from Kansas to New Mexico, where he again pursued publishing at a number of newspapers, including the *Santa Fe New Mexican*, which was owned by the Atchison, Topeka and Santa Fe Railway (AT&SF). Greene worked at the *New Mexican* under Max Frost, who also served as chief of New Mexico Territory's Immigration Bureau.[81]

In 1886, Garrett approached rancher Charles Eddy with his vision of irrigating the Pecos Valley, introducing him to Charles Greene at Eddy's ranch headquarters.[82] The meeting between Garrett and Eddy heralded the beginnings of corporate irrigation in the valley. Eddy and Garrett recognized that the two small irrigation ditches that Eddy had been working on—one to irrigate his ranch at Halagüeño, the other at Double Crossing—were inadequate for their vision of an irrigation empire.[83] Consequently, what began as an idea to capture water for Eddy's cattle evolved into an ambitious project to build two large canals, each thirty feet wide and five feet deep. The Northern Canal would draw water from the Hondo River and stretch forty miles across Garrett's property. The Southern Canal would take water out of the Pecos close to Eddy's property and reclaim 300,000 acres. Garrett and Eddy set up their first company, Pecos Valley Land and Ditch, in 1887.[84] Garrett and Eddy hoped to make money by selling land in the new town of Eddy, which they laid out in 1888, and charging an annual rental fee for water.[85]

The two men authorized Greene, who gave up his career in journalism, to begin promoting their plans, first in America and then in Europe.[86] In the late 1880s, Greene served as chief promoter of both the Pecos Valley in New Mexico and Bear Valley in California. He was a busy man, traveling between the Pecos Valley, Chicago, New York, and Europe. On October 25, 1890, the *Eddy Argus* stated: "Charles W. Greene, who has been nearly all over the world during the past two years and inspected hundreds of irrigation works, declares that the works of the Pecos Irrigation and Improvement Company are [the] largest and most substantial . . . seen."[87] Greene, who initially maintained an office in Chicago, the home of early investors, set up his main office in New York and traveled to Europe at least twice between 1888 and 1891. Greene placed a great deal of faith in the project. Expecting to get rich off land promotion, Greene sank a good deal of his personal earnings into lands near Eddy.[88]

In the late 1880s, Eddy and Garrett had engineers draw up plans and specifications for the irrigation project in the Pecos Valley.[89] Yet Eddy, Garrett, and Greene could not have anticipated the myriad challenges. Together, they faced the initial task of raising capital for what would become an impressive irrigation scheme to water half a million acres of creosote and sagebrush.

Even as the trio began putting together investors, other irrigation companies sprang up throughout New Mexico and the West. There were no incorporated irrigation works in New Mexico until 1888, when nineteen companies incorporated. Another thirty-two formed in 1889, and by 1891

entrepreneurs had established eighty-eight companies and had built ditches to water 40 percent of the irrigable land in New Mexico.[90] As other irrigation companies sprang up across New Mexico, Charles Eddy scrambled to get needed capital for his own expanding project to irrigate the desert. To raise more money for the venture, Eddy looked north.

Eddy frequented Colorado Springs, Colorado, in the late 1870s and early 1880s, when Amos Bissell bankrolled his cattle drives from New Mexico to South Park. Bissell, who had funded the Eddy brothers' Colorado ranch, may have introduced Charles Eddy to some of the businessmen and vacationers to the city.[91] The Eddys, while not cash-laden themselves, were a prominent family from New York State. They hailed from Milford, Otsego County, a scenic village ten miles south of Cooperstown and some 150 miles northwest of New York City, where members of the family owned a grist mill and a tannery and served in local government.[92] John Eddy, the brothers' father, was a banker and delegate to New York's 1867 Constitutional Convention.[93] Part of Milford was named Eddyville, and the patriarch of the Eddy family was called "Squire." Sometime in the 1870s, John Eddy went to New York, got into financial trouble, lost almost all of his vast properties, and died insolvent.[94]

Not by choice then, but because of the financial straits of their father in New York City, John Arthur and Charles Bishop Eddy headed west in the

1870s, like hundreds and thousands before them, to seek their fortunes. The Eddys had undoubtedly spent time with men of means. Consequently, they moved with ease among those connected to old eastern money, those "eager to invest in the West . . . knowing and trusting 'one of their own.'"[95]

Long before ski resorts catered to millions, many parts of the West were the playgrounds and investment arenas of rich easterners and their sons. In the nineteenth century, Colorado Springs, Colorado, became the playground of, if not the famous, then certainly the rich. Before the mines of Cripple Creek yielded any riches, the cities at the base of Pikes Peak became famous worldwide for their beauty and health-restoring climate. Colorado Springs and its neighbor, Manitou, were fashionable summer resorts in the Rocky Mountains, attracting thousands who came west to improve their health, including many in the last stages of consumption (tuberculosis) and asthma.[96]

On October 9, 1875, President Ulysses Grant stopped to give a speech on the porch of the Colorado Springs Hotel and was surprised to see so many guests at the resort. Because of some good promotion by two resident doctors, one from England and the other from Virginia, and because of actual results, Colorado Springs became the mecca for those suffering from consumption. Good food, rest, exercise, and air at high altitude seemed to slow lung degeneration. Medical facilities sprang up to treat easterners with deep pockets. Experts recommended bathing in and drinking waters from the series of springs near the town, and soon a thriving business in treating consumptive invalids developed. One such patient, James J. Hagerman, arriving in 1884 from Paris, began a remarkable recovery.[97] Hagerman, a short, feisty millionaire, would later dump a good portion of his money into the irrigation ditches of New Mexico.

The Philadelphians who established the Springs in 1871 sought to create a model city against the majestic backdrop of the Rocky Mountains. General William Palmer bought up the water rights along Fountain Creek in the early 1870s. Realizing that the town could not survive, and he could not sell real estate or operate a resort, without an adequate irrigation system to green things up properly for easterners who were accustomed to such things, Palmer built the El Paso Canal. The ditch stretched eleven miles to feed laterals running next to home lots in what is now the downtown section of Colorado Springs.[98]

Palmer's plans for making the Springs attractive to easterners included railroad connections, paved streets, streetcars, parks, trees, and irrigation. In 1885, the city was three miles long and two miles wide, with broad streets and avenues 140 feet wide and city blocks 400 feet square. Palmer's irrigation system watered hundreds of cottonwood trees, large green lawns,

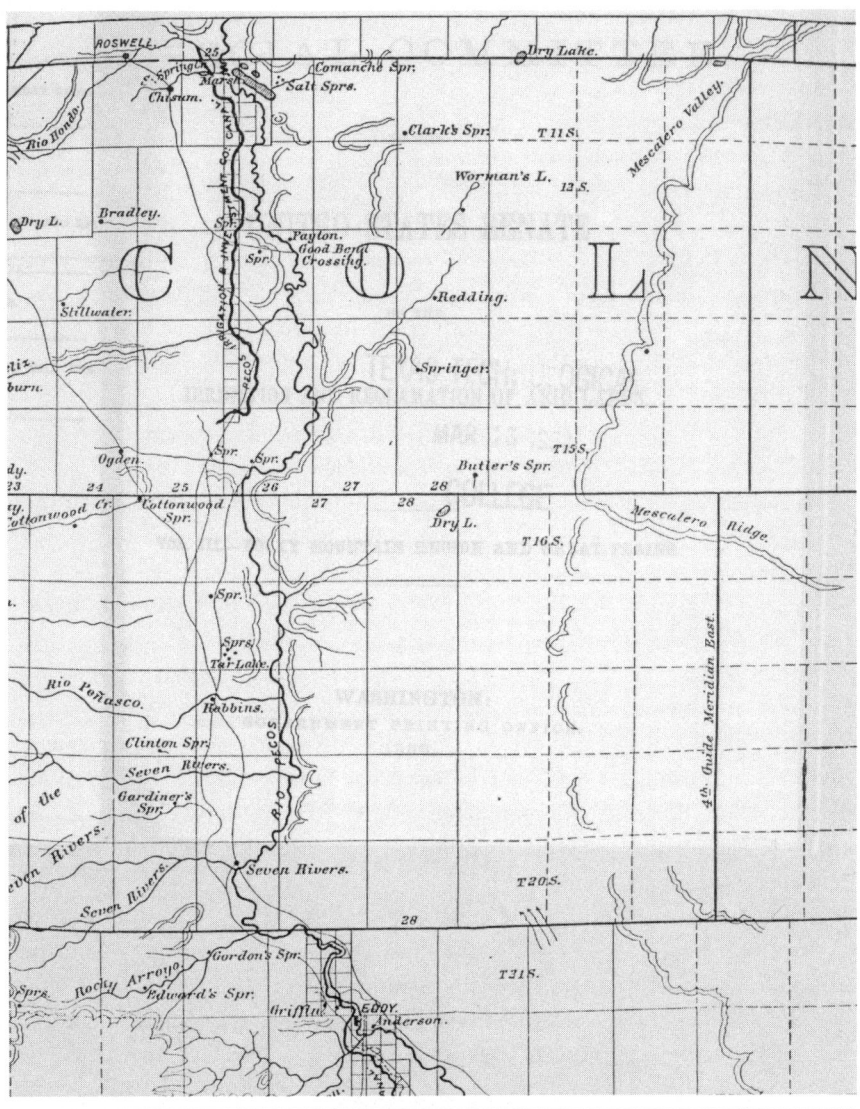

Canals of the Pecos Irrigation and Improvement Company and Pecos Valley Land and
Ditch Company, by H. Hartmann, Santa Fe.
(From Report of the Special Committee of the United State Senate on the Irrigation and
Reclamation of Arid Lands, *vol. 3, Rocky Mountain Region and Great Plains,*
51st Cong., 1st sess., 1890, S.R. 928, pt. 4).

29

and gardens. Before the turn of the century, Colorado Springs boasted 30,000 inhabitants, numerous churches, schools, business blocks, a sewage system, electric railways, light and power, a water system, and the leading mining exchange center in the world. In order to keep out what was considered to be the vulgar element of society, city fathers outlawed the "vicious influences" of liquor and gambling.[99]

In Colorado Springs, city of health seekers and millionaires—especially within the exclusive suburb of Broadmoor—Charles Eddy recruited investors for his dream city and his plans to irrigate the desert of New Mexico. In the 1880s, Broadmoor served as the private getaway for wealthy gentlemen pursuing various types of leisure activities. Surrounded by handsome estates, the Broadmoor Hotel and Casino offered recreations unavailable in Colorado Springs proper. Broadmoor offered a country club, one of the first in the country and certainly the first in the West.[100]

Although sporting activities were commonplace at Broadmoor throughout the 1880s, in April 1889 members organized an official country club. Here, scions of large fortunes indulged in golf and polo, cricket, bowling, riding, shooting, bicycling, and horse racing, and fashioned investment deals in their spare time. Two years later, after securing a site near the Broadmoor Hotel and Casino from William A. Otis, organizers incorporated as the Cheyenne Mountain Country Club Association.[101] Membership consisted of "leading business and professional men . . . as well as men of leisure." In June 1901 the club reorganized as the Cheyenne Mountain Country Club, its members known as "The Grizzlies."[102] By all indications, the Grizzlies were a closely knit group of businessmen and socialites primarily from the East Coast and Europe.

The Cheyenne Mountain Country Club, the El Paso Club (another prominent social club in Colorado Springs), and the mining industry attracted prominent families from both sides of the Atlantic. In 1903, then president Theodore Roosevelt joined his former Rough Riders, including country club member Joseph Stevens, in a pickup game of polo. Chester Alan Arthur II, the former president's son, was a popular member who served as president of the club from 1905 to 1908. Besides making additions to the club's facilities, Arthur, according to at least one member, devoted his life "to the art of having a good time, in which he succeeded supremely well."[103] So many prominent men from England spent at least some part of the year in Colorado Springs that for a time newspapers referred to the city as "Little London."

It is not surprising that members of the club consummated business deals between rounds of polo. Of more significance is the number of members and their relatives who invested in the irrigation schemes of Charles

Promoter and town builder Charles Bishop Eddy. Eddy came west with his brother, John Arthur, in 1879 and started a Colorado cattle operation. In the early 1880s Eddy started his Halagüeño Ranch in southeastern New Mexico near present-day Carlsbad. From here he drove cattle to his Colorado VVN Ranch near Salida. Intrigued by Pat Garrett's vision of irrigated agriculture, Eddy bought water rights in 1887 on the Pecos River and began recruiting investors. *(Courtesy Southeastern New Mexico Historical Society, Carlsbad, New Mexico.)*

Eddy, and later James Hagerman, in the Pecos Valley. Among the investors who had connections to the club were Richard J. Bolles, F. E. Bryant, William P. Bonbright, James H. McMillan, Percy Hagerman, Herbert J. Hagerman, and Joseph Stevens.[104]

Charles and John Eddy recruited fellow New Yorker Joseph Sampson Stevens to the Pecos Valley as early as 1887. Stevens was renowned as one of the world's greatest polo players, establishing a reputation for his prowess at Broadmoor and Denver, as well as in clubs on the East Coast. From 1887 to at least 1907, Stevens played competitively against fellow club members, the Tenth Cavalry, and other teams of national prominence.[105]

Stevens came from a distinguished family. His father, attorney Frederic William Stevens, served on the board of directors of five of the nation's leading banks and was a leader in public and civic affairs. Frederic Stevens's mother was the daughter of Albert Gallatin, secretary of the U.S. Treasury from 1801 to 1813. In October 1862, Frederic Stevens married Adele Livingston Sampson, daughter of Joseph Sampson, one of the founders of the Chemical National Bank of New York City. The elder Stevens served as a director of the bank from 1871 until his death in 1928.[106]

Suffering from tuberculosis, Joseph Stevens went west to Colorado Springs in the mid-1880s. There, Stevens, who was not yet twenty-one, became fast friends with the two Eddy brothers. The Eddys had family and business connections to the Chemical Bank, the same bank where Stevens's father served for fifty-seven years as director.[107] In 1886, just prior to his twenty-first birthday, when Stevens stood to inherit considerable wealth, he traveled to the Eddys' Halagüeño Ranch, just outside present-day Carlsbad. According to Stevens's cousin, Francis Tracy, writing in the 1950s, Halagüeño was the only settlement between the ranches lying along Seven Rivers and those along Black River to the south. The federal government owned the land between these settlements, and Stevens, perhaps noting the recent land boom in California, saw in southeastern New Mexico a place to invest his money.[108] That Stevens and Eddy had a mutual friend in Amos Bissell could not have hurt Stevens's growing enchantment with New Mexico. Joseph Stevens lured others to the Valley. Already intrigued by the Eddy brothers' grandiose designs for irrigating the Pecos Valley, Stevens jumped into an investment scheme that ultimately recruited a number of investors from Switzerland, Chicago, New Bedford (Massachusetts), Colorado Springs, and New York City.[109] Impressed during a trip to the wide-open spaces of the territory, Stevens immediately set about recruiting his father, Frederic, and cousins Francis G. Tracy and John Adams, to Eddy's irrigation scheme.[110]

In consideration for their investors' filing on desert lands[111] sight unseen, Stevens and the Eddys arranged to have them travel to New Mexico Territory. Stevens first told his cousin Francis Tracy about the Pecos Valley of New Mexico in 1886 or 1887. By then Tracy had been a gentleman farmer for two years at Long Island, New York. Tracy farmed one mile east of the Meadow Brook Hunt Club, an international polo ground. According to Tracy, reminiscing about his early years in New York and later in New Mexico, he was a close friend of Elliot Roosevelt, Theodore Roosevelt's younger brother, who lived at the hunt club.[112] It is quite possible that Elliot Roosevelt whetted Tracy's appetite for adventure by telling him of his own adventures hunting buffalo with a cousin on the High Plains of Texas.[113] At any rate, when Joseph Stevens revealed his investment concept, the idea of going West took hold.

In November 1889 Tracy and Frederic Stevens made the trip west via St. Louis and Toyah, Texas. Tracy and his uncle stayed with Charles Eddy in the new town that would bear his name until being renamed Carlsbad some years later. Traveling north toward Roswell, Tracy and Stevens saw their lands for the first time.[114] After seeing water in the ditches leading to the property, both men decided to invest further in the Pecos Valley irrigation venture.[115] Tracy filed a Homestead entry[116] in Las Cruces before heading back east with his uncle. He was sufficiently impressed with the prospects in southeastern New Mexico that he returned to Long Island, sold his property there, and headed west permanently. Arriving in Eddy in late January of 1890 on the stage from Texas with a pair of collies, the twenty-six-year-old Tracy had already filed both a Desert Land claim and a Homestead entry for land along the Northern Canal and expressed interest in acquiring additional lands. Tracy immediately started working as surveyor and foreman for Pecos Irrigation and Investment Company during 1890 and 1891 and then began selling real estate.[117] Others, including cigar king Robert Tansill, followed Stevens to New Mexico, providing initial capital for irrigation.

Robert Weems Tansill had made a fortune manufacturing cigars in Chicago. Born August 20, 1844, in Prince William County, Virginia, Tansill was the only child of a successful businessman. Tansill was a direct descendant of Mason Locke Weems, historian of the American Revolution and clergyman at the church where George Washington worshiped. In the spring of 1861, Tansill went to Illinois and stayed. Six years later, Tansill began jobbing cigars and selling confectionery. In 1868 he moved to Chicago where he continued selling cigars and candies. In 1871 the great fire swept the city, burning local cigar manufacturing plants. Tansill borrowed some capital and rebuilt one of the dilapidated plants into the largest cigar factory in Chicago. He made his name manufacturing "Tansill's Punch" cigar, sold across the United States through direct marketing. By 1886, Tansill manufactured his product both in Chicago and New York, promoting sales from both locations. Each month Tansill manufactured more than one million of his Tansill's Punch brand alone, and he built his family a stately house on Chicago's prestigious Dearborn Avenue.[118] By 1888, Tansill suffered from tuberculosis. Like many others of his financial standing, Tansill heeded his doctor's advice and "retired," leaving his cigar empire to his sons and heading west to Colorado Springs. According to his younger son, Robert Weems Tansill II, his father's doctor told Tansill that he only had "30 days to live if he didn't get to an arid region such as the Southwest." Tansill suffered from debilitating lung hemorrhages, at one point experiencing eleven in ten days.[119]

After moving to Colorado Springs in the late 1880s, Tansill met Charles Eddy and Charles Greene.[120] Eddy and Greene convinced Tansill that he should visit the Pecos Valley and encouraged him to recruit investors from Chicago. Greene traveled to Chicago and organized a trip for Tansill and twenty other potential investors from the city. The party, including Greene, his daughter Lillian, Tansill, Tansill's brother-in-law Ed Motter, and others, left Chicago in the spring of 1889 by rail for Toyah, Texas, some sixty miles south of the Texas state line. There, Pat Garrett met the group, providing hacks, buggies, and buckboards to take the party to Eddy's ranch.[121] Three years after the trip, Tansill described the stretch from Toyah to Eddy's new town in the desert:

> At that point [Toyah, Texas] we took private conveyance for more than a desert. Our train consisting of twelve wagons, loaded with people, provisions and camp equipage, rolled out from Toyah one bright September morning, forcibly reminding one of a caravan entering the desert. Our first camp was at Screwbean, where we pitched our tents and spent the night. The next night we reached the ranch of C. B. Eddy, known as Halaguen[o] just across the river, north of this city [Eddy], where we were most hospitably entertained. From there we went on to Roswell. We spent about two weeks in all in looking over the Valley. All of the party was delighted with the trip, the country and its prospects.[122]

Eddy's sister, Mary Fox of New York City, along with her friend Mrs. Arthur A. Mermod Jr. of St. Louis, were visiting the ranch at the time and helped entertain the guests from Chicago.[123] Mermod's husband had since left the Stirrup Ranch in Colorado. Although Mermod resided in St. Louis, he maintained ties to the Eddy brothers in New Mexico and joined E. G. Shields, land commissioner from El Paso, in the real estate business. Tansill was so delighted with the trip that he returned to Chicago to bring his family to see the Pecos country. According to Tansill, his family liked the country as much as he did. Originally, the Tansills planned to build a home and reside in Colorado Springs at least part of the year and in the Pecos Valley during the winters. The altitude of Colorado Springs proved to be too rarified for Tansill's wife, and the family took up more or less permanent residence in New Mexico.[124]

Tansill proceeded to recruit more investors to Eddy's burgeoning irrigation schemes.[125] Tansill's son maintains that his father recruited William McMillan, a manufacturer of train car wheels from St. Louis. The storage dam above the town of Eddy (Carlsbad) was named after McMillan. McMillan's brother, James McMillan, born May 12, 1838, in Hamilton, Ontario, Canada, moved to Detroit in 1855. In 1860 James McMillan became manager of the Michigan Car Company to build freight cars. McMillan was successful in railroads, shipbuilding, and steamship line investments. He and John S. Newberry established car plants at Hamilton, Ontario, Cambridge, Indiana, and St. Louis. Between 1889 and 1902 he served as a U.S. senator. He met regularly at his home with a group of conservative Republicans that called themselves the SOPC, or School of Philosophy Club. Over the years the group included Senators William B. Allison of Iowa, Nelson Aldrich of Rhode Island, Eugene Hale of Maine,

Francis G. Tracy, about 1890. Trained as a lawyer, Tracy came to the Pecos Valley from Long Island, New York, in 1889. He immediately began working for the Pecos Irrigation and Investment Company. A tireless promoter of irrigation in the valley long after private initiative had failed, Tracy was a true crusader, badgering federal reclamationists to provide additional storage and improvements to the irrigation system around Carlsbad well into the twentieth century. *(Courtesy Southeastern New Mexico Historical Society, Carlsbad, New Mexico.)*

Representative Willis C. Hawley of Oregon, politico Marcus Hanna, and presidents Benjamin Harrison and William McKinley.[126]

By far, Tansill and the Eddys' most brilliant stroke was the recruitment of mining capitalist James Hagerman, by telling him of the canal work already commenced by Eddy, Pat Garrett, Charles Greene, and partially financed by Joseph Stevens.[127] By 1889, Tansill and the Eddys convinced Hagerman to lend his enormous financial muscle to the venture.[128] Perhaps what linked Tansill and Hagerman, besides their enormous wealth and vision for the future of the Pecos country, was tuberculosis. Both men experienced serious bouts of the disease, and both moved west to slow its progress.

The recruitment of investors by Tansill and later by James Hagerman reflects important economic and political ties—in effect, the network of connections between nineteenth-century American captains of industry, especially the iron and steel industries, and politicians. These connections proved valuable in recruiting investment for irrigation of New Mexico's Pecos Valley.

James John Hagerman was born near Port Hope, Ontario, on March 23, 1838, and grew up along the St. Clair River near Detroit. There, Hagerman worked at a number of jobs in shipping, carpentry, milling, and farming. Hagerman attended the University of Michigan in 1857 and 1858 but could not afford to finish school. He returned to work in the steam shipping

Charles Eddy's house in La Huerta, 1892. Seated on the porch is Arthur Mermod, son of the owner of the Mermod-Jaccard jewelry firm of St. Louis. Arthur and his brother Eli had been shipped off to Colorado to dabble in the cattle industry on a ranch near the Eddy brothers' VVN Ranch. Arthur later invested monies in irrigation and real estate ventures of the Pecos Valley. The two women are Mrs. Mermod and Charles Eddy's sister, Mary Fox.
(Courtesy Sacramento Mountains Historical Society, Cloudcroft, New Mexico.)

business, working for E. B. Ward, who became Hagerman's mentor in the early days of the iron and steel industry.[129] Hagerman graduated from Michigan in 1861 and went on to manage Ward's Milwaukee Iron Company until 1873 when he and a Milwaukee syndicate opened the Menominee iron district in Michigan's Northern Peninsula. Hagerman was the principal owner of the Chapin Mine, the largest producer of iron in the country. He became wealthy from the steel business, supplying iron for steel rails to various railroad interests.[130] While amassing a fortune, Hagerman met and became friends with a number of businessmen and politicians, among them Andrew Carnegie, Marcus Hanna, William McKinley, and Charles Otis.[131] Hagerman sold the Chapin Mine to the Cambria Iron Company in 1883, making himself a fortune. As Hagerman's fortune increased, his health declined. Having contracted tuberculosis in 1873, he finally left Wisconsin in 1881, traveling throughout the United

James J. Hagerman. Chief investor in the Pecos Valley irrigation and railroad enterprises, Hagerman had made his fortune in iron, railroading, and Colorado mining. Suffering from tuberculosis, Hagerman recuperated in Colorado Springs and Europe, where he used his political and business ties to recruit both investors and settlers to the Pecos Valley. *(Courtesy Special Collections, Tutt Library, Colorado College, Colorado Springs, Colorado.)*

States seeking medical treatment. In 1882, he left for an extended tour of Europe.[132] When he returned to the United States in 1884, he settled in Colorado Springs because of its dry, healthful climate.[133]

In the Springs, Hagerman quickly immersed himself in new projects, buying choice property while railing at his constantly hovering doctors and nurses, and referring to the Springs as a town "as dead as Julius Caesar with no business worth mentioning and little hope for the future." Nonetheless, Hagerman built a mansion worth $110,000 on Cascade Avenue.[134] In 1886, Hagerman along with other investors incorporated the Colorado Springs and Manitou Street Railroad Company. The first cars, which were drawn by eighteen draft horses, began operation in 1887 and ran through the business district every fifteen minutes. Hagerman also involved himself in banking. He and others managed the first national bank in the region. In 1894, Hagerman donated two city lots to St. Stephen's Church.[135] For someone who had little confidence in the city, Hagerman played a major part in shaping its future.

In the mid-1880s, besides buying property in the Springs, Hagerman began buying claims in the infant silver camp at Aspen. He also acquired a number of stone quarries and coalfields on the Western Slope. Several wealthy eastern transplants, including Jerome Wheeler, who owned considerable mining property in Aspen and half interest in Macy's Department

Former home of Mr. and Mrs. James Hagerman, Cascade Avenue, Colorado Springs, Colorado, as photographed in 1999 by the author. After an extended stay in Europe, Hagerman moved to the high, arid Springs in 1884 to help his tubercular lungs. Despite his declining health, Hagerman embarked on a number of business ventures, including building the Colorado Midland Railroad and investing in Aspen's Mollie Gibson Mine.

Store, intrigued Hagerman with tales of building a railroad over the Rocky Mountains. Wheeler and others, including Irving Howbert, a Colorado Springs pioneer and president of the First National Bank of Colorado Springs, wanted to take advantage of the weakened condition of the Denver and Rio Grande Railway by building west past Leadville. Hagerman joined Wheeler and Howbert, who encouraged Orlando Metcalf, a steelmaker from Pittsburgh who had also gone west for his health, Hagerman's steelmaker friend Charles Otis, and J. R. Rusk, a New York financier, to join the trio. Jay Gould's manipulation of Denver and Rio Grande Railroad Company stock forced out Colorado Springs builder and railroad magnate General William Palmer in 1883. In 1884 Palmer's company went bankrupt and fell into receivership.[136]

Because of its circuitous route, the Denver and Rio Grande ran 196 miles to Leadville, still 108 miles from the newer silver fields in Aspen. Hagerman saw the advantages to building a direct route 204 miles up Ute Pass west of Colorado Springs to Leadville, 100 miles shorter than its narrow-gauge competition. Hagerman became president of the Colorado Midland in 1885 and started construction, raising $7 million in capital and ironically bringing materials and crews into the area on the Denver and Rio Grande. Hagerman's crews completed their task in late November 1887, having built the railroad over some of the most spectacular and tortuous terrain in the world.[137]

Conveniently, the Midland skirted all of Hagerman's property in Aspen and along the Western Slope, a major bargaining point in getting Hagerman to build the Midland.[138] The line connected the mining operations of Leadville and Cripple Creek with Aspen's Mollie Gibson Mine, where Hagerman later made a fortune in silver. When Hagerman finished the line, the Midland's directors urged him to continue as president, but he was exhausted. Sick again, Hagerman resigned, but he reappeared in 1890 to negotiate the sale of the Colorado Midland to the Atchison, Topeka and Santa Fe Railway.[139]

In February 1891, Hagerman contracted pneumonia and nearly died. In July of that year, miners in Aspen struck rich ore in the Mollie Gibson Mine.[140] As Hagerman prepared to travel to Europe to recuperate, he left his business affairs in the hands of a trusted friend—the secretary/treasurer of the Mollie Gibson Mine, Henry C. Lowe. Lowe's family had long been involved in Colorado mining, and apparently Hagerman had been a friend of the family. Hagerman gave Lowe the power of attorney to conduct all of Hagerman's affairs in his absence. Lowe apparently double-crossed him, or so Hagerman thought—either in regard to the Mollie Gibson or in some other business matter.[141]

Hagerman, who realized the huge potential of the Mollie Gibson before other investors, sought to keep its wealth a secret so as to buy out major shareholders before the price of stock soared. Part of the arrangement to keep the potentially rich strikes a secret involved paying off the chief engineer of the project, C. E. Palmer, with a substantial salary and an option on stock if Jerome Wheeler, Hagerman's partner in the Colorado Midland venture, were convinced to sell out his shares. Wheeler, who was vice president of Mollie Gibson Consolidated Mining and Milling Company, filed suit against Hagerman, then president of the company, alleging that Hagerman conspired with other major investors to deceive him about the mine's potential.[142]

Wheeler, who invested heavily in Colorado mining activities and owned a bank in Aspen as well as other properties, sold his shares in Mollie Gibson Consolidated Mining and Milling to become a major shareholder in the Bushwhacker Mining Company. Although Wheeler withdrew his suit

Mollie Gibson Mine, Aspen, Colorado. James Hagerman and Richard Bolles were major shareholders in the operation. Hagerman's Colorado Midland Railroad conveniently skirted much of the mining activity in Aspen.
(Courtesy Denver Public Library, Western History Collection, W. H. Jackson, WHJ-763.)

against Hagerman, legal wrangling and an anonymous forty-four page pamphlet supporting Hagerman revealed the manipulation behind capital investment schemes of the nineteenth century.[143]

Despite legal wrangling, Hagerman positioned himself to reap a fortune from the Mollie Gibson, much of which he poured into irrigating New Mexico's Pecos Valley.[144] Initially, Hagerman intended to invest modestly in the Pecos irrigation project, but eventually the project consumed his resources. Recalling those early days when he invested his first $40,000 in the project, Hagerman admitted that his dreams got the best of his judgment: "One false step in business, as well as in morals, is often the precursor of others."[145] Of the $5,000,000 fortune he made at the Mollie Gibson, Hagerman ultimately invested some $2,500,000 in New Mexico's Pecos River Valley.[146]

Sometime between 1887, when he finished the Midland to Aspen railroad, and 1890, when he sold it, Hagerman traveled to the Pecos Valley at the request of Robert Weems Tansill, who lived in Colorado Springs at the time. Henry Lowe, who had known Tansill for several years, introduced Hagerman to Charles Eddy, whom Hagerman described years later as "surely as persuasive a scamp who ever lived."[147] But "Eddy [also] came highly recommended by the president of the Denver National Bank who knew him well and by [George] Williams, Pres[ident] of the Chemical Nat[ional] Bank of N[ew] Y[ork] and they endorsed him in the highest terms."[148] Tansill, who had already seen the country, and Hagerman joined Charles Eddy in Toyah, Texas, for the trip north along the Pecos River to Eddy, the name given to Charles Eddy's community in the desert.[149]

Back in Colorado Springs after visiting the Pecos Valley, Hagerman convinced others, including Richard Bolles, to invest in Charles Eddy's irrigation project. Richard J. Bolles was born in New York City on August 1, 1843, the son of a doctor.[150] His business career began in New York, where he was a member of the New York Stock Exchange. In 1883 he moved west to invest in Colorado mining enterprises, especially in the Aspen area. Bolles later joined James Hagerman as one of the Mollie Gibson Mine's principal investors. He amassed a fortune from his properties in Aspen, Glenwood Springs, Carbondale, and Grand Junction. The investor apparently became a regular in the social circles of Colorado Springs, where he "was a member of every congenial group of successful business men who forgathered in the hotels and clubs of that period [and] . . . [h]is ebullient humor and fund of anecdotes made him a social favorite."[151] Bolles served as president of the Cheyenne Mountain Country Club for six years, longer than any other member at the time.[152] Bolles's business interests in the Springs included membership on the board of directors of the First National Bank and various other business activities.

Besides recruiting Richard Bolles, Hagerman convinced William P. Bonbright to invest in the Pecos Valley. The Bonbright family served as financial and civic leaders of Colorado Springs during the 1890s, speculating in the stock market and local mining exchange. The family owned one of the oldest brokerage offices in the city, specializing in Cripple Creek mining stocks, organizing and promoting mining companies and western securities, and acting as financial agents for nonresidents. Bonbright's experience in stocks and fund-raising for western investments made him a crucial part of the unfolding irrigation scheme in the Pecos Valley. Bonbright's firm operated an office in London and was involved in various corporate reorganizations in the Colorado Springs area, including the Colorado Springs Electric Company and the Colorado Springs Light and Power Company.[153]

Relatively untouched by a succession of Native Americans, Spaniards, and Mexican Americans, the Pecos Valley changed beginning with the U.S. military presence in the 1860s, allowing John Chisum and other cattlemen to monopolize the scant water holes and vast cattle range on either side of the Pecos. Cattlemen Charles Eddy and promoter Charles Greene took Pat

Garrett's concept of irrigating the valley, expanded it, and recruited much-needed capital. After gaining initial seed money from New Yorker Joseph Stevens, Eddy and Greene turned to Chicago and Colorado Springs, a playground mecca for some of the nation's industrial and political elite. James Hagerman, a millionaire made rich from iron ore in northern Michigan, used his ties to the steel, railroad, and mining industries to attract more capital for irrigating the New Mexico desert. As Eddy began to pour investment capital into dams and canals, the tools providing the impetus for corporate irrigation of the valley were the nation's flexible land laws—especially the Desert Land Act of 1877.

2

CREOSOTE AND SAGEBRUSH

MANIPULATION of the nation's land laws, particularly the Desert Land Act, provided a means for speculators to accumulate vast holdings across the West. Federal investigations revealed widespread abuse in the 1880s. Investigators in New Mexico indicated that such abuse was condoned by politicians and land officers alike. In the Pecos Valley, land accumulation began with cattlemen. Corporate irrigation companies expanded the practice, relying extensively on the time-honored practice of using dummy entrymen, friends, and relatives as the basis for speculative land development. Sidestepping competition for land by the older established community of Seven Rivers, Charles Eddy added thousands of acres to the holdings of his irrigation companies in the Pecos Valley.

Pecos Valley settlers and, more important, speculators, acquired land the same way many other western settlers did—through grants from the U.S. government. There were a number of ways to obtain land in the West; many took advantage of liberal forms of entry on government land. Francis Tracy, who would become a major player in the development of the Pecos Valley, claimed that most of the lands taken to develop the valley were acquired under the Desert Land Act. Tracy estimated that such lands encompassed at least 100,000 acres.[1]

The U.S. government first began to help the "landless" after the Panic of 1837 with the Preemption Act of 1841, which allowed those living on unsurveyed land to purchase up to 160 acres of land at $1.25 per acre prior to government survey without bidding on it. Lawmakers designed the law to help settlers save their land from speculators.[2] Preemptors had to be head of a household (any age), a widow, or a single man over the age of twenty-one, and be a citizen or have filed for citizenship before he or she could get land. An applicant could not own more than 320 acres anywhere and could not abandon his home in order to claim public lands.[3]

In 1862 Congress passed the Homestead Act to allow Americans to scatter throughout the West and help "develop the nation's resources and character." Settlers could get 160 acres of land free of charge, except for filing fees of ten cents per acre, but had to build a home and live on the land. Claims could be commuted and title finalized after eighteen months by paying $1.25 an acre. The government required that the land be surveyed with legal subdivisions before it could be homesteaded.[4]

During the 1860s, the Office of Surveyor General attempted surveys in New Mexico, but Indian hostilities and lack of military protection limited work along the Pecos River. Consequently, the office confined its studies to the vicinities of Fort Stanton, near Lincoln, the Hondo River, and the Mesilla Valley along the Rio Grande. As late as the 1860s, then, irrigation in the Pecos Valley centered on the Hondo River region, just west of Roswell. By the 1870s the Office of Surveyor General had plenty of unoccupied land to survey in New Mexico. Indian attacks had largely ceased, and the office expanded to new areas, including the Pecos Valley.[5]

The late 1870s brought not only surveys but other ways to gain government land. President Ulysses Grant, traveling in the West, recommended that lands in larger quantities be given to settlers to induce them to improve their property with water. On March 3, 1877, Congress passed the Desert Land Act—designed for California, Oregon, Nevada, Washington, Idaho, Montana, Utah, Wyoming, Arizona, Dakota, and New Mexico. Settlers on "desert" lands could get one section (640 acres) for twenty-five cents per acre down and one dollar per acre on final proof of compliance. The law stipulated one entry per person. The applicant had to be a citizen of the United States or had to have applied for citizenship, and the government gave no rights to the land until the applicant proved up the property by irrigating. While the law was intended to encourage irrigation by family farms, 95 percent of all claims were fraudulent and were made for the benefit of land speculators.[6] One renowned western water authority speculates that entrymen simply got a hogshead of water and a witness, went to the claim, and poured water on the land in the presence of the witness. The entryman then paid the witness, and the two went to the land office where the witness swore he saw irrigation take place.[7]

The Desert Land Act tended toward corporate reclamation projects because of the nature of the lands included and the enormous cost of irrigating them. In New Mexico and the seven other territories and three states of the West where the law was first applied, lands available under the Desert Land Act had to be devoid of timber and minerals and could not be productive agriculturally without irrigation. Ordinary settlers found irrigation a costly venture, but land speculators, pooling their resources and those of eastern and foreign investors, took advantage of such liberal land laws.[8]

All told, a crafty land seeker might accumulate 1,120 acres of land by using the Preemption, Homestead, Desert Land, Timber and Stone, and Timber Culture Acts in combination. In addition, cattlemen and others acquired large chunks of land using dummy entrymen—that is, relatives, friends, and compensated third parties all filing claims for each other. Often such applicants used fake names, and many claims with the same handwriting and different names acquired substantial properties in the West.[9] In the Pecos Valley, the vast majority of lands were taken under the Desert Land Act, some 100,000 acres in all.[10]

Under the Desert Land Act, an applicant only had to spend twenty-five cents an acre at the time of entry, then had three years to make proof of reclamation and complete the paperwork.[11] Using dummy entrymen, cattlemen and land speculators bypassed the law's intent. Cattlemen obtained thousands of acres of land in New Mexico along streams and rivers for virtual payment of twenty-five cents an acre. Evading laws was easy. Dummy entrymen slept on a claim two or three nights over a fourteen-month period. Some houses purportedly used to show improvement were birdhouses.[12] Land officers made no attempt to identify entrymen.

Although U.S. government policy required land acquired for homestead or preemption entries to be capable of growing crops, in practice this requirement was ignored in the Pecos Valley, which most saw as a grazing region. By the 1870s, members of the Office of Surveyor General openly propagandized for the cattle industry in New Mexico. Large cattle ranchers such as John Chisum and later Charles Eddy benefited from the process. Chisum and Eddy were two of the biggest landowners, but none of the large cattle companies owning large amounts of land depended solely on private lands. All used the public domain as though it were their own, and gradually extended their holdings. The only practical place to homestead in southeastern New Mexico was along the Pecos. That reality, combined with well-financed cattle companies, meant big business controlled much of the land.[13]

The domination of big cattle operators was in keeping with the spirit of the Gilded Age and concentration of wealth. During the 1870s and 1880s, there was general acceptance in the public mind of the notion of a man quickly enriching himself, even at the cost of his neighbors.[14] By the mid-1870s, when Henry M. Atkinson of Nebraska took over the Office of Surveyor General of New Mexico, surveyors had almost completed their work. By and large, Atkinson's tenure in office was dominated by cattlemen's demands for land and for his falsifying surveys to include nonarable land so cattlemen could claim it.[15] Large portions of land were unsuitable for anything except grazing. The 1880s saw cattle as king, at least until the middle part of the decade.[16]

Papers, periodicals, and livestock journals pointed out the large profits to be made in the cattle business. Some estimated that an investment of $5,000 could net the investor a profit of $40,000 or $50,000 within four years. Within that four years, a calf worth five dollars could be fed at very little cost on the public domain and then sold for forty or fifty dollars. Such rhetoric caused the cattle business to take off.[17] Contrary to wishful thinking at the time, eastern concepts of the yeoman farmer and the homestead entry on 160 acres of land simply did not work in the West and certainly not in New Mexico, where cattlemen obtained the choicest pieces of land along springs, creeks, and rivers. Homestead entries for New Mexico far outnumbered existing population figures or the number engaged in farming. Many of the excess homestead entries went to cattlemen, and in 1885 all of southeastern New Mexico was devoted exclusively to cattle ranching.[18]

Many landholders in the Pecos Valley possessed at least 10,000 acres. For years the General Land Office tried to figure out how landholders had obtained such large holdings in the valley and elsewhere. According to some, local land officers knew full well how the land was attained, but found insurmountable difficulties in proving it. Probably the largest difficulty was that testimony from a landholder would convict not only his neighbor, but himself as well. One man owned more than 10,000 acres, and except for about 400 acres the owner made no improvement on any of it. At regular intervals, small roofless stone houses of the specified preemption size stood entirely deserted. The rancher hired cowboys to live the required time in one of the houses, sign a statement to that effect before authorities, and then deed the land to the rancher. Cowboys knew that to show a more permanent interest in the land meant tough circumstances, including death, and they readily agreed to such contracts. Such was regular practice in the Pecos Valley.[19]

Fraud in the land laws, especially in the West after 1880, became widespread through abuse of the Preemption, Timber and Stone, Timber Culture, and Desert Land Acts. Besides the issue of fraudulent land entries, the accumulation of such large parcels of property led many in Washington to fear speculation across the West. In the 1880s, a reform movement gained new impetus with the discovery of fraud and huge land acquisitions involving foreign corporations, noblemen, and alien ownership. The uproar over large holdings centered on alien landlordism by British and Scottish interests. A Senate investigation discovered vast holdings in well-financed land and cattle companies.[20] Beginning with the census of 1880, reformers including William Sparks, Henry George, and George W. Julian began clamoring for changes in land laws, which appeared to favor speculation. Statistics showed another trend. In 1880, tenancy, mortgages, and the size of landholding were on the increase. Twenty-five percent of the nation's farmers were tenants. The Homestead, Preemption, and Desert Land Acts and assorted other laws allowed speculators to "emasculate" the intent of the democratic laws through fraudulent land accumulation, leading to monopolization and tenancy.[21] By 1895 the figure rose to 34 percent, with the largest increase coming in the West.[22]

Beginning in 1882 the federal government launched a number of investigations into land fraud in the western states. Although these so-called land frauds were a major issue in 1882, 1883, and 1884, they largely concerned the abuse of land in northern and central New Mexico owned through Spanish land grants. A number of individuals, including Max Frost, publisher of the *Santa Fe New Mexican* and head of the New Mexico Immigration Bureau, became embroiled in issues concerning land fraud and Spanish grants.[23] As no Spanish claims existed near the middle or lower Pecos, other instances of land fraud prevailed there. In 1883, special agent H. H. Eddy (no relation to Charles Eddy) investigated some two hundred homestead claims in New Mexico, and found that only 32 percent complied with homestead regulations. Eddy was paid extra for the danger surrounding his investigations in the Pecos Valley in 1884, where he found that most homestead claims were false. Only one in fourteen proved legitimate. Eddy showed Henry Atkinson's involvement in fraud and connected him with two murders of entrymen on preempted land.[24]

Most homestead entries lacked any semblance of settlement or improvement. In some cases the land had been abandoned for several years; in others the applicant was under legal age, or the required house had been built by a party other than the applicant. In some cases the resident on the land had never filed a claim and did not know that one had been filed.[25] From 1882 to 1886, seven special agents for the General Land Office investigated land fraud in the West, including New Mexico. One of the recurring problems, according to investigators at the time, was that native New Mexicans were unreliable witnesses who would swear to anything—and Mexican American juries seldom returned a guilty verdict even though the evidence was conclusive. The government viewed natives as people ignorant of the law who could be deceived into signing virtually anything. Many times investigators discovered native New Mexicans who had settled on a piece of land they thought they had filed on, only to find out that, deceived by unscrupulous land speculators, they had filed on some other piece of land that was worthless. Their home in the meantime was filed on by a person representing the party who had given the New Mexicans the false descriptions in the first place. Consequently, many native settlers did not get the valuable land they wanted but rather worthless acreage. If locals complained about the situation, they were told by officials that they had committed perjury by entering on land that they had never lived on. They were warned that the consequences of speaking out would be arrest and prosecution.[26] Investigations into land fraud across the West revealed widespread deception and prompted others to accuse New Mexico political groups of collusion.

New Mexico was no exception when it came to land fraud. All across the West there were numerous examples of cattlemen securing the choicest lands. Reports from land office agents throughout the 1880s indicated land fraud in Colorado, the Dakotas, Montana, Nebraska, and Kansas as well. Most agreed that the choicest lands in these areas had been controlled and then attained though questionable legal practices by powerful groups of cattlemen in the regions to keep out small-time settlers.[27] In many cases, on the surface, the legal papers were so technically correct that local land offices could hardly determine the fraud. In other cases, the land offices themselves were co-conspirators.

In late 1884, Secretary of the Interior Henry Teller realized that a large percentage of preemption claims in the West were filed with no intention of perfecting entries. Teller noted that the proportion of entries to perfected titles was less than one-half, and he called for stopping such abuse.[28] He sent an army of special agents from the land office into western states and territories. Agents unearthed a multitude of cases where land interest had been made through false affidavits. G. W. Pritchard, U.S. district attorney for New Mexico, investigated such land theft and secured information by 1884 that allowed him to prosecute a long docket of cases. Pritchard stated in 1884 that "the practice [of stealing land] was carried on to an alarming extent in New Mexico and in many instances the guilty parties [had] been apprehended and [Pritchard meant] to prosecute [them all]."[29] U.S. Land Commissioner William Andrew Jackson Sparks, appointed in 1885 by President Grover Cleveland, accused General Land Office workers of participating in fraud or looking the other way. He suspended final entry on land in the West except for those who made the final payment with cash or script.[30]

This had the effect of stimulating speculators and monopolists to feverish activity, grabbing lands before the public domain closed. Land sales and entries under the various land grant laws reached a high point during Cleveland's administration. Others, including Surveyor General of New Mexico George W. Julian, another Cleveland appointee, lamented that various political groups within the territory aided and abetted such actions.[31] Julian noted that in 1884 when the Democratic territorial convention met, it adopted a resolution decrying Washington's claims of land fraud in New Mexico and denied that such frauds existed to any considerable extent. Julian contended that members of the convention knew full well—from the records of the General Land Office, reports from special agents, actions of courts and grand juries, and situations arising in the Office of Surveyor General—that land fraud in New Mexico was rampant. Julian considered taking a man's home a crime second only to murder. "To rob a nation of its public domain and steal the opportunity of landless men to gain homes was not only a crime against society, but a mockery of the poor."[32] Julian repeatedly lambasted the land rings and political groups that perpetuated such fraud. In 1887 he denounced the various political powers in the Territory, claiming that they should have "sounded the true keynote and battle cry of reform in New Mexico, while rebuking the ravenous conclave of land-grabbers, whose hidden hand made it the foot-ball of their purposes, and led astray the honest and confiding rank and file . . . who would gladly have responded to a brave and honest leadership."[33] Julian added: "[T]he grinding oligarchy of land sharks, whose operations have so long been the blight and paralysis of the Territory, should be completely routed and overthrown."[34]

By 1885 the various government investigations and consequent shake-ups in the U.S. land offices across the West were being felt in the Pecos Valley. In October 1885, the registrar in the U.S. land office in Las Cruces, John R. McFie, was ousted in favor of E. G. Shields, who would become a major promoter of irrigation and later joined Arthur Mermod, former neighbor of Charles and John Eddy in Colorado, as a real estate broker in the town of Eddy. But in 1885 observers hoped Shields would enforce every rule of landholding to the letter. Residents hoped that Shields would "not set himself up to know more than his superiors," alluding to perceived improprieties by McFie in helping cattlemen amass large holdings. The *White Oaks Golden Era* newspaper, considering McFie's replacement by Shields, commented that the president of the United States "probably went to church last Sabbath and heard an eloquent discourse on office holders ignoring the law, for the good deed [the ousting of McFie] was done early Monday morning."[35] Furthermore, the paper commented that many had been "certain for the past six weeks that it would only be a matter of a little time until McFie was ousted and the land ring [operating in southeastern New Mexico] broken up."[36]

Reflecting the views of small ranching interests, the *Golden Era* said: "The appointment of Mr. Shields as registrar of the U.S. Land Office is received with satisfaction by the small rancheros [ranchers] in this locality, and they think they will now receive that justice which they say has been so long denied. . . . If Mr. Shields will demonstrate that he intends to be the servant of the people, and not the cringing, obsequious tool of a hybrid land grabbing gang, all good citizens will rally to his support."[37]

The territorial governor of New Mexico, Edmund Ribson Ross, in 1885 commented: "There's a general belief that considerable areas of good agricultural land have been illegitimately entered and included in great cattle ranches."[38] The governor went on to say that he believed that large quantities of public lands had been added using the preemption laws through "the boldest perjury, forgery, and false pretense, and that in at least some instances this has been done, if not with the connivance, at least through the inadvertence and carelessness of officials." The governor also noted that in many cases land had been absorbed into great cattle ranches merely for the purpose of gaining control of water supplies, keeping out settlers and their small herds, and some ranchers accumulated lands for purely speculative purposes.[39]

Cattlemen, beginning in the mid-1880s, by incorporating ditch companies and appropriating available water, could keep settlers from using it. Consequently, they kept competitors off even the public land. This had the effect of making land that others wished to enter valueless unless they were supplied with water from irrigation companies created by cattlemen.[40] In fact, cattlemen from the older, established community of Seven Rivers, twelve miles north of Charles Eddy's new community, argued that Eddy and other ranchers had obtained their land improperly. In 1885, M. J. Denman charged Charles Eddy, who ran cattle near Seven Rivers, with having obtained large acreage through fraud.[41] Denman came from Texas and lived in Lincoln County near Seven Rivers as a land agent. Supporters of Eddy, many of whom were cattlemen themselves, quickly denounced Denman and supported their friend. One supporter, Marshall Ash Upson, a colorful character in his own right, lived in Roswell and had lived in the territory since 1864, and in Lincoln County since 1875.[42] Upson was a newspaperman who had established and edited the *Albuquerque Press* in 1867–68 and the *Las Vegas Mail* (later known as the *Las Vegas Gazette*) in 1870. Upson also served as acting adjutant general of the territory of New Mexico under Governor Robert B. Mitchell. He served as postmaster in Roswell and as justice of the peace. Upson made a living in 1885 as a land surveyor and notary public and had an office at Seven Rivers. He had known Charles Eddy since 1880 when Eddy purchased a herd of cattle and improved a ranch near Seven Rivers.[43]

Denman's charges revolved around Eddy procuring land from several Mexican Americans near Seven Rivers. Denman and government investigating agencies accused Eddy of acquiring lands from settlers through coercion. Many Mexican Americans had taken up residence along streams and close to springs in Lincoln County. They were unfamiliar with U.S. land laws. Upson, in his affidavit, testified that Charles Eddy acquired lands from Mexican Americans living close to the Pecos River, but explained that Eddy bought the land from these settlers who preempted the property.[44]

Eddy's supporters tried to turn the tables on Denman and the government investigators by painting Eddy as the friend of Mexican Americans. Upson argued that strong class hostility existed between the "less intelligent" Americans, many of whom had apparently drifted in from Texas, and Mexican Americans who lived along the Pecos for one hundred miles. Upson maintained that for several years Mexican Americans had not been allowed to live in these areas peacefully, that many of them had been killed or wounded without provocation, and some had been driven from their homes by intimidation and threats.[45]

Upson and his supporters contended that rather than stealing land from such citizens, Charles Eddy had taken them under his wing and tried to protect them in their rights concerning land laws. Upson claimed that in so doing, Eddy had incurred the wrath and hostility of Denman and others like him who were intent on obtaining title to the Mexicans' land for themselves. Eddy's friends said he had never entered into a land purchase without scrupulous and careful attention to avoid conflicting claims. Upson claimed that Denman and his followers were "a class of people in our midst that were led by men of unprincipled character who persistently incited their followers to deeds of vice and crime; that newcomers, ignorant of our laws were induced by these leaders to enter upon and occupy lands, legally held and occupied by bona-fide actual settlers; that many of these immigrants are outlaws from Texas and fugitives from justice, who eagerly attach themselves to these leaders for purposes of plunder, blackmail, and kindred crimes."[46] He gave numerous examples whereby he claimed Denman and his band threatened and intimidated Mexican Americans along the Pecos River, and he contended that Denman often had Mexican Americans arrested simply for the sake of obtaining their land. Upson further claimed that after Denman had them arrested on frivolous charges, on their release his thugs drove them from their homes, forcing them to leave behind all their worldly possessions to be plundered or destroyed.[47]

Denman responded by attacking the character and conduct of Quay Taylor, special agent for the General Land Office. Denman, Robert M. Gilbert, and others hoped to be named district land inspectors for the area of southeastern New Mexico, with headquarters at Las Cruces. Denman organized a petition to the General Land Office in Washington, D.C., charging the officers at Las Cruces with fraud and blasting some of the most powerful and substantial citizens of the region.

Upson minced no words in condemning Denman as a "communist of the most dangerous type, and a common barrater in the most offensive sense of the term; . . . nine-tenths of the litigation and quarrels in this precinct are acknowledged to be incited by him, and that he bears the reputation, where he is best known, of being a liar, a thief, and a perjurer."[48] Denman represented smaller ranchers and settlers near Seven Rivers, an area notorious for its violence, especially connected with the Lincoln County War of the 1870s. Many Seven Rivers settlers were former Texans, including Denman. Texas history suggests an inherent hatred toward those of Mexican heritage. Upson portrayed Denman and his followers as Texas outlaws and fugitives from justice who were jealous of Eddy's relationship with the local Mexican American populations.[49]

In a similar reproach of Denman, Charles H. Slaughter, also a large cattleman from Seven Rivers, praised Eddy's reputation to the fullest and denounced Denman as a blackmailer who circulated malicious slander relative to Eddy's land titles. Like Upson, Slaughter did not mince words in condemning Denman, whom he had known for twelve years and whom he considered to be a thief, a swindler, and blackmailer. Slaughter insisted that Eddy had properly obtained title to his lands. Denman, according to Slaughter, was simply jealous of Eddy's relationship and the kindness that he showed all Mexican Americans in the region. Slaughter referred to Denman and his ilk as "border ruffians" and claimed that Mexican American settlers had always feared such Texans. Slaughter also related several incidents where such settlers had been shot down in cold blood or driven out of the country.[50]

Following the *Golden Era*'s publication of Upson and Slaughter's affidavits, thirty citizens, most cattle raisers themselves, signed a deposition describing Denman as a "depraved and reckless man, of evil reputation, and charged with many crimes; that no faith or credence can be placed in what he says, nor the statements of his followers and tools who do his bidding."[51]

That Charles Eddy had obtained numerous acres from those of Mexican heritage near Seven Rivers and along the Pecos River was never denied by his supporters. What was in question was the method by which Eddy and others obtained the land. Eddy County deed records from 1887 and 1889 attest to Anglos obtaining land from persons with Spanish surnames. Many were illiterate and could not sign their own names, and transferred ownership of the land for less than one dollar per acre.[52]

Certainly one of the easiest ways to obtain large amounts of land was through the Desert Land Act. The General Land Office had reservations about the application of the Desert Land Act in New Mexico Territory. Eight months after it passed in March 1877, the General Land Office suspended all Desert Land entries in New Mexico and ordered hearings to determine their legality. The order for suspension was revoked within the month at the insistence of Secretary of Interior Carl Schurz, and General Land Office officials instructed land officers in New Mexico to report immediately any cases of suspected fraud in the territory.[53] Officials, including local federal land commissioner Shields, publicly regarded Desert Land entries as a source of "a great deal of crookedness," but that did not stop Shields from working closely with Charles Eddy in securing lands for the developer's irrigation scheme.[54]

During the fifteen years that the Desert Land Act was in effect, there was a parallel growth in the cattle industry. Across the West, cattle companies utilized cheap grazing lands made available through the Desert Land Act. In fact an almost nationwide decrease in Desert Land entries by 1887 correlates with a falling demand for grazing land as the cattle industry reached its maximum development. By the mid-1880s the ranges were overstocked, and ranching profits declined. After the "Big Die" in 1886 when so many cattle died across the Great Plains and the western part of the United States, most cattle operations across the country were reduced in size.[55] With the decline of the cattle industry by the mid to late 1880s, investors turned from ranching to land development schemes through irrigation.[56]

From 1888 to 1891, eighty-eight irrigation companies were incorporated in New Mexico. At least 40 percent of the land brought under irrigation during the next decade was irrigated by these companies. Outright entries by companies under the Desert Land Act were not allowed, as federal laws aimed at establishing individual setters and their families on the land. But companies acquired thousands of acres by circumventing the law. Companies such as Pecos Irrigation and Investment acquired much of their land from individuals who had filed Desert Land entries and then sold the land for next to nothing to company officials.[57]

Company officials often registered numerous consecutive land purchases and deed changes on the same day. This was the case in Lincoln and later Eddy County along the southern stretches of New Mexico's Pecos

River. Pecos Irrigation and Investment in 1888 and 1889, and Pecos Irrigation and Improvement Company thereafter, added significant acreage to their holdings. Following the required three-year period, applicants proved up on entries and then deeded their properties to the company in return for a small fee or, in the case of Pecos Irrigation and Improvement, a free trip to the desert Southwest.[58]

The Pecos Irrigation and Improvement Company's land schemes were exposed in part by the report of Ralph Tarr, part of John Wesley Powell's survey team in 1888 and 1889. The Tarr Report was a comprehensive study of the topographical and geological components of the valley.[59] Tarr noted that landholders in many instances obtained their property through bribery of government officials, notary publics, and witnesses. At the time of his report, many of the leading citizens of Lincoln County were under indictment by a grand jury for fraudulent land entry. He doubted that any of them would be convicted. Tarr saved some of his most condemning rhetoric for the Pecos Irrigation and Investment Company, soon to be reorganized as Pecos Irrigation and Improvement Company. He noted that before the company declared the line of its ditch, it gave information concerning that ditch to friends and relatives. Observing Charles Greene's success in attracting Chicago investors to the valley in September 1888, Tarr counted a party of thirty people from Chicago who filed on nearly 25,000 acres of land along the company's ditch. Not one of the entries was a homestead or preemption claim; all were Desert Land entries. It was clear that this group of rich urbanites were not planning to settle on the land but were simply purchasing it for speculative purposes. Many in the party were related to company officials. Members of the party who entered claims under the Desert Land Act later deeded their property over to Pecos Irrigation and Investment (later Irrigation and Improvement) officers, thus adding to the land and assets of that company.[60]

The records in Eddy County's deed books confirm Tarr's observations. Frederic Stevens and Francis Tracy filed Desert Land entries numbers 562 and 584 with the land office at Las Cruces on November 25, 1889.[61] Thomas Fennessey, John Arthur Eddy, Charles B. Eddy, G. W. Williams, Percy Hagerman, Anna Hagerman, and C. H. McLenathen, along with many others, filed Desert Land entries only to deed them to the company. One example is Thomas Fennessey.[62] Fennessey filed a Desert Land claim on section 27 of township 22, which he proved up in 1889. He immediately sold the property to John A. Eddy. E. G. Shields, the federal land officer and real estate dealer, also filed a Desert Land entry. Shields proved it up in 1889 and sold to Charles Eddy within days. Almost everything the Eddys bought went to the company.[63]

The same thing happened near Roswell. Within a week of the announcement that Pecos Irrigation would furnish water to the area, applicants filed Desert Land entries on 10,000 acres. Anticipating fraud and abuse in the valley, Tarr suggested that the U.S. Geological Survey (USGS) declare a strip of land twenty miles wide and extending from five miles north of Roswell south to the Texas line off limits and removed from entry. Tarr accurately predicted that with a team of government surveyors in the area, unless the government acted quickly, speculators would immediately follow and file Desert Land entries on the basis of the government's presence.[64]

Ironically, in 1889, Tarr found that there was a popular sentiment in the Pecos Valley for repeal of the Desert Land and Preemption Acts. He soon discovered that those who objected most strongly to the laws were people who had no money to take advantage of them. A second group, those who actually had the wherewithal to buy land, enthusiastically supported ending the laws because they had already obtained all the land they needed and did not want competition from other large landholders. On the contrary, Tarr thought the second group would rather see "people of limited means who [sic] they could fleece."[65] By the time James Hagerman joined Charles Eddy and Pecos Irrigation and Investment Company, Eddy and his friends and relatives had entered most of the land to be irrigated in the Pecos Valley under the Desert Land Act. Hagerman later claimed that all this activity transpired before he had anything to do with the company. Although the company had accumulated vast acreage by 1890, Hagerman claimed that the company's sole purpose at first was to sell "water rights" and collect "water rentals."[66] Technically speaking this was true.

Charles W. Greene, the promoter for Pecos Irrigation and Investment Company, responding in 1890 to questions about water rates and rents in the Pecos Valley, said that the company did not own nor had it attempted to control in any way the promotion of government lands, but he said this after the bulk of federal lands had passed from dummy entrymen, friends, and relatives to company officials.[67] Greene claimed that lands planted in fruits within the irrigated area produced anywhere from $100 to $500 per acre annually. Greene did promote the desirability of government lands in the valley and counted on settlers quickly filling the region.[68]

According to Greene, within thirteen months after the irrigation project in the valley began in 1887, applicants had entered on 200,000 acres of land not worth taking before the prospect of irrigation. Greene said, "From the fact that nearly 200,000 acres must be finally proved up under the requirements of the Desert Land Law during the year 1891, the success of the enterprise is assured beyond question."[69] In response to queries about who owned lands near the irrigation ditch—the government or the company—Pecos Irrigation officials said they owned no land in the valley except for their ditch right-of-way.[70]

Pecos Irrigation and Investment Company utilized the flexibility in western land laws to accumulate not only ditch rights-of-way, but land as well. Technically, the company itself owned only the access and right to deliver water to the land. But by 1891 much of the choicest government land had been snatched up and conveyed to individual company officials. Charles Greene and company were not the only ones eager to promote the Pecos Valley. E. G. Shields, registrar for the General Land Office in Las Cruces, quoted prices of fifty dollars per acre for the best lands with water rights, ditches, fences, and other improvements. Shields was not bashful about promoting his own real estate firm, Shields and Mermod, which was selling land from twenty-five dollars to fifty dollars an acre, ready to plow. Shields, perhaps conscious of a conflict of interest between his own real estate and the government's, quickly pointed out that in 1891 a considerable amount of government land still existed for settlement under the various laws of the United States.[71]

Deed record books for Eddy County during the 1880s and early 1890s are riddled with examples of irrigation company officials obtaining individuals' land and water-right deeds following proof of reclamation in exchange for a few dollars.[72] In many cases, a series of deed exchanges took place between individuals who filed Desert Land entries sight unseen from thousands of miles away and early company investors like Joseph Stevens, Henry C. Lowe, and Robert Tansill. After a series of these transactions, the lands eventually became the property of the irrigation company or subsidiary valley companies.[73]

In following township 22 east, range 27 south, one can trace lands that passed from the federal government to various individuals and finally to a director of the company. For example, Arthur Mermod was granted through the Desert Land Act all of section 26. He sold twenty acres in the northern half of the southwest quarter to J. M. Hibner, and the north one-half of the southeast quarter and south one-half of the southwest quarter to Pecos Irrigation and Improvement. His wife was also granted a section; she deeded part of the land to Pecos Irrigation and Improvement. Schoolteacher Edith Ohls deeded her grant to William P. Bonbright, James Hagerman's Colorado Springs mining partner, in 1891. Joseph Stevens's name appears often in the deed books. He received a Desert Land grant and then sold parts of it to Charles Eddy, William Bonbright, John Eddy, and others. He also bought land from H. L. Potter, John Eddy, Charles Eddy, and others.[74]

Transactions dating from the early to mid-1880s involved a number of land transfers to Charles Eddy, some to Amos Bissell, and others to John Eddy, Charles Eddy's older brother. Several deed transactions for 160 or 320 acres involved those with Mexican surnames. Many were illiterate and could not sign their own names, and all of them transferred lands to Anglos for less than one dollar per acre.[75] Many of the Desert Land entries by Mexican Americans involved transfer of property to Pecos Irrigation and Investment Company between 1887 and 1889 for at most a few hundred dollars.[76] Not only did individuals transfer their deeds of property to various company officials, but they also transferred their water rights. Mortgage records for Eddy County during 1891 and 1892 read like a who's who of early company officials in the Pecos Valley. Many early investors tied to the company, such as Robert Tansill; M. A. "Ash" Upson, Charles Eddy's supporter against the Seven Rivers cattlemen; Charles McLenathen; B. A. Nymeyer; and J. O. Cameron, held mortgages for property originally obtained through the Desert Land Act and sold to subsequent investors.[77] While the Eddy County clerk stayed busy recording the many transactions taking place, the federal land offices stayed busy as well.

The year after Congress passed the Desert Land Act, John Wesley Powell offered his findings and viewpoint concerning the use of arid lands in the West. To prevent the sort of rampant land law abuse occurring in southeastern New Mexico, Powell recommended an alternative to speculative control in the West. Powell realized that the 1,120 acres accumulated through the Preemption, Desert Land, and Timber Culture Acts was not enough land to graze cattle, and 160 acres through the Homestead Act was too much for one man to cultivate as an irrigated farm. Powell suggested 40 acres for farmers on irrigated land.[78] He believed 2,560 acres might be adequate for unirrigated parcels, with adequate water for emergencies.[79] He advocated planned reclamation and settlement, creating enemies in both the War Department and General Land Office. He alienated land, cattle, mining, and timber speculators, many of whom were getting rich through land fraud.

Although Powell was vilified by many, ten years after he released his report he was vindicated by drought in 1887. Everything Powell warned against—recurrent aridity of the western lands and indiscriminate settlement—came true. Senators William M. Stewart of Nevada and Henry M. Teller of Colorado led a campaign to locate and designate reservoir and irrigation sites for future development.[80] A short-lived alliance between Powell and Stewart resulted in a March 20, 1888, joint resolution of Congress, calling for an irrigation survey to locate reservoir sites and lands susceptible to irrigation. An amendment to the joint resolution in October

1888 temporarily withdrew from settlement all lands made susceptible to irrigation by reservoirs and canals that the survey might designate. Lawmakers wanted the amendment to prevent speculative companies from rushing in to buy up potentially profitable lands.[81] President Cleveland signed the bill on October 2, 1888, with $100,000 in appropriations. Lawmakers added $250,000 in a sundry civil appropriations act during the second session.[82] Powell's crews simultaneously began work in Montana, Nevada, Colorado, and New Mexico in 1888, dividing the workforce into two divisions: one for topography and one for engineering and cost estimates. For hydrographic work, Powell had his protégé, Frederick Newell, set up an instruction camp on the Rio Grande in New Mexico.[83] As part of the soon-to-be-aborted survey, Powell's foot soldiers trudged across the United States conducting surveys in various western valleys. Carrying out instructions issued by Captain C. E. Dutton, USGS, Ralph S. Tarr, as part of Powell's team, conducted a preliminary reconnaissance of the Pecos Valley in late winter 1888 and early spring 1889.[84]

The General Land Office was seemingly oblivious to the March 20, 1888, joint resolution allowing John Wesley Powell to start work. The Land Office gave no official notice of the law to local offices until August 5, 1889, a year after Powell's surveys began. In the meantime, speculators followed geological survey teams across the West, filing on lands wherever they worked. The secretary of the interior then ordered lands filed on since October 1888 removed from filing lists for all areas withdrawn by Powell's surveyors. But in the fall of 1889, land officers did not know which lands had been withdrawn—and therefore where legal filings could take place.[85] By April 1890 the land commissioner, concerned with increasing land patents that might fall into government land withdrawal areas, ordered local land officers to accept no more entries in the arid West. This action brought an immediate flood of protest by western senators.[86]

By June of 1890 Powell's lieutenants had surveyed 200 reservoir sites, many approved by the Interior Department,[87] but pressure was building against the survey. Powell's engineers in New Mexico selected thirty-nine reservoir sites, totaling 40,170 acres of land.[88] Powell's withdrawal of the land in New Mexico threatened to limit land accumulation by Pecos Irrigation and Improvement Company. If the act withdrawing the lands were not repealed, Pecos Irrigation and Investment's projects, on which $700,000 had already been spent, would have to be abandoned.[89] Charles Eddy's concern was reflected in a letter that he sent to the governor of New Mexico Territory, L. Bradford Prince. On July 7, 1890, Eddy said,

I am of course aware of the ruling of the Secretary of the Interior and the Attorney General regarding the law of October 2, 1888, which withdraws from settlement all of the arid lands of the West, and the serious results which will follow if same is not repealed this session of Congress. Cannot you head the movement on the part of the people of New Mexico in urging action by Congress and will [you] suggest to us our best method of assistance. No matter what they may include for the future, all claims made since October 2nd of any nature whatever . . . upon reservoir sites actually surveyed and reserved by the government *should be reinstated.* Nothing else would be just. I have advice from reliable quarters in Washington that very active measures are needed to be taken by the entire west country. And if it is possible to defeat [the measure] Powell is prepared to put a larger force in the field this summer to cover all the west and most available government lands in the west. . . . I appeal to you on this subject believing you can do more for this relief than anyone else. . . . I would be greatly pleased to learn your views.[90]

In July of 1890 James Hagerman, who began investing in the Pecos Valley during the previous year, traveled to Washington to address the interpretation of the attorney general on the withdrawal of irrigable lands from entry. Hagerman recalled that in the fall of 1890, "[t]he Attorney General had interpreted a law of Congress in a way which would have ruined our Pecos Enterprise unless we had secured legislative relief speedily—in this we were completely successful and we were left in better shape than before the row began."[91]

An amendment to the Sundry Civil Appropriations Bill, signed on August 30, 1890, effectively killed the Powell survey, although operations continued using meager USGS funds. The government restored all entries made in good faith after the survey began.[92] The defeat of Powell's removal program was a boost to western land and irrigation companies.

Powell's aborted work, if not stopping speculation in the West, certainly gave insight into irrigation potential and progress in western valleys. Although Powell's vision proved prophetic, speculators and western promoters won the day.[93] However, Powell's efforts and those of reformers like George Julian and William Sparks led to the General Revision Act of 1891. The act repealed the Timber Culture and Preemption Acts, extended the period after original homestead entry several months before it could be commuted to a cash entry, and modified requirements for the completion of Desert Land entries. Now applicants for Desert Land entry had to submit plans for irrigation and spend at least one dollar per acre in each of the three years on irrigation. The applicant had to live in the state, and the acreage one could acquire was now limited to 320 acres. The government released no title to the land unless one-eighth of it was cultivated in three years. These safeguards made fraud more difficult.[94] Such reforms may have come too late, because speculators and railroads had already grabbed the choicest lands.

Using the Desert Land Act and other laws to accumulate land in the Pecos Valley was indicative of the general land and irrigation boom across the country in the 1880s. Together the Desert Land Act and the Homestead Act opened 4 million acres in New Mexico. The early ease of obtaining lands under desert entries and the subsequent transfer of such lands led to accumulations of vast acreage in the Pecos Valley. Subsequent settlers in southeastern New Mexico could also buy irrigable land for between $25 and $100 an acre, at 6 percent interest from land developers.[95]

Early settlers under the Desert Land Act paid only twenty-five cents per acre on as much as 640 acres of land, the balance of one dollar per acre to be paid on proof of reclamation within three years.[96] In New Mexico and elsewhere investors filed on unseen lands since the law did not require residency or cultivation. Purchasers simply applied water to some part of each forty-acre tract within a three-year period, as witnessed by the applicant and two others. On final proof of reclamation three years later, New Yorkers and Chicagoans deeded the lands over to Stevens, the Eddys, Tansill, friends, relatives, and parties connected to the valley's irrigation company(s). These lands and others secured in a like manner formed the basis of Eddy, Garrett, and Stevens's design for accumulating vast acreage in the Pecos Valley.[97] Many irrigation companies formed between 1888 and 1890—thirty-three in New Mexico[98]—and the lure of big profits caused rapid development. However, the supply of irrigable land was far greater than the number of settlers who wanted to live on it.[99]

3

JACKRABBITS, CANALS, AND COTTONWOODS

CHARLES EDDY's vision for irrigation and land development in the Pecos Valley soon outstripped Pat Garrett's early notions. Working from the foundation of lands accumulated under the Desert Land Act and with capital from New York, Chicago, and Colorado, Eddy sought additional money, water, land, and settlers for a model community in the desert. Eddy built his town twelve miles south of the established village of Seven Rivers, the community of Texas cattlemen that now fought Eddy not only over control of grazing lands but also for political control of the region.

When he visited Colorado Springs in 1887 to raise money for irrigation, Charles Eddy watched the activity there with much interest. He began to formulate in his mind the idea of replicating General Palmer's city in southeastern New Mexico. As soon as crews completed work on Avalon Dam to the north in 1888, work began on the new townsite in the desert. B. A. Nymeyer, a transplanted Dutchman who surveyed and practiced law with J. O. Cameron, laid out the townsite,[1] and Charles Eddy immediately began platting streets, digging irrigation ditches, and erecting buildings. The similarities to Colorado Springs in town design, if not scenery, are striking. Eddy and his new partner, James Hagerman, even hired Palmer's engineer, E. S. Nettleton, who acted as an advisor in constructing the Pecos Valley's irrigation systems.[2] Eddy, New Mexico, was laid out with broad streets and large city blocks. Deeds to city property came with the stipulation that alcohol was not to be sold or consumed under any circumstances. And like Colorado Springs, Eddy and other cities in southeastern New Mexico became havens for health seekers—especially consumptives. Citizens christened the town of Eddy in the fall of 1888—with a bottle of champagne in a dry town.[3]

Eddy established the Pecos Valley Town Company,[4] organized April 6,

1889, "to acquire, hold, and subdivide tracts of land in the territory of New Mexico and to turn such tracts of land into suitable town lots and to sell them to settlers."[5] In addition to improving lots and selling real estate, the Town Company was responsible for public improvements such as gasworks, electric light works, waterworks, and streetcars, all of which Eddy proposed for the new town in the early 1890s.[6]

In 1888, Eddy platted the original townsite to include eighteen blocks. To accommodate the growing population of almost 600, in late fall of 1889 city fathers added a nine-block area on the northern end of town. Eddy sold lots in the original townsite and those in the new addition at prices ranging from $58 to $400 each. Company officials now planned a twenty-seven-block addition. Many of the early additions to the town bore the names of early investors, such as the Stevens addition, the Greene addition, and an addition named after Charles McLenathen and his partner Franklin Campbell, who operated the first real estate firm in Eddy.[7]

Eddy drew up business lots 25 feet wide and 150 feet deep from the street. Residential lots stretched 50 feet across and 150 feet deep. Streets in the new town were 80 feet wide, with double rows of shade trees on each side, much like those in Colorado Springs. Charles Eddy even planted cottonwoods throughout his town. Early settlers working for the irrigation company traveled eighty-nine miles north to Salt Spring to secure thousands of young trees, which they planted during the winter of 1889 and 1890. Townspeople watered them from irrigation ditches built the year before.[8]

The Pecos Irrigation and Investment Company let contracts for the construction of nine new buildings in the desert community of Eddy.[9] By October 1889 the irrigation company had constructed a $2,000 schoolhouse, where the first teacher, Coloradan Edith Ohls, taught thirty-five students. The irrigation company also planned to build a $20,000 hotel in 1890.[10] One of the first businesses in the new town was the company-owned lumberyard, which W. A. Finlay later bought and operated.[11] Other buildings included the Pecos Irrigation and Investment Company's headquarters, the Pecos Valley Town Company's office, offices for Pecos Valley Land and Ditch, and the company-owned *Argus* newspaper office.[12] W. A. Hawkins and Albert B. Fall, who would later achieve national notoriety for his involvement in the Teapot Dome Scandal, established a law office in the new town during its first year.[13]

Although somewhat extravagant and biased in their claims, irrigation company land promoters Charles Greene and E. G. Shields wrote numerous "progress reports" to the *Eddy Argus* in the early 1890s. Greene, who visited the valley in December of 1890, had not been there since February

Eddy (Carlsbad), New Mexico, 1890. Notice the cottonwood trees and canals to water them. *(Courtesy Sacramento Mountains Historical Society, Cloudcroft, New Mexico.)*

1889. He noted that in 1889 crews had made only preliminary surveys of company ditch lines. The town of Eddy had been located and the land secured, though only one building, the headquarters for the irrigation company, stood on the townsite. Greene noted that transportation had improved considerably in one year. In 1889 it took him two days each way from Pecos, Texas, to Eddy. In late 1890 Greene made the trip by leaving Pecos at 9:00 A.M. and arriving in Eddy at 10:00 P.M. the same day. Greene found a town that had almost doubled in size and stayed in a hotel that he said "would be a credit to any western city, [in] a town beautifully laid out."[14] E. G. Shields commented that as late as 1890, the valley was more or less a complete desert without inhabitants, but by 1891 the population exceeded one thousand, and according to him there were hundreds of houses throughout the valley.[15]

Observers noted in the spring and summer of 1892 that work crews had graded twelve miles of road and had dug twenty-four miles of ditch to water town lots. Rows of small cottonwoods stood along each side of the street,[16] and during the winter of 1892 and early 1893 crews continued to plant trees, some four boxcar loads in all. Not only in Eddy, but also in La Huerta, a private suburb across the Pecos River, the planting of cottonwoods lent a kind of civilization to this desert community, perhaps impressing eastern investors used to seeing trees over four feet tall.[17]

Boosters with the company-owned newspaper, the *Eddy Argus*, noted with exuberance the rapid progress being made in the town. In 1892, the company set aside entire blocks for a courthouse, public school buildings,

Eddy (Carlsbad), New Mexico, 1892, with Hotel Hagerman at the end of the street.
(Courtesy Sacramento Mountains Historical Society, Cloudcroft, New Mexico.)

playgrounds, and a public park,[18] and the entire riverfront for the railroad depot and sidetracks. Eddy boasted a two-story hotel, the "Hagerman," of sixty rooms with wide verandas, the structure having cost some $65,000.[19] Visitors noted a well-stocked livery stable, owned by Pat Garrett and James Brent, both former sheriffs of Lincoln County,[20] two sale and feed corrals, a blacksmith and wagon shop, three construction companies, three real estate and insurance companies, a brickyard with live kiln, a lumberyard, four grocery stores, two drugstores, a doctor, a dentist, two barbershops, four Chinese laundries, a butcher shop, a billiard hall, restaurants, and a few scattered cottages throughout the town. Many businesses such as laundries and restaurants catered to the large forces of transient workers and company men in the town. Early residents also constructed churches for congregations of various faiths, including Episcopalians, Baptists, Methodists, and, somewhat later, Roman Catholics.[21]

Despite the *Argus*'s attempt to portray the town's sophistication, the community still displayed some frontier characteristics. One of the problems townspeople experienced was that loose cattle had a penchant for running through the streets and knocking over newly planted cottonwood trees. Some residents wanted to invoke an 1884 law against loose stock. At one point, H. S. Church, Eddy's irrigation and town superintendent, decided to protect the cottonwoods by planting ocotillo cacti beside the trees and by hiring a mounted night watchman to chase the cattle out of town. Church was concerned not only with the trees, but with the welfare of horses, who were often gored by bulls running free through the streets. According to the local newspaper, "One bull was driven out and killed, and there was another that comes in town and feels a bit lonesome [and] if he finds a horse or mule tied, he gores them for recreation." Like many frontier settlements, Eddy also had a problem with cattle rustlers. The *Argus* noted that a party of men left early in 1890 to round up the thieves.[22]

With the establishment of several businesses, waterworks and sewage systems, a national bank,[23] and several other businesses, settlers and company officials alike realized the importance of a railroad to the new town. James Hagerman, who had joined the company in 1889, especially focused on transportation as the key to the community's success. The town depended on wagons and freighters to deliver produce and other goods to and from Las Vegas to the northwest. Hagerman believed a railroad would encourage settlement, thereby increasing profits for the irrigation company. He raised the needed funds and construction began. By the middle of July 1890, construction crews had laid five miles of Pecos Valley Railway track from Pecos, Texas, north to Eddy.[24] Charles Eddy kept close tabs both on crews building north from Pecos and on other enterprises in the Pecos Valley. In 1891, telegraph service allowed Eddy to stay in close contact with the company's general office, a two-story brick building in downtown Eddy; with the Hotel Hagerman, which had a thriving business housing prospective investors from the east; with Hagerman's farm south of Eddy; and with the residence of his attorney, W. A. Hawkins.[25] By the spring of 1893 the Pecos Valley had a telegraph line extending from Eddy to Roswell.[26]

As Eddy grew from village to town, politics played a key role. In 1889, the territorial government of New Mexico divided Lincoln County into three parts: Eddy County, which encompassed the town of Eddy and a large part of southeastern New Mexico; Chaves County, which included Roswell and an even larger area; and Lincoln County to the west, which included the Hondo River and some of the more mountainous terrain of the region. With the restructuring of county jurisdictions, political maneuvering enabled some factions and locales to rise in power at the expense of others. Political fever ran high in November of 1890 in the campaigns for new Eddy County commissioners, judgeships, and other elected offices.[27]

The largest political battle occurred between supporters of the new town of Eddy as county seat and those who saw the older Seven Rivers community as the logical choice. Many former Texans in the Seven Rivers

Hotel Hagerman. Many potential investors and business partners stayed at the hotel.
(Courtesy Sacramento Mountains Historical Society, Cloudcroft, New Mexico.)

area were bent on controlling the Pecos Valley regardless of the activities of "intruders" like Charles Eddy and his backers. As they had tried under the leadership of M. J. Denman to stop Eddy's accumulation and control of land in the valley, the Seven Rivers cattlemen now tried to stop him from taking control of county political machinery as well. Eddy was acutely aware of how important it was to secure positions in the new county, and in July he wrote a personal letter to L. Bradford Prince, territorial governor in Santa Fe: "Dear Sir, According to the bill creating our county . . . the governor is to appoint . . . county commissioners in August. In a county [as] progressive as our's [sic] . . . there [are] always . . . obstructionists. . . . [I]t may be that they will ask you to appoint commissioners who would prove very unfortunate to our enormous interest. And before you take any action, let us suggest the names of men who are first class businessmen and thoroughly reliable citizens. . . . Charles B. Eddy, Eddy, New Mexico."[28]

Shortly after Eddy County came into existence, at a mass meeting in December 1890 townsfolk selected Francis Tracy and James Cameron to represent the financial interests of the newly established county. Tracy and Cameron proceeded to Lincoln County for a hearing scheduled to determine the allotment of resources and liabilities of the former Lincoln County. These resources were to be distributed among the three new counties—Lincoln, Chaves, and Eddy.[29] Seven Rivers lost its bid to become the seat of Eddy County, and the community gradually faded away. But as Seven Rivers receded in importance, the burgeoning town of Roswell, ninety miles upstream, grew as a rival to the town of Eddy.[30] When in 1890 Roswell adopted the name "The Pride of the Pecos," Eddy decided to call itself "The Pearl of the Pecos."[31] This rivalry reflected itself not only through competition in local newspapers, but also in a building boom that took place in both cities in the early 1890s. In March of 1890, the Eddy newspaper announced that it would bet ten to one that the Eddy Hotel would be built before Roswell's. Both cities took bets on which hotel would be built first, and the *Eddy Argus*, confident of Eddy's success, said, "[this] causes great glee in Eddy. Hundreds of people are collecting their wads, and if Roswell does not suffer a financial panic in less than three months we will be greatly surprised."[32]

Roswell became the county seat of Chaves County, named for Colonel J. Francisco Chavez, a prominent citizen. Anglo residents, offended that the county was named for a Mexican American, lobbied to have the name changed. The territorial legislature eventually changed the spelling of the county to Chaves. By 1889 optimistic reports showed Chaves County with 10,000 people, 300,000 cattle, 50,000 sheep, and over $5 million worth of property.[33]

Historians consider Captain Joseph Lea the "father" of Roswell, although Van C. Smith founded the town in 1869. Smith was a gambler, lover of fine horses, and poker player. He built two adobe buildings, a blacksmith shop, stable, chicken house, and two corrals. He named the new settlement Roswell in honor of his father. Lea came to the area in 1877. The former Confederate soldier raised cattle near Roswell. By 1885 the Roswell area boasted a population of 100, with nine houses and two stores in the town proper. The *Pecos Valley Register* began publication in November 1888. By 1889 Granville Richardson opened a law office; a bank opened in 1890, as did the Pauly Hotel. As in Eddy, much of Roswell's growth after 1889 was generated by the activities of Charles Eddy's irrigation company and land development in the valley. Townspeople built churches, established a volunteer fire department, and encouraged other businesses to move to town. Roswell incorporated as a village in 1891 with a population of 400. Construction of a courthouse and school began in 1891, telegraph service reached the area in 1894 and electricity in 1901. By 1900 Roswell had a population of 2,049.[34] Most of its houses were one-story adobes, fronted by wooden sidewalks. The town became known as a healing place for consumptives, and several more hotels were constructed. To provide entertainment the town built an opera house.[35]

One reason the town of Roswell prospered was because Nathan Jaffa, owner of the town's first department store, suffered stomach ailments, worsened by Pecos River water. Jaffa sought to assuage his problems by drilling for water. His first well, dug by William Hale, flowed pure clean water. By 1900 the Roswell area had 153 such wells. Town boosters touted the pure waters through advertisements, brochures, and newspapers and succeeded in attracting doctors, laborers, merchants, and orchard farmers.[36]

Almost forty years before the founding of Eddy in southeastern New Mexico, Secretary of War Jefferson Davis had assigned Brevet Captain John Pope to examine the possibilities of artesian well development near the thirty-second parallel to encourage possible railroad development across the region. In the 1850s, however, survey and settlement of the Pecos Valley was out of the question because of its inaccessibility and the danger of Indian attack.[37] Pope's expedition, however, held the promise that artesian aquifers existed beneath southeastern New Mexico. The city of Eddy provided pure drinking water from wells three miles from town, tapping an underground flow from mountain summits to the west,[38] and as early as 1891 settlers near Seven Rivers drilled artesian wells, one only 180 feet deep.[39]

In the 1890s James Hagerman and Charles Eddy concerned themselves with providing irrigation waters from the surface water of the Pecos River, not from artesian waters. James Hagerman, who examined the potential of using artesian water for irrigation, thought that although artesian water would be a financial boon to southeastern New Mexico, in the long run "they [artesian wells] will never figure very extensively in irrigation but they afford a cheap and convenient supply of water for town and domestic purposes."[40]

In the late nineteenth century, both Roswell and Eddy grew simultaneously with Eddy's and Hagerman's expanding vision for irrigation in the valley. During the last week of 1892, citizens of Eddy prepared to vote on incorporating their town. To incorporate under the laws of New Mexico Territory, the town had to have a bona fide population of 1,500. According to a census taken in August 1892, the town had more than 1,500 residents, almost 2,000—not counting the suburbs of La Huerta and Phenix. The local newspaper commented: "[Eddy] is entirely too large a town to twaddle along any farther. She has outgrown her baby clothes and now demands adult garments in the shape of a complete city government." The town incorporated on January 16, 1893.[41]

Part of the success for either Eddy or Roswell depended on immigration to the valley. James Hagerman believed that all the country needed was the right sort of people to settle it. "You want steady, industrious, frugal people—people who understand farming and fruit growing and are not afraid to work and do not expect to make a fortune in one or two years."[42] In the early days, Eddy and Hagerman brought investors, not farmers, to the area, but by 1891 the emphasis shifted to farmers.

Throughout the late 1880s and 1890s a number of visitors traveled to

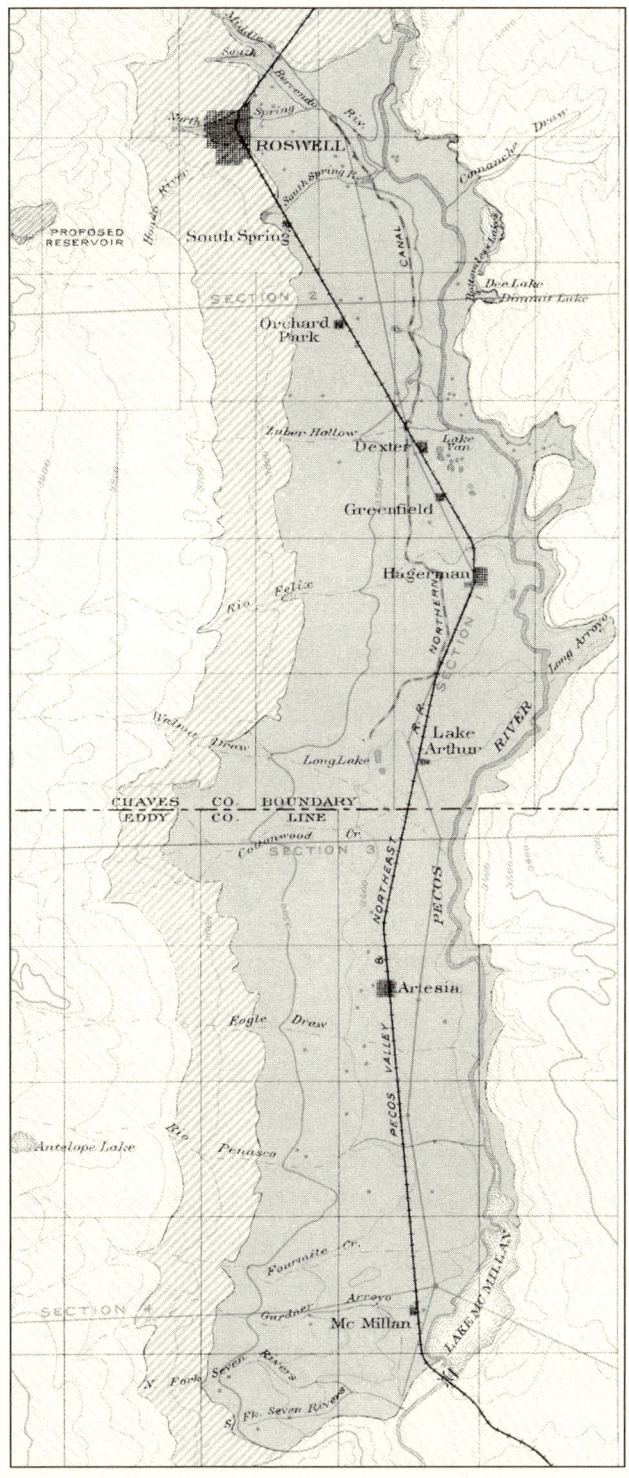

Roswell Artesian Basin
of southeastern New
Mexico, by C. A.
Fisher.
(From Cassius A. Fisher,
Preliminary Report on
the Geology and
Underground Waters of
the Roswell Artesian
Area, New Mexico
[Washington, D.C.:
Department of the
Interior, U.S. Geological
Survey, 1906].)

69

Charles Eddy's house in La Huerta.
(Courtesy Sacramento Mountains Historical Society, Cloudcroft, New Mexico.)

the Pecos Valley. Charles Eddy and other company officials arranged trips as compensation for applicants filing Desert Land entries, the land often being turned over to the company once the applicant, aided by company canals, showed proof of reclamation. Charles Eddy's lower ranch property some eighteen miles southeast of Eddy was the scene of constant visitation by investors and speculators from Chicago, New York, and elsewhere. Eddy's four-room house sat on a slight hill, with shade trees, barns, sheds, corrals, and eight to twelve acres of peach, pear, apricot, nectarine, apple, and olive orchards nearby. His hack left regularly to pick up eastern investors and tourists.[43]

Eddy used his ties to banking in New York to attract Franklin G. Campbell, a guest of Eddy's from Cherry Village, Otsego County, New York, who planned to settle in the valley.[44] Another investor and later the town's superintendent, H. S. Church, was encouraged by the Eddy-Hagerman interests to come west when the town was founded. Church invested in various enterprises in the valley, including the town of Eddy and irrigation ventures, and was very active in community affairs.[45]

Many early settlers, like Francis Tracy, moved to the valley to work for the irrigation company in some capacity. Still others moved to the town to set up businesses. Many Coloradans, enticed by Eddy and Hagerman, purchased land near Eddy and in the townsite proper. A number of original settlers in the desert community sold real estate privately as well as for the company. Charles McLenathen, for example, initially teamed up with Franklin Campbell, and later joined Francis Tracy in business.

E. G. Shields, who had been appointed registrar in the General Land Office at Las Cruces largely because his predecessor was accused of fraud, actively pursued his own career in the real estate business with partner Arthur Mermod. Appointed to his position as registrar by President Cleveland, Shields was a civil engineer who not only sold real estate and insurance, but also worked as a civil engineer for the Pecos Valley Railway. Shields ran the survey line for the company as it laid track from Pecos to Eddy in the early 1890s.[46]

Conveniently, while Arthur Mermod never maintained a permanent residence in Eddy, he served as a U.S. land commissioner, recording entries and final proofs in his Eddy office as early as August 1890.[47] Mermod was typical of many investors during the early days of the Pecos Valley enterprises. He stayed in Eddy a couple of months each winter, but spent most of his time in other places, including his home in St. Louis and his ranch in Colorado. The *Argus* eagerly welcomed Mermod and others of his financial standing and encouraged them to stay in Eddy for longer periods during the year. The paper's editor reasoned: "We wish he [Mermod] would stay permanently, for he is one of the kind of men Eddy needs and must have. He is the steady, energetic and enterprising son of a wealthy father. His father is the senior member of the Mermod and Jacquard [*sic*] Jewelry Company, which is famous all over the continent."[48]

Unlike Mermod, cigar maker Robert Tansill actually planned to farm and spend more time on his land. Tansill placed Ed Motter, his brother-in-law and former cashier/secretary, in charge of his investments and farming activities in the valley.[49] By 1892 Motter had planted fields of alfalfa on Tansill's farm south of Eddy. From the earliest days Tansill spent much of his time in the lower valley—especially during winter months. Having improved his sections of land, he sold them and built the two-story brick Tansill block in the new town. In 1896, after selling his Chicago and Colorado Springs homes, Tansill and his wife, Mary, settled in La Huerta, across the Pecos from Eddy.[50]

A number of investors, largely from the Colorado Springs area, planned winter residences and improvements in the valley. In 1893 James Hagerman's old partner, Richard Bolles, joined the mix of investors improving land and planning residences in the valley. Bolles purchased 640 acres near Otis, a settlement not far from Lookout, five miles below Eddy, and hired E. G. Shields to improve his lands. Shields, in turn, hired a force of seventy-five men to work on the Bolles property. The work involved

Mary and Robert Weems Tansill, and son R. W. Tansill Jr., Colorado Springs, Colorado. Tansill had made a fortune manufacturing "Punch" cigars in Chicago, and "retired" following his doctor's advice to move west to slow down his tuberculosis. Tansill, who met Charles Eddy in the late 1880s, energetically recruited Chicago investors for Eddy's New Mexico irrigation project. Tansill faithfully served as receiver for the bankrupt Pecos Irrigation and Investment Company, hoping to interest the federal government in rehabilitating the irrigation project.
(Courtesy Special Collections, Tutt Library, Colorado College, Colorado Springs, Colorado.)

converting it into a state-of-the-art horse racing stock farm. Shields laid out sections in forty-acre tracts for grazing and alfalfa. He planted other fields in grass, rye, corn, oats, sorghum, and millet. Shields then set aside one portion of the Bolles property exclusively as training grounds for Bolles's racehorses. In the spring of 1893 Bolles contemplated moving a large part of his Colorado stable of thoroughbreds, trotting, and racehorses to the Pecos Valley. He built a racing track west of Eddy and transported a number of horses to the facility.[51] Bolles built a residence to serve as his winter headquarters.[52]

The indispensable E. G. Shields also made a number of improvements on property owned by William Bonbright. Shields contracted to improve the property for the Colorado Springs millionaire who resided in the Springs and elsewhere most of the year. William McMillan of St. Louis, Charles Otis of Cleveland, and Bonbright all had erected, or were planning to erect, winter residences in the Eddy area.[53]

The residence and farm of Richard J. Bolles, not far from Eddy (Carlsbad), about 1896.
Bolles was Hagerman's partner at the Mollie Gibson in Aspen, Colorado, and later invested
heavily in irrigation lands in southeastern New Mexico. When fortunes turned in the
Pecos Valley, Bolles sued the irrigation company and moved to Florida where
he invested in Everglades properties.
*(Tracy Family Album, Courtesy Southeastern New Mexico Historical Society,
Carlsbad, New Mexico.)*

Early in January 1891 Hagerman announced that he would live at least
part of the year in Eddy and would begin construction of a residence in two
months.[54] Attending to business matters and experiencing acute lung problems, however, Hagerman let a year pass before hiring contractors to build
a place on "Hagerman Heights,"[55] for which he planned a garden and
orchard. Since his home was located on an elevated area east of the river,
Hagerman had to build some sort of conveyance for water from the river to
the Heights; so engineers designed a flume. Before water could reach the
powerhouse at the end of the flume (designed to provide electricity for
Hagerman Heights and the town of Eddy), the whole fabrication collapsed,
and H. S. Church, the supervising engineer, temporarily abandoned the
entire project. The only way Church could get water to the Heights was by
pumping it, and the only way to get pumps to work was to provide power.
The flow of the river did not furnish enough power, and late in 1892
Hagerman decided to build a power dam to pump water to the Heights.

The millionaire thought Eddy could use extra kilowatts for manufacturing and lighting the city, and for a proposed street railway like the one in Colorado Springs.[56]

The dam built principally for supplying water and electricity to Hagerman Heights was known appropriately as the Hagerman Dam. In 1892 workers built the first dam of timber and brush on a sandstone base, which they later replaced with concrete. In later years workers built a small hydroelectric power plant at the east end of the dam, which became the parent of Southwestern Public Service Company, now Xcel Energy, which supplies electricity to customers in seventeen states.[57] Hagerman Dam, from end to end, measured 150 feet. At the mouth of a 4,200-foot flume leading to Hagerman's property, Church placed three turbines forty-four inches in diameter, each representing some fifty-five horsepower.[58]

In December 1892 Church completed plans for a power and pump house and a reservoir on the Heights. The improvements supplied Hagerman's residence, orchard, garden, and other areas with water and electricity. The reservoir had a capacity of 1.5 million gallons for domestic and irrigation needs and for fire emergencies in nearby Eddy. Church's pump supplied the Heights with 250,000 gallons of water every twenty-four hours.[59] Following the successful implementation of the pump system, Hagerman himself turned the switch on a Westinghouse 700 incandescent light system, illuminating the town of Eddy in the spring of 1894.[60]

Hagerman also spent a small fortune on fruit trees, shade trees, and the like. Expenses for his Pecos Valley homesite included forty barrels of Portland cement, some twenty tons of feed, several pounds of alfalfa, bluegrass, Egyptian corn, and sorghum seed, a brick residence on Hagerman Heights, various pieces of furniture, a stone stable with stalls for eight horses, room for a coachman, a carriage room, a harness room, a hay room, a grain room, and closet. In the spring of 1893, Hagerman planted 12,316 grapevines and 2,950 fruit trees, including 37 fig trees. The area around his residence was covered with 5 English walnut trees, 58 olive trees, 12 orange trees, 26 palm trees, a number of yuccas, pomegranates, and oleanders, 99 China umbrellas, 20 elm trees, and 46 Russian mulberries. Hagerman added black locust trees, 790 Carolina poplar trees, 350 cottonwoods, 18 weeping willows, 354 Russian mulberries, and 40,000 Osage hedge plants. He spent thousands of dollars on Hagerman Heights, including the dam ($18,000) and flume ($26,894) to get water from the Main Canal into the ditch ($3,775) leading to his residence, a pipeline ($5,038), pumps and waterwheels ($4,067), a pump house ($8,580), and other accessories totaling over $100,000. While H. S. Church wrestled with the problem of supplying electricity and water to the Heights, and to the city proper, Hagerman hoped to

James Hagerman's residence on Hagerman Heights, a bluff just east of the Pecos River, across from Eddy (Carlsbad), about 1896. Built in 1892, the residence featured twelve rooms, including a library, billiards room, and smoking chambers.
(Tracy Family Album, Courtesy Southeastern New Mexico Historical Society, Carlsbad, New Mexico.)

grow additional fruit trees on his farm eighteen miles south of town.[61]

Like Hagerman, Francis Tracy, L. Wallace Holt, local cattlemen, and others who hoped to make money raising fruit, organized the Pecos Orchard Company in the fall of 1892, which, according to the Bureau of Immigration, planted 250,000 fruit trees that year and another 300,000 in 1893. Charles Greene supervised the interests of the orchard company in Chicago, trying to arrange a financial base for its operations. Besides his work for the irrigation company, Greene had personal interests in the valley. He owned 640 acres of vineyards south of Eddy near Malaga. He owned another three sections of land through the Pecos Irrigated Farm Company, which was capitalized at $250,000. Greene planned to sell the land in forty-acre lots—all planted and ready for settlers. He offered the land for sixty dollars per acre, or $2,400 for the forty-acre lot, with payments spread over four years. During the fall of 1894 the company expected to plant 2,000 acres of apple orchards in the valley, and continue planting until the total acreage approached 10,000 acres.[62]

At the territorial level, one of the staunchest supporters of the Pecos Valley, and Eddy in particular, was Max Frost, editor of the leading newspaper in the territory, the *Santa Fe New Mexican*.[63] Frost also served as the leader of the New Mexico Immigration Bureau, established in 1880 with the goal of producing a predominantly white population.[64] Frost, who made a living promoting the territory and had substantial landholdings himself, was duly impressed with progress in the valley. Whereas four years before Chaves and Eddy Counties had not existed, Frost noted that in 1893 each county had a population of between 6,000 and 7,000. While Eddy was still a new town, Frost saw a community that already had 2,500 citizens, a national bank, a large hotel, newspapers, a $30,000 courthouse, a $10,000 school building, some 20,000 shade trees, lights, streets, several churches, a public park, a social club, many houses, and business blocks. He estimated that by the end of the year Eddy would have an electric railroad, extending to Hagerman Heights, La Huerta, and Otis, six miles south of town.[65]

Locally, in November 1892 organizers held the first formal meeting of the Eddy Club, although the group had met unofficially for eight months. The club listed thirty-five to forty members, including Charles Eddy, Charles Greene, and others who wished, among other things, to establish a public library and social club in Eddy.[66] Comprising irrigation company officials and local town boosters, the club actively promoted Eddy and the lower Pecos Valley through advertisements at meetings, conferences, conventions, state fairs, and at the World's Fair in Chicago, Illinois. On September 16, 1892, the World's Fair celebrated New Mexico Day. Governor Thornton, former governor Bradford Prince, future governor Miguel Otero, and Charles Eddy addressed audiences on the grounds of the Territorial Building.[67]

The Eddy Club and the men who attended the World's Fair voiced the typical theme of western promoters and irrigationists during the late nineteenth century: "The great need of the country at the present time is a class of energetic and enterprising farmers. As a rule the pioneer of the west has to rough it and await capital; in this case capital is already here and every inducement is offered to farmers of every class. When irrigation is more thoroughly understood and results therefrom generally known, there is no question but what the lands of the Pecos Valley will become very valuable."[68]

Due to tremendous boosterism by New Mexico territorial governors, the immigration bureau, Eddy townspeople, and investors and officers of Pecos Irrigation and Improvement, the town and Eddy County grew. Responding to numerous questions concerning availability of public lands and employment, the local company-owned newspaper encouraged settlers to take advantage of the public domain, which would allow the company to dispose of the water in its canals. To those who recognized the local paper as the voice of the company, the *Argus* responded that references to the paper as a boom sheet did great injustice. "They [give] people at a distance the idea that the *Argus* indulges in wild, wicked, sinful exaggeration, all this which is not the case."[69] But in fact the *Argus* was the company mouthpiece, and its primary objective was to entice settlers to the region and make the community of Eddy alluring.

Local booster Charles McLenathen, one of the community's first real estate brokers and land agent for Pecos Irrigation and Improvement, listed 100 reasons in the local paper why prospective settlers should own a home in the valley. Among the reasons enumerated was the rich soil of the Pecos Valley, which McLenathen claimed was the richest and most extensive agricultural land in the Southwest. Touting high production of cereals, fruits, and other crops, McLenathen focused on valley produce, whose success farmers demonstrated time and time again. They produced harvests of wheat, rye, barley, corn, vegetables, sweet potatoes, and alfalfa. Another advantage, according to McLenathen, was the valley's location, 1,200 miles closer to eastern markets than California fruit farms. Local markets, adjacent mining camps, and stock ranges to the east assured a great demand for all products grown in the valley. McLenathen claimed twenty acres of well-cultivated land would support a family, and in a few short years yield a decent living. The claim that sixteen families per section was more efficient than one family per section, of course, enhanced land sales of twenty- to sixty-acre blocks. Describing the valley's climate as "matchless salubrity," McLenathen asserted that consumption (tuberculosis) never originated in the Pecos Valley, and living in the valley's air and sunshine from September to July cured the illness and totally rejuvenated patients.[70]

McLenathen and others may have been right about the climate, and they used it as a major draw to get settlers to the valley. The arid environment drew thousands to the Southwest, and settlers moved to southeastern New Mexico specifically for tuberculosis and respiratory illnesses. Doctors encouraged moderate to vigorous exercise in the arid regions of the West to slow down the lung disease. In 1890 Mr. and Mrs. Lucious Anderson built a two-story adobe hospital in Eddy that became known as the Anderson Sanatorium, one of the earliest buildings in town. Mrs. Anderson, a nurse and humanitarian often referred to as the "Angel of the Valley," kept the sanatorium open from 1890 until 1920, when the building was torn down.[71] Because of its mild climate, the town of Eddy became somewhat famous as a health resort, especially for tuberculosis patients.[72] By the turn of the century, health officials estimated that tuberculosis caused more deaths—8 to 10 million of the 75 million persons living in the country in 1905—and greater economic loss to the United States than any other disease. Some writers place tuberculosis among the most important factors in the settlement of New Mexico, at least as important as ranching, mining, and railroads to the territory. Thousands of health seekers flocked to all parts of the territory during the mid to late 1880s,[73] and some physicians called for a national sanatorium.[74] By 1908 New Mexico ranked fifth in the nation for the number of beds available for tubercular care.[75]

Pecos Valley promoters, and promoters of New Mexico in general, knew the value of the climate and touted the valley as "the great sanatorium of the world."[76] In the southeastern portion of the state, physicians built a number of facilities, including two in Las Vegas and one each in Alamogordo, Deming, Valmora, Las Cruces, Lincoln, Roswell, Tucumcari, and Carlsbad, and a twelve-bed baby sanatorium in Cloudcroft.[77] Many doctors who came to southeastern New Mexico in the nineteenth century were themselves tuberculosis patients.[78] Those patients with deep pockets like James Hagerman, Robert Tansill, and Joseph Stevens could afford the finest physicians and moved wherever they needed to, including New Mexico's Pecos Valley. Though of smaller means than these men, farmers and townspeople also sought the wide-open spaces and sunshine of the valley.[79]

The *Eddy Weekly Current*, the independent Democratic paper in Eddy, tried to explain the benefits of living in the Southwest and claimed that the prevalence of tuberculosis patients limited the amount of crime in the community. The paper's editor commented: "[M]en and women [in the valley] have concluded not to die like dogs, but instead seek a climate where life may be prolonged. It is safe to say that one in every three who come here to make a home is suffering from lung or throat troubles. They are people who, above all others are agreeable to live among and certainly all law abiding."[80] Although evidence abounds that plenty of violence existed in southeastern New Mexico in the late 1880s and early 1890s, the paper quickly downplayed the more flamboyant instances and touted the region as a healthy, peaceful place to live.[81]

Pecos Valley companies and investors also tried to portray the valley as a progressive, modern English-speaking community to attract the "right" type of people. Arthur Mermod, Charles and John Eddy's former neighbor in Colorado, promoted the valley in Colorado, St. Louis, and New Mexico. Handbills printed in local newspapers indicated that Mermod was a U.S. commissioner, notary public, and general agent for Pecos Valley lands.[82] Real estate agents, including Mermod and company officials, sent circulars, promotional ads, and newspaper articles across the country touting the advantages of life in the Pecos Valley, and they often recruited local testimonials. Many of these "unbiased" valley citizens who had grown fruit trees and vines[83] were intimately connected with land sales and irrigation in the valley.

County officials registered 108 real estate transfers during December 1892, and valley supporters commented that many counties in western

Nebraska and Kansas probably recorded twice as many mortgage foreclosures during the same time. Taking the comparison further, boosters proclaimed, "if the people in the arid parts of those states could get just one look at the beautiful and fertile Pecos Valley, they would never want to stick a flower in the ground in Kansas or Nebraska again."[84] To demonstrate the valley's advantages to Nebraskans, Kansans, and the rest of New Mexico, boosters compiled a list that in their estimation made their newly incorporated town a full-fledged city. Highly interpretive at best, the promoters' list extolled the amount of business transacted, the "excellence" of nearby irrigation lands, and hotel accommodations. But perhaps most important, the list gave the impression to those east of the hundredth meridian that Eddy, a town without saloons, a town with schools and churches, was not a town of desperadoes and ruffians as depicted in so many dime novels. Here, echoing the tone of Max Frost, was a town of "progressive English speaking people."[85]

In January 1893 the Eddy county clerk filed seventy-one real estate transfers on a single day. Using such figures to prod settlers before the choicest land was gone, local promoters also tried to convince easterners and midwesterners that irrigation farmers produced more per acre than farmers in the East. Continuing McLenathen's previous claim that twenty acres sufficed for a properly managed farm, promoters asserted that Pecos Valley land produced many times the value of farm products raised on a 160-acre tract in Iowa or Illinois. Pecos Valley promoters promised the eastern farmers that less land and more work could be advantageous. They recommended that newcomers plant ten acres of alfalfa and devote remaining acreage to fruit trees. Promoters pointed out that fruit lands in other irrigated sections of the United States produced $300 to $1,000 an acre based on quality and quantity of the fruit. Boosters claimed every ten-acre farmer near Eddy who grew orchards would realize in a ten-year period property worth $10,000, besides a comfortable living.[86]

The Pecos Irrigation and Improvement Company did not own the local *Weekly Current,* and the *Current* warned prospective settlers about selecting government land in areas with little chance of ever getting water. The paper reported that a number of unscrupulous parties engaged in getting people to file on government land that would never be serviced by the local irrigation company. "Unsuspecting parties fall into the hands of shyster locators, and when they awaken to the fact that their land is worthless, they raise a great howl against the country."[87] Messages like this had the effect of steering land seekers not to government lands but into the arms of company land agents and local real estate dealers.

Boosted by unscrupulous promotion and false promises, Eddy continued to grow. Optimism abounded that settlers from the United States would be successful in the valley. By December 1892, several families arrived in the Pecos Valley from Louisiana. Under a banner headline, "Here They Are: The Louisiana Colony Arrives Today: Seventy-five heads of families arrive on today's train," the local newspaper described this latest group of colonists as those that are "mostly desired in the Pecos Valley."[88] The Louisiana colonists chartered a special train to carry their household effects and some 200,000 board feet of lumber and shingles to erect houses and businesses at a point on the river called Lookout, southeast of Eddy.[89] Also in the spring of 1893, German representatives from San Antonio, Texas, arrived in Eddy to survey the valley's irrigated lands. The company made plans for relatives and friends near San Antonio to sell their homes quickly and move to the Pecos Valley.[90]

Although irrigation company officials sent circulars to communities across the United States and Europe and set up elaborate displays at state and world fairs, the largest contingent of settlers in Eddy and vicinity came from neighboring Texas. According to promoters this was no surprise, because it demonstrated that those who had the best opportunity to judge the merits of the country by frequent trips had the most confidence in the valley.[91] E. G. Shields addressed questions from Texans living in the more easterly portion of that state, which received far more rainfall than eastern New Mexico. But, according to Shields, anyone who went to the Dallas State Fair and saw the Pecos Valley display could understand that the Rain Belt country could never compete with irrigated cotton, wheat, or fruit.[92] He estimated that a settler could build a good four-wire fence using cedar posts for $120 a mile. Clearing and grubbing expenses varied, depending on the number of mesquite trees, ranging from $1.00 to $5.00 per acre. Plowing could be contracted from $2.50 to $3.50 an acre. According to Shields, "it [was] not necessary to break and then plow in this soil but one plowing fits the land perfectly for cultivation."[93] Taking top-end figures, a new settler improving forty acres of land might pay $500, exclusive of "water rights," rental fees, tools, and seed. Paying $1,200 for a forty-acre tract of irrigated land, a settler might spend $1,700 just to get started.

Promoters assured potential settlers that farmers irrigated and cultivated 69,000 acres of land between Eddy and Malaga. The Department of Agriculture analyzed the water and pronounced it "wholesome to the soil and good in every respect." The area was purportedly perfect for settlement and irrigated farming.[94] Some supporters optimistically estimated that Eddy and Chaves Counties would soon be home to 25,000 to 30,000 people, and as Max Frost put it, "if this thing keeps on, the Eddy country tail will wag the New Mexico dog financially, politically, and in every other way."[95]

The irrigation company viewed foreign investment and immigration, like domestic immigration, as vital to the success of the valley enterprises, and officials tried to entice Europeans to the area. Early in 1891 a quasi-governmental syndicate from Switzerland sent Henri Gaullieur to the United States. Gaullieur's goal was to avoid unscrupulous land agents while locating a colony where Swiss immigrants could move, settle, and prosper. Well educated and wealthy, Gaullieur had lived in the United States where he had developed numerous friendships and acquaintances, including the first director of the U.S. Geological Survey, Clarence King.[96] Gaullieur visited a number of irrigation projects in the western United States, including two or three in Wyoming. He traveled to the Pecos Valley without the solicitation of anyone connected with Pecos Irrigation and Improvement, according to a company circular issued in December 1891.[97] But according to Percy Hagerman, after examining his father's notes and files in the early 1930s, Charles Eddy was well aware that Gaullieur was in the country and in fact sent his friend Henry VanCleef of Denver to Cheyenne, Wyoming, to encourage Gaullieur to investigate the Pecos.[98]

Gaullieur traveled to the Pecos country, became an enthusiastic disciple of Charles Eddy, and fell in love with the desert surrounding the Pecos River. On his return to Switzerland, he reported that the Pecos Valley was the best location for settlement he had found in the United States.[99] Gaullieur met Eddy, but not Hagerman. During the summer of 1891, Hagerman met Gaullieur through a letter of introduction from Eddy, dated June 1.[100] In July Hagerman and his family went to Europe primarily to regenerate the patriarch's failing lungs. While they were there, they visited the Gaullieurs at the Chauteu de Kiesen near Interlachen. The Gaullieurs entertained the Hagermans for several days, during which time Hagerman raised the question of promoting Swiss immigration to the Pecos Valley. The two men reached an arrangement by which Gaullieur became a paid immigration agent of Pecos Irrigation and Improvement.[101]

After another visit to the Pecos Valley, Gaullieur became confident that fruit growers from the lower valleys in Switzerland would do remarkably well in the Pecos Valley, although a number of obstacles needed to be overcome: "This is a new country and very different from their's [sic] where there are hills and mountains and trees and clear running streams. You have plenty of water here for all purposes, but other things are different and a stranger in a strange land, with no knowledge of the language, and where the habits and ways are so unlike is liable to find the first year or two uphill work."[102] Gaullieur was concerned that his people have the protection guaranteed by law and was assured that the company made their welfare a priority.[103] Gaullieur believed that given the proper moral support, the European immigrants would succeed in New Mexico. Pecos Irrigation and Improvement's operations satisfied Gaullieur. He told locals: "[I] was thoroughly satisfied from the examination that [I] made that this valley will excel in raising all agricultural products. I have seen no place where I believe immigrant farmers from my country could do so well as they can here, and I shall so report to the syndicate upon my return to Berne, and advise them to concentrate a portion of our immigration to this valley."[104]

Charles Eddy convinced Gaullieur that forty to eighty acres of land in the valley would provide plenty of revenue for any farmer, promising that five acres devoted to fruits and grape vines would provide the farmer $200 to $300 per year. Gaullieur suggested that farmers might use the remaining land for farm produce and vegetables.[105] The company made plans favorable to the Swiss colonists, arranging for land, houses, equipment, and virtually everything they might need once they arrived in the Pecos Valley. Interestingly, Gaullieur told local officials that the immigrants needed no financial help since his country did not permit immigration of the poorer classes who depended solely on their labor to get by. Prospective immigrants to the valley, according to Gaullieur, gained permission to leave Switzerland only after proving ownership of enough capital to allow farmers and stockmen to succeed without going into debt.[106]

Gaullieur gave the committee responsible for settling colonists in the United States a very favorable report that the committee printed in both German and French. Application requests to leave Switzerland for the Pecos Valley were apparently so great that Gaullieur decided to travel with the group and see them settle in before returning to his homeland.[107] The first contingent of Swiss colonists included eighty-five settlers; by the end of 1891 still a larger number had crossed the Atlantic. Each of the eighty-five original colonists had at least $1,000, and many brought $10,000 or more. The *El Paso Herald* called them "intelligent, wide awake, and thrifty people."[108] The new colonists purchased, on average, eighty acres apiece. Eddy County Deeds of Record Book 4 lists more or less consecutively the Swiss immigrants and their warranty deeds, all from Pecos Irrigation and Improvement Company, starting on page one with Charles Auguste Bremond (who married into the Hagerman family). The Swiss entries all show Vaud, New Mexico, as the place of residence. Witnesses to the transactions included Charles Eddy and Arthur Mermod. All of the land was in township 23 south, ranges 27 and 28 east; the Swiss paid $1,000 or more for forty acres of land. For example, Adrian Sutler paid $1,200 for forty acres (the northeast quarter of the northeast quarter of section 23, township 23 south, range 28 east, tract number 94) near Vaud on April 23, 1892. Frederick Auguste Dardel on April 28, 1892, paid $2,400 for eighty

Schoolhouse near Vaud, later Florence, New Mexico, about 1896. Most of the students attending the school were Swiss immigrants who arrived in the Pecos Valley in 1892. *(Courtesy Southeastern New Mexico Historical Society, Carlsbad, New Mexico.)*

acres in the southeast quarter of the northeast quarter of section 23, township 23 south, range 28 east, tract number 90, and the southwest quarter of the northwest quarter of section 24, township 23 south, range 28 east, tract number 91. While Gaullieur served as colonizing agent, his colleague, Frederick Dominice, relocated to the Pecos Valley to ensure the success of early colonists.[109]

According to the *Eddy Argus*, the colonists liked the desert country, its rich soil, abundance of water, and climate. Swiss colonists located south of Eddy, in Vaud, which they named for their homeland canton, a wine-producing area bordering France. Gaullieur and Dominice secured lands alongside the colonists, whom Gaullieur praised as good agriculturists and fruit growers.[110]

With the influx of American and foreign settlers, the area in and around Eddy grew. By August 1892 sixty-five Italian families arrived at Malaga Station southeast of Eddy, taking possession of lands purchased nearby. Contractors built eight or nine buildings that fall. Company officials expected between fifty and sixty additional families by spring 1893.[111] New Italian immigrants settled many of the lands near Malaga, almost exclusively in forty-acre tracts. The Italian settlers of southeastern New Mexico are also listed in Eddy County Deeds of Record Book 4. Entries run more or less consecutively, with warranty deeds from Pecos Irrigation and Improvement. All of the settlers, except Dominice who purchased eighty acres, bought forty acres in townships 23 and 24 south, range 28 east. Angelo Grofita, on October 1, 1892, paid $1,800 for forty acres in section 4 of township 23 south, range 28 east. Giuseppe Nastasi also bought forty acres for $1,800 near Malaga.

The influx of eastern capital and foreign settlers into the Pecos Valley aided the development of towns like Eddy and Roswell and assured the emergence of southeastern New Mexico as a viable agricultural region. Promoters first targeted investors with deep pockets. Then they launched an aggressive promotional campaign designed to convince American farmers that smaller irrigated farms would yield more than rain-fed lands in the East. Promoters also tried to disabuse prospective settlers from the notion of southeastern New Mexico as a rough, uncivilized place. Rather it was progressive, civilized, and English speaking. Promoters also attracted foreign colonists to the Pecos Valley. Convinced by Charles Eddy that their welfare was uppermost in the minds of irrigation company officers, Swiss and Italian immigrants arrived in 1891 with hopes of building a bright future in the New Mexico desert. Early success and momentum attracted more capital, settlers, and townspeople, making the region economically and culturally unique in New Mexico Territory.

4

DAMS AND VIOLENCE IN THE PECOS VALLEY

T HE construction of dams, flumes, canals, laterals, and railroads in the lower Pecos Valley gave the region much of its flavor during the halcyon days of the projects. A flume just above town and a dam six miles upstream generated hundreds of construction jobs for mostly Mexican labor. As total investment increased, projects got bigger and more capital-intensive, measured by the flurry of activity by investors, laborers, farmers, and townspeople. Violence fed by alcohol and racism also increased commensurate with the amount of construction activity in the valley. Charles Eddy tried to present an image of progressiveness and sobriety in this new town for the benefit of investors, settlers, and towns-people, who converged on the Pecos Valley in the early 1890s.

Construction projects in the valley began in earnest when Eddy hired irrigation engineer Edwin S. Nettleton[1] in June 1887. The engineer meas-ured the river's flow using an elaborate system of wooden floats made by G. W. Witt, who assisted in the preliminary examinations. Based on their measurements, engineers designed the dam and canals to accommodate the river's flow.[2]

In July 1887 laborers began working on the first canal and headgates nine miles above Eddy's new town. Dennis W. Duncan plowed the first fur-rows for the Halagüeño Ditch, originally used to irrigate Eddy's holdings in La Huerta.[3] After work began, Eddy incorporated his first company, Pecos Valley Land and Ditch, with Pat Garrett. At this point the men projected an elaborate plan to buy and enlarge the Harroun Ditch[4] to the northeast. Eddy and Garret planned to establish a main canal that would divide and extend by flume to the west side and south along the river.[5]

Workers also began what became the Avalon diversion dam a few miles north of the Eddy townsite. When James Hagerman came to the valley in 1889, neither the dam nor the canals were complete. From the diversion

Main Canal.
(Courtesy Sacramento Mountains Historical Society, Cloudcroft, New Mexico.)

dam, workers had only completed a small ten-foot ditch, which returned water south to the banks of the river, and they had constructed a small wooden flume to carry the stream over the river.[6] To hurry completion, Hagerman enlisted the advice of several irrigation engineers, although in the late nineteenth century "irrigation engineers" could more accurately be described as mining or railroad engineers.[7] Hagerman brought to New Mexico H. H. Cloud, a railroad engineer who had worked on his Colorado Midland rail system. Assisting Cloud was Louis D. Blauvelt, another engineer from Colorado, who in the early 1900s worked as chief civil engineer on the Denver, Northwestern and Pacific Railroad's Moffat Road moving up the Front Range. As late as the 1920s Blauvelt worked in Colorado building roads for the state highway department.[8]

Hagerman also relied on the expert advice of "Colonel" E. S. Nettleton of Colorado, whom Eddy had brought to the valley in 1887. Hagerman described Nettleton as the best irrigation engineer in the United States at the time. Unlike Cloud and Blauvelt, Nettleton had an impressive resume as a water expert. He was a railroad engineer, but had experience in reclamation engineering. Graduating from Oberlin College with degrees in civil and mechanical engineering, Nettleton joined the Union Colony and moved to Greeley, Colorado, in 1870. In Greeley he platted the townsite and irrigation system and designed a water-flow meter to gauge stream flow. He later laid out the plans for Manitou Springs and created General William Palmer's plat and irrigation system for Colorado Springs. He worked as Colorado state engineer from 1883 until 1887 and joined John Wesley Powell's abbreviated irrigation survey during 1888 and 1889.[9] Nettleton served Hagerman in an advisory capacity, but his approval of the initial irrigation concept in New Mexico's Pecos Valley clearly encouraged the millionaire to invest more than a modest fortune in the enterprise.[10]

With Eddy as manager, the expertise of irrigation engineers available, and Hagerman bankrolling the operation, the newly organized Pecos Irrigation and Investment Company expanded the proposed irrigation scheme, merging the current project with one proposed in 1886 by Pat Garrett on the Hondo. The company laid plans for completion of the Avalon Dam above Eddy and two large canals. The proposed Northern Canal would take water out of the Hondo River below its junction with the Spring Rivers and use it to irrigate 75,000 acres near Roswell. In 1888 Leslie M. Long, an engineer educated at Troy University, had developed a plan to build a reservoir west of Roswell, near the ruins of Missouri Plaza. He and his backers founded the First New Mexico Reservoir and Irrigation Company in 1888. First New Mexico sold its water rights to Pecos Irrigation and Improvement, a company Hagerman formed in 1891 to supersede the Irrigation and Investment Company. Pecos Irrigation and Improvement planned to build a canal thirty feet wide, six feet deep, and fifty miles long.[11]

The Southern Canal would take water into a short "main" canal on the east side of the Pecos near Seven Rivers. The company planned to enlarge the Main Canal to forty-five feet wide and seven feet deep for three miles. The canal would then divide, with part going east to an east side canal and part going by flume across the river to the west side, seven miles below the dam, to feed the new Southern Canal. The Southern Canal was to be forty-five feet wide at the bottom, sixty-three feet wide at the top, and six feet deep, and would stretch forty miles south to the Texas line. Company officials estimated the Southern Canal could reclaim 125,000 acres of land.[12] Altogether, Hagerman projected in 1889, the expanded system would irrigate a total of 143,000 acres.[13]

In 1889 Hagerman hired the Denver, Colorado, construction firm of W. C. Bradbury and Company to build and oversee work on the diversion dam and the Northern and Southern Canals. Bradbury and Company had experience in both railroad and canal construction.[14] Bradbury hired a number of subcontractors for various parts of the project, all under the direction of Bradbury's superintendent of construction, "Captain Mann," of Colorado Springs.[15] H. V. Clark served as local project secretary and financial manager, and J. L. (Jim) Warren directed a company commissary that kept the workforce fed, clothed, and housed.[16] Warren had his hands full serving the needs of the various subcontractors and construction camps. Hundreds of men filtered into Warren's commissary to buy tobacco, candy, and the like. He estimated that men from the various work camps ate half a ton of candy each month.[17]

Initially work progressed concurrently on all irrigation works under the project—on the giant flume that crossed to the west side of the Pecos just north of Eddy and furnished water to the Southern Canal;[18] on the diversion dam and reservoir six miles above Eddy; and on the Northern Canal, which took its water directly from the Hondo six miles east of Roswell.[19]

Bradbury hired 1,000 men and 700 teams of horses to dig some eighty-five miles of ditch, six feet deep and thirty-five feet wide. To get the job done the company used a giant ditching machine, the New Era Road Machine, which plowed the earth and threw it on a rubber conveyor belt, which constantly carried it over the side of the ditch. Roswell newspaperman and writer Ash Upson described the contraption in December 1888:

> Imagine a large wagon frame, with a tread of eight feet, enormous wheels, with tires six inches wide. Underneath, and at one side, is a large plow, of the kind known as a "prairie breaker." An endless band of gum elastic about three feet wide, revolving over rollers, extends from the plow at right angles with the furrows. A cog wheel attachment to the wagon wheel gives motion to this endless band. Now, when the machine is in motion the plow turns the sod or ground just like any plow, the dirt so turned over falls on the band, which in motion, carries it off to one side and deposits it a distance of 22 feet from the furrow. The machine in an up and down trip will excavate two furrows and deposit the dirt 44 feet apart. . . . Fourteen horses and three men work the machine, doing the work of thirty men and horses with ordinary plows and scrapers.[20]

The machine plowed a furrow twelve inches wide and six inches deep. The company ran the machine ten miles per day, working as quickly as possible, throwing some 1,000 cubic yards of dirt in the process. The company used two of these machines on the projects. [21]

As canal construction progressed, work crews just north of Eddy toiled on the wooden flume, reportedly the largest in the United States, that would carry water across the Pecos. The irrigation company hired local contractor G. W. Witt of Witt Brothers Construction to complete this structure.[22] The flume drew much attention. Stretching 475 feet, the structure

W. C. Bradbury's construction camp and commissary, 1889. Bradbury's crews built Avalon
Dam and the early irrigation system near Eddy (Carlsbad).
(Courtesy Southeastern New Mexico Historical Society, Carlsbad, New Mexico.)

stood atop a network of wooden trestles and was 25 feet wide and 8 feet
deep. The flume stood 38 feet above the channel of the Pecos River.[23] A
200-foot-long terre plein at the west end of the flume conducted water
from the flume to the canal on the west side of the river.[24] The company
newspaper crowed: "The great ditch at the terre-plein is so wide that a
buggy can turn in it without the wheel rubbing the box. . . . [I]t is twice as
wide as an ordinary bridge, three teams being able to walk abreast in it
without being crowded. The sides are so high that a medium-sized man
had to tiptoe to look over them. It is truly a great flume—the greatest, perhaps, in the country."[25]

By March 29, 1890, the Witt Brothers' construction crews had finished
work on the flume and moved to the already thriving camp next to Avalon
Dam. Workers still had much to do on the dam. Crews were busily filling a
100-foot gap in the structure with 25 feet of stone. Workers followed
placing another five feet of rock on top of the fill, followed by sandbags and
dirt, while still others worked on the dam's headgates and spillways.[26] Engineers Blauvelt and Cloud supervised work at the dam site. Two hundred
and sixty men worked on the dam by late March,[27] and over 300 by early
April 1890.[28] The engineers designed Avalon Dam as a "prism of rock"
facing upstream, built on bedrock. Blauvelt and Cloud used the river to

Pecos Irrigation and Investment Company's wooden flume, 1890. Twice destroyed by floods, this wooden version was replaced by a concrete version built in 1903. *(Courtesy Sacramento Mountains Historical Society, Cloudcroft, New Mexico.)*

scour away the earth resting on top of the bedrock as they built the dam.[29] The rockfill dam was built to raise the water level, provide storage, divert water into canals, and equalize flow. The outer slope of the dam had rockfill one and one-half feet thick. The dam was 270 feet thick at its base, 48 feet high, and 1,135 feet long. The outlet to the Main Canal was 30 feet wide and 25 feet deep. The dam and canal headworks cost $90,000.[30]

As construction crews labored on the dam, engineers and promoters touted the strength and security of the structure. H. H. Cloud estimated that the reservoir created by the dam would stretch six full miles and in places be three-quarters of a mile across, backing up over a billion cubic feet of water. Cloud was sure that "the security of the dam is beyond all question, as the greater portion of it is rock, with foundations and sides of solid limestone, and the overflow being entirely distinct from the face of the dam, no water action will be able to destroy it."[31] Significantly, at the time of the dam's construction, the stream flow of the Pecos was lower than it had been for some time, flowing 1,200 cubic feet of water per second, or roughly 9,000 gallons per second. Yet, company officials did not seem concerned that an increase in the volume of water flowing down the Pecos might materially affect the dam. E. G. Shields, using Nettleton's claims as gospel, thought the dam's safety was unquestionable: "Expert

Construction of Avalon Dam (originally Rock Dam), 1890.
*(Courtesy Sacramento Mountains Historical Society,
Cloudcroft, New Mexico.)*

engineers, of which General Nettleton of Colorado is one, have pronounced it [the diversion dam] absolutely safe, every precaution having
been taken by the construction of spillways to carry in themselves, without
calling upon the canal or scourgates, twelve times the capacity of the river
at flood season."[32]

Irrigation officials were confident that the dam six miles north of Eddy
would stand any storm that came down the Pecos. In August 1890, five
months after the dam was completed, that confidence was put to the test
when a man came riding into town yelling at the top of his voice that the
dam had burst and that everyone should flee to the hills. The natural reaction of the citizens of Eddy was to scramble to the highest land available on
Hagerman Heights or some other elevated place. When word finally came
that the dam had not broken, that storm waters had only risen rapidly in
the reservoir, those who had fled returned to town and purchased tents and
supplies in case the worst should happen. The dam had not broken, and
according to the *Argus*, it never would: "It wasn't put up to burst, it was
built expressly to hold water for irrigating purposes."[33]

By spring 1891 the reservoir at the diversion dam was almost filled with
water, and irrigation officials tested the dam by closing the headgates and
forcing water to run down the spillways. To many, the test justified the

belief that the dam was "one of the strongest structures of [its] kind in the United States."[34] Engineers used the best technology of the time, but the safety of the dam depended solely on the ability of the spillways to keep water off the crest of the dam, which was not built to withstand an overflow.[35]

In early 1891 the dam and thirty miles of canals were complete, twenty-four miles of canal were in operation, and over thirty miles of laterals were built that could deliver water to 15,000 acres. The local newspaper lamented that Bradbury and Company would soon finish their work on the "great [Southern] canal" and that E. E. Clark, superintendent of water services for the company, "Captain" Mann, and Jim Warren would soon depart the valley.[36]

In 1892 the irrigation company planned to spend another $60,000 to extend the Southern Canal from Black River to the Delaware River to water another 82,000 acres of land. By the end of 1892, near the southern end of the project, workers had constructed a second flume, 1,040 feet long, 14 feet wide, and 6 feet deep, to carry waters in the Southern Canal system across the Black River.[37]

The Northern Canal followed a mostly southerly route some eight miles west of the Pecos and roughly parallel to it.[38] W. C. Bradbury subcontracted with the railroad construction firm of Ward and Courtney to build twenty-five miles of ditch. Using 300 teams of horses and men, the company finished its work in early February 1890.[39] Project promoter and immigration agent Charles Greene, visiting the worksites in December 1890, noted that workers had completed the Northern Canal past the Felix River.[40] By February 1891 crews had placed thirty miles of the canal into operation.[41] By the end of 1891 workers completed the Northern Canal as far south as Tar Lake. Company officials estimated that the canal could irrigate 68,000 acres of land. By 1896 the Northern Canal stretched thirty-eight miles and, according to promoters, could water 200,000 acres, although only 12,500 were under cultivation by 1897 and only 4,000 were irrigated.[42]

After construction of the initial projects—the diversion dam, the Southern and Northern Canals, and the giant flume and terre plein—company officials focused on other works in the valley. The East Side Canal ran parallel to and east of the Pecos River and had a bottom width of twenty feet and was nineteen miles long, but only four miles were used. By 1897 it irrigated only 2,500 acres. This canal, like the Southern Canal, received water from the Main Canal stretching from the east side of the diversion reservoir south three miles along the east side of the Pecos before reaching the Pecos River flume. At that point, irrigation structures turned much of

Water over the main spillway at Rock (Avalon) Dam, about 1891.
(Courtesy Sacramento Mountains Historical Society, Cloudcroft, New Mexico.)

the water west across the river via flume and terre plein, while a lesser amount stayed on the east side and was diverted into the East Side Canal. The East Side Canal, which watered Hagerman Heights, was expected to irrigate 10,000 acres, and later to extend twenty-five miles to water nearly 50,000 acres of land.[43]

Other projects included the sixteen-foot-wide Hagerman Canal, which took water from the east side of the Pecos sixteen miles south of Eddy, where it ran for eight miles into a storage reservoir. At the reservoir's lower end, water was taken out and carried by canal another six miles to irrigate lands below the reservoir.[44] Expecting the canal to water 15,000 acres of land, the company planned to extend it another twenty-five miles. Ditch crews constructed thirty-five miles of major laterals on the Southern and Hagerman Canals and some fifty miles of sublateral ditches to allow individual property owners to "reclaim" and irrigate the desert.[45]

The company also planned to expand several other canals. Based on company figures, an extension to the Northern Canal would cost the company $30,000 and irrigate an additional 30,000 acres of land, while an extension on the East Side Canal, also estimated at $30,000, would provide water for an additional 25,000 acres of land. By the end of 1891,

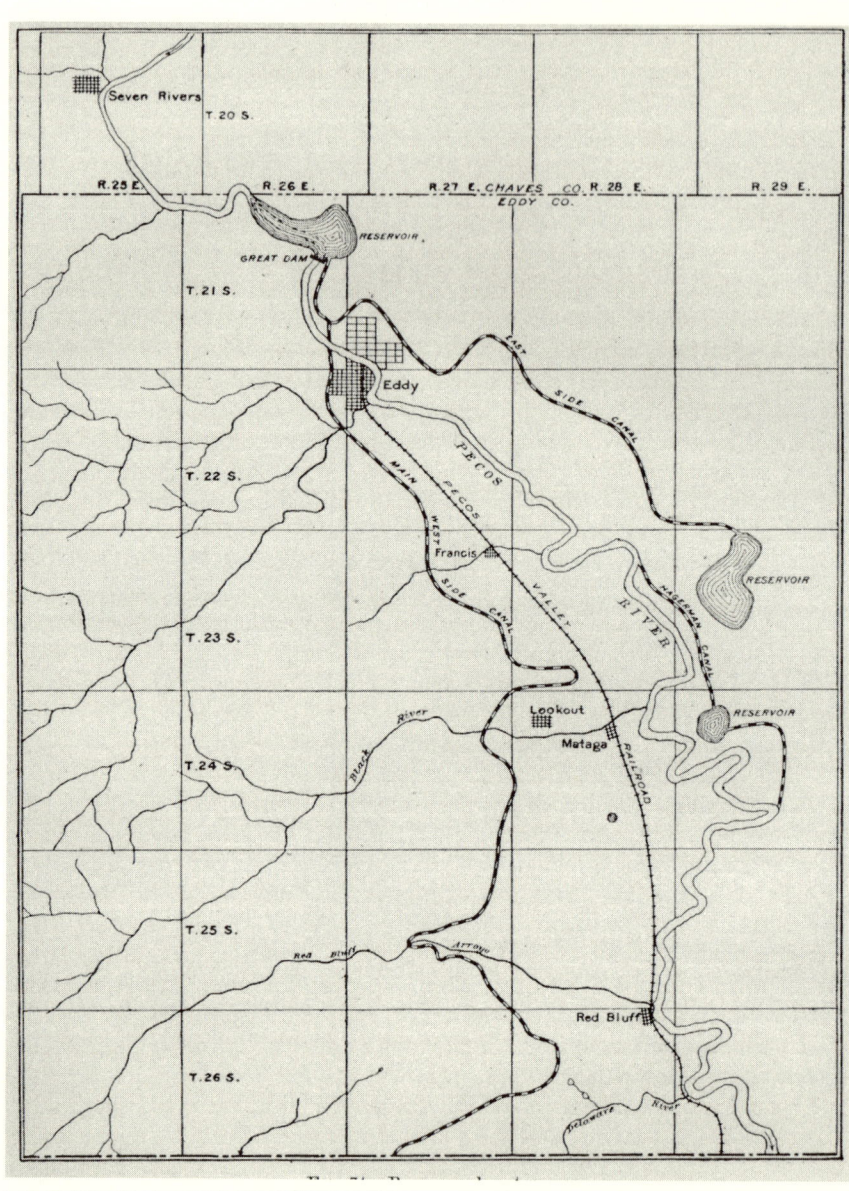

Pecos canal system.
(From Report of the Secretary of the Interior, *vol. 4, pt. 3,*
52d Cong., 2d sess., 1892, Ex. Doc. 1, pt. 5.)

construction crews had laid or planned to lay canals and laterals to irrigate 280,000 acres of land in New Mexico's Pecos Valley.[46] The company contemplated a total of 127 miles for all the canal works, not counting laterals dug to individual sections of land.[47]

Realizing the need for more water storage to fulfill its plans, Pecos Irrigation and Improvement began planning another, larger reservoir north of Avalon. In October 1892 laborers began building a second major reservoir in the valley. Seventy teams began excavation work, with another 100 due to come on board by the spring of 1893. Referred to as the "Seven Rivers Dam" because of its relative location, or Reservoir Number One, it soon became McMillan Dam, named after William H. McMillan, the president of a railway equipment manufacturing company in St. Louis, and a prominent investor. Louis Blauvelt worked as the chief engineer and chose the same rockfill design used on the Avalon diversion dam, but McMillan would hold eight times the water, enough to supply farmers with irrigation water for seventy-two days. Officials expected to spend close to $200,000 on the structure.[48] Completed in 1893–94, the dam was 1,686 feet long, 52 feet high, and 20 feet wide at the top.[49] A thirty-foot-wide canal extended from the east end of the reservoir. McMillan Dam had six massive wooden headgates, each four by eight feet, with iron gates six inches thick. The headworks alone cost $20,000.[50]

Working with tons of rock using nineteenth-century technology and horsepower was irksome, backbreaking work. Accidents occurred. In March 1890 an accidental explosion of "giant powder" caused the death of Ignacio Himenez and the injury of Meliton Gonzalez.[51] Some problems were minor in comparison. At one point during 1890, Bradbury and Pecos Irrigation and Improvement Company officials became concerned with numerous holes that appeared spontaneously in the canal system. Observers noted a number of prairie dogs burrowing into the banks of the canal and lateral network. Workers soon dispatched the rodents by applying shovels of sand and tamping the area.[52]

Expense proved to be the major problem. The combined irrigation works cost the company half a million dollars.[53] Preparing irrigated company lands for settlers and investors, railroad plans, and material and transportation costs caused financial woes throughout the life of the projects.[54] By the spring of 1891 James Hagerman and other investors had spent close to a million dollars on various works in the valley, and by the fall of 1893 that figure had doubled.[55] But because of the belief of Hagerman and others in the valley and its agricultural potential, by 1900 Eddy County's population had increased to 3,227. Population figures continued to increase

Headgate at Lake McMillan Reservoir. McMillan served as the storage facility for Pecos River waters released into Avalon Reservoir, the distribution reservoir downstream near Eddy (Carlsbad). McMillan suffered severe leakage and siltation over the course of its ninety-year history. In 1987, Brantley Reservoir inundated McMillan Dam and Reservoir.

(Record Group 115, Records of the Bureau of Reclamation, Public Relations Photograph Collection, Series 3—Historic, Photographs by Project Sites, Box 004, National Archives and Records Administration, Rocky Mountain Region, Denver Colorado.)

into the twentieth century.[56] Many people moved to southeastern New Mexico from Texas, and their parents often from somewhere else in the South, for example, Kentucky, Tennessee, Alabama, the Carolinas, Mississippi, Georgia, Arkansas, Missouri, or Virginia.[57]

Many of the people who entered the valley did so as part of Charles Eddy's plan to irrigate New Mexico's Pecos Valley, which required a cheap, massive labor supply. The flurry of activity brought workers to the valley that provided the skills and services necessary for growth. Progressive ideals often did not extend to minority groups such as Mexican Americans. The introduction of property taxes after 1848 caused many New Mexicans to lose land. Land taxes were based on the presumption that land was used for commercial enterprises, such as those pursued by Anglos, but the traditional economy of Mexican Americans was based on subsistence agriculture and stock raising, in which cash profits were scant. During the Spanish and Mexican eras, taxes were levied not on lands but on products of land, and were paid in kind. The demand for taxes in cash resulted in many cash-

Raising the headgates at McMillan Reservoir, about 1895.
(Courtesy Southeastern New Mexico Historical Society, Carlsbad, New Mexico.)

poor Mexican Americans falling into delinquency and losing land through tax sales.[58]

In the 1880s when commercial cattle operators entered the territory, Anglos took full advantage of tax laws to wrest land from Mexican Americans. By the late 1880s and 1890s, Mexican pastoral agriculture near Missouri Plaza and elsewhere had been replaced, and displaced Mexican American farmers ironically became the workforce that made possible the newcomers' prosperity.[59] Competition for water to feed nearly one million cattle in New Mexico by 1900 increased after the Newlands Act of 1902 went into effect. Irrigation unlocked the resources of the region, which led to economic growth made possible in large part by Mexican labor.[60]

Charles Eddy, charged with forcing Mexican Americans off their land, also became their major employer, recruiting Mexican nationals and Mexican American workers to valley projects for extremely low wages. Some of his workers were boys, making two and a half to five cents per hour, working ten-hour days, or $15 per month, to complete Avalon Dam and clear a site for the new town.[61] By comparison, Francis Tracy, who began working for the company as a crew foreman supervising Mexican laborers planting orchards and digging ditches, made $75 a month for his efforts.[62] A Roswell schoolteacher taught twenty students in a one-room schoolhouse and made $45 a month, $13 going for room and board.[63] Trail bosses hired to herd the Eddys' cattle in the early to mid-1880s made $100 a month.[64] Cowboys made much less.

Those promoting the growth of the valley during the 1890s tried to recruit a "better class" of laborers. Trying to lure prospective workers to the valley, the local newspaper reported that farmers needed laborers, presumably white, and paid $25 to $30 a month plus board. Pecos Irrigation and Improvement, Witt Brothers, who built the original Pecos River flume, and W. C. Bradbury were building large dams and railheads that required massive labor. Regarding the availability of work, the paper commented: "No sober, industrious man ever fails to find work here, his wages are of course regulated by what he does. If he is a laborer he gets $25.00 a month and board, if a carpenter $2.40 to $4.00 per day, if a brick mason $4.00 to $5.00 per day, and other trades and occupations in proportion. No man with pluck, energy and a little money need have any hesitation in coming to the Pecos Valley. He can't help doing well."[65]

In October of 1889 Bradbury advertised for "American men" to build canals, offering $1.75 a day, or $25 a month and board.[66] Too few men applied, so company official Jim Warren went to Pecos, Texas, to hire 200 laborers. He also went to Las Cruces, New Mexico, where he recruited 90 men, adding to the Mexican and Mexican American populations in Eddy County.[67] Most Mexican immigrants were imported under contract for particular industries or tasks.[68] Laborers often worked in undesirable locations like smelters and in agriculture in the deserts of the Southwest. In many cases Mexican workers had no machinery, nothing to work with but rudimentary tools.

Employers in the Southwest often segregated Mexican and Mexican American workers into their own camps, company towns, or *colonias* (colonies). Such was the case in the Pecos Valley. Low wages, cheap rent, low land values, and undesirability determined the site. This had the effect of reinforcing prejudice and stereotypes and prevented acculturation into the now-dominant Anglo culture. Eddy's work crews lived in tent cities close to the construction projects. So many workers had erected tents and shacks in the southwest part of Eddy that it became known as "Ragtown."[69] Local newspapermen, visibly impressed with the flood of workers into the valley, urged readers to visit the camps themselves: "The camp at the dam is a veritable city of tents. It is inhabited at present by one hundred and eighty laborers, besides a competent force of mechanics. There is a blacksmith shop, a harness shop, a commissary store, etc., and the immense dining tent for the men. Here and there and everywhere are sleeping tents. This canvas city is well worth seeing and the *Argus* suggests that Eddy people who have not seen it pay a visit."[70] One group of Mexican Americans lived in a settlement called Chihuahua, originally located on the west side of the Pecos River. Employers often used natural or artificial barriers to separate Mexicans from the Anglo residential areas. In August 1890 the general manager of the railroad company forced the settlement to move to the east side of the river in the northeast part of town. The entire population, some twelve families, abandoned their jacales and moved into dugouts on the eastern banks of the Pecos River, separated from Eddy by the river and the railroad tracks.[71]

Despite harsh living conditions, Mexican nationals were willing to move to southeastern New Mexico. Following the Mexican War, the Mexican economy became an integral part of the U.S. economy, especially as a labor resource: "A worker from central Mexico could move northward, from town to town, through the western United States, engaged in mining, farming and maintenance work for investor and production linked companies." According to some, in New Mexico railroads accelerated the redistribution and concentration of lands into Anglo hands.[72]

Following the first railroad built into El Paso in the 1880s, subsequent lines employed thousands of Mexican workers to lay track in the Southwest.[73] El Paso became a labor supply center, because by the turn of the century the Southwest relied primarily on Mexicans for hired hands. Many Mexicans moved north looking for work because they had lost their lands and therefore their livelihoods. The policies of Porfirio Díaz, who served as the president of Mexico from 1877 to 1880 and 1884 to 1911, prompted large landowners to dispossess the *peones*, or peasants, of their lands. Some five million families became landless. Men needing jobs soon learned the farther north they went, the more money they earned. In Chihuahua, hacienda (large ranching and farming interests) workers earned thirty-five centavos to one peso (between seventeen and seventy-five cents) per day, which was among the highest salaries in Mexico. But in Texas and New Mexico, they could often earn one dollar or more per day.[74] Until 1917 U.S. immigration policies allowed free immigration for anyone who could work and was not insane, a convicted criminal, or a prostitute. No passports were needed, and some men brought their families with them.[75]

While the companies recruited Mexicans for labor, the Anglo population looked down on them, regarding Mexicans as "indolent, undependable, degenerate, dishonest, impoverished, addicted to vice and gambling."[76] To help control the growing "undesirable" populations, Eddy's founding fathers placed great emphasis on morality. They thought their new town should stand out as a community of churches, little vice, no drinking, and as essentially upright in every way. This attitude in no small way reflected the desire of Hagerman and Eddy to attract hardworking family men to enhance the success of their irrigation project. It also reflected the national temperance movement to prohibit the manufacture, sale, and consumption of alcohol. From early on, founders made sure that settlers within Eddy stayed sober. Eddy, both the town and its namesake, fought hard to give the appearance of temperance in the Pecos Valley.[77] Irrigation company officials pursued this from the start: "It is pretty generally understood . . . that saloons, gambling dens, houses of ill repute, etc., will not be tolerated in Eddy. This place is going to be a temperance town, a decent town, a chaste town. It is going to be a place where good people can locate to bring up good children. It is patterned after Colorado Springs, Colorado, a town that has a national reputation as a strictly temperenced place."[78] They estimated that a town like Eddy without "a lot of beer kegs, dirty spittoons, bleary-eyed bums in front" would attract families to the region.[79] Pecos Irrigation officials made sure that Eddy stayed dry. "Eddy," crowed the local newspaper in 1893, "is the first city in the territory without saloons."[80] To encourage morality and sobriety, settlers built churches for many denominations, belatedly including Catholics.[81]

Despite all the promotional efforts to paint the area as progressive, sober, religious, and civilized, southeastern New Mexico was still a wide-open territory. Recently a haven for desperadoes and renegade Apaches, southeastern New Mexico experienced the last days of unbridled adolescence encouraged by the presence of large labor camps of single men. Although city father Charles Eddy, and later James Hagerman, forbade drinking in Eddy proper, liquor was readily available in locations nearby. At either end of the new town of Eddy, one mile north in Lone Wolf and two miles south in Phenix, men found any number of diversions to escape the labor of town building and watering the desert. Prostitutes, gamblers, and those with an affinity for hard liquor regularly spent their nights at either end of town.[82] Lone Wolf, or "Wolftown," located just west and slightly north of Eddy, was the closest area to buy liquor and other diversions before Phenix became a growing concern in 1892. Businesses in Wolftown intercepted laborers as they came into town from construction camps. Wolftown "druggists," close at hand, sold massive quantities of beer, stout, and whiskey.[83] While Lone Wolf saloon proprietors laid in a stock of liquor "for medicinal purposes," the Pecos Valley Town Company considered opening a road to the wooden flume that bypassed Wolftown altogether.[84]

Phenix, at the other end of town, with about 900 residents, served as a magnet for thirsty construction crews, outlaws, and Texans.[85] Saloon and brothel interests catered to workers and townspeople alike. Laid out in April 1892 by county surveyor B. A. Nymeyer, specifically to attract saloon men who had unsuccessfully tried to settle in the town of Eddy, Phenix lay approximately two miles south of town. Given the number of advertisements for liquor and beer appearing in Charles Eddy's company newspaper,[86] one has to wonder about his true motives in outlawing liquor sales in town. To placate workers and recoup some of their wages, Eddy worked with Nymeyer to supply them with alcohol. At the same time, promoters and company officials maintained Eddy's air of respectability for prospective investors and settlers. Curiously, Nymeyer and County Judge J. O. Cameron were business partners.[87]

Known by the euphemism "Jagville," Phenix attracted violence as a regular feature. The violence almost always involved minorities and workers constructing dams, rail lines, and canals in the area.[88] In the late 1880s and early 1890s, the violence that plagued the Pecos Valley was usually perpetrated within Phenix, Wolftown, Mexican, or other minority communities. Mexicans, Mexican Americans, Italians, and Chinese all found themselves in violent circumstances in the early 1890s. Racial bias ran deep in the Pecos Valley, and the general attitude of newspapers and officials seemed to be that although the Mexican workforce was important for building dams and canals, such workers were expendable.[89]

Both of Eddy's newspapers reflected the racist rhetoric and attitudes of the time in covering Phenix activities. Observers in Eddy had little sympathy for Phenix, and especially its patrons of Mexican origin. In January of 1893 the *Eddy Weekly Current*'s story "One Mexican Less" discussed the recent stabbing of one Pedro Hernandez by a Mexican countryman, following an evening spent at nearby Phenix. The incident occurred as the two made their way back to Chihuahua, the Mexican encampment on the banks of the Pecos.[90] The *Argus* added that someone disemboweled Hernandez with a knife, that an Eddy doctor was summoned to attend the injuries, but that Hernandez died anyway. The murderer apparently fled the country on a horse stolen from a black man in Eddy.[91] Later that month, under the headline "Another Mexican shot," the *Weekly Current* gave coverage to a Mexican national shot four or five times and found lying dead in the road north of Phenix.[92] Two months later, in March, the *Weekly Current* proclaimed "Another Mexican Killed!" and reported the shooting of one Martin Barias by Teofilo Trejo after the men left a Phenix saloon together.[93]

Local reporters gleefully related graphic and often bizarre details about violence almost every week. For example, Celzo Zammaron, a Mexican national, in what the local press called an apparent suicide, shot himself in the left eye with a .44 carbine Winchester. Witnesses found Zammaron's right finger still resting on the trigger, the muzzle against his face under his left eye where a bullet had torn out the eye and made an aperture extending all the way to the mouth. The victim's brains splattered against the walls of the building. The suicide was said to be the result of a refusal by a Mexican prostitute to receive his attentions any longer.[94]

A quarrel between two Mexican men in Chihuahua resulted in one getting a revolver and shooting the other in the leg. The shooter escaped before officers could capture him.[95] While some locals believed that escapes were fortunate as they saved the county useless expenses,[96] others believed that "jurymen have no right to say: they are Mexican; let them kill one another unmolested. Here was a chance to hang a foreign outlaw, who will probably again be charged to the county. The proper disposal of a fellow of his stripe might keep other outlaws from the neighboring republic from coming to this locality, and drive away some of the Mexican vagabonds who compose the population of Phoenix [sic]."[97]

Such statements printed by the company newspaper again demonstrate the hypocrisy and racism exhibited by the irrigation company. While the company had no qualms about hiring Mexican laborers to build dams and canals at fifty cents a day, officials publicly lamented Mexican activities in Phenix and Wolftown. Compounding the irony here is the fact that some saloons in Wolftown and Phenix were company drinking holes designed to soak up the meager wages paid to Mexicans by Eddy and the irrigation company.

Women openly plied their trade as prostitutes in Phenix and Wolftown. In March 1894 a fifteen-year-old Mexican girl named Theodora Guerra died at Phenix from an overdose of morphine. The *Argus* surmised that Guerra, suffering from pneumonia, probably ingested the drug with the intent of committing suicide. Lured away from her parents in Texas, the girl had been placed in a Phenix bordello.[98]

Phenix attracted spectacle as well. In May 1893 Phenix hosted a boxing match between William "The Kid" Davis and C. E. Welch, "the Dallas painter," at the King Theater stage for a ten-dollar purse.[99]

Fires were common in Phenix. On June 2, 1893, the *Weekly Current* reported that a great fire burned all but two Phenix establishments when a barrel of whiskey ignited spontaneously. Within three minutes, buildings went up in flames. A northern wind spread the fire. Since the only water available was a well behind the saloon, patrons saved some buildings by pouring bottled beer on the flames. The sheriff, attempting to put out a fire on the roof of one building, fell and sprained his leg. The *Argus* reported that about "one hundred Mexicans stood upon the railroad and watched the fire without offering to assist in saving the property." Despite Charles Eddy's attempt to recover wages paid to Mexican laborers, many of the businesses that burned refused service to non-Anglos. The image of Mexican workers standing on the railroad while Anglo saloons burned down symbolized the frustration and resistance by workers keenly aware of the valley's double standard.[100]

As was the case in Phenix, in 1890 and 1891 Wolftown saw much violence. Local observers noted, "the boys occasionally shoot up the Wolf in the early hours of the morning. The *Argus* advises them to be careful. The penalty for such amusement is very heavy, and they will certainly be arrested."[101] Oftentimes, if someone was shot, particularly a Mexican or Mexican American, few witnesses came forward to report what had taken place. In one particular situation, a Mexican had been shot and killed near "the Wolf." Although the shooting occurred in a crowd, no one volunteered any information until two days later, when it was reported that an Anglo Monte dealer fired the fatal shot, mounted a horse, and started either to Lincoln to give himself up or south to escape. The Mexican allegedly insulted the dealer, who pulled out a revolver and shot him between the eyes.[102]

Not all area violence occurred in the watering holes. Some occurred in construction areas. One worker, reminiscing on his earlier days building irrigation projects, remembered that the territory was full of bandits: "They ha[d] long hair and beards and would ride through the countryside stopping wagons . . . killing the people and taking their horses."[103] Pointing out the violence of the times, young Frank Pompa, a worker at both dam sites, recalled that a train came through Eddy bringing mail and payrolls for the work crews. Riders would go to the train and pick up the mail and payroll, and on one occasion Pompa came across a scene moments after bandits stole the payroll and shot a number of men and their horses. Pompa still recalled seeing blood running down the road.[104]

Racism and competition between the Anglo "Americans" for whom Bradbury had advertised and the Mexican crews sometimes erupted in violence. On April 17, 1891, Mexican workers apparently involved themselves in some sort of "conspiracy to do up the Americans," but were restrained by some of their fellow countrymen.[105]

Few minority groups escaped racial violence. In December 1892 an Italian named Pompais Bozake violated the law against carrying deadly weapons when he was found with "a small cannon buckled around his body, at Phoenix [*sic*]. . . . [T]he court after investigating the case assessed a fine of $50.00 in costs which the Italian promptly produced from the recesses of a pair of baggy pantaloons and departed a wiser man."[106] Cases such as this where someone fired weapons either accidentally or with intentional damage multiplied during the hectic period of dam building.

A small number of Chinese lived in Eddy in the early 1890s. In another instance of violence, Wing Lee, whose laundry was next to the offices of the *Eddy Weekly Current*, found himself a potential victim because of his dog. When five shots rang out, news reporters and others gathered around the laundry where they peered inside a window and witnessed a white man holding a revolver and swearing at the top of his lungs that he would kill every Chinaman in Eddy if Wing Lee's dog, chained within ten feet of the laundry door, ever snapped at him again. The man, whose name was Horton, apparently ventured too close to the dog, which snapped at him but did not bite him. Horton went to his house, procured a revolver, then fired five shots at the dog, but was so upset (or inebriated) that he hit nothing except the side of the building. Wing Lee had secured the dog to guard laundry hanging near his establishment where someone had stolen several articles in the past.[107] That the townspeople held the Chinese in low esteem is apparent from an *Argus* article. It states that at the Chinese New Year's celebration the men were taking "draught like Mexican [men]," but they were not drunk. The author supposed "that will come tonight after they hit a pipe [of opium] or two."[108]

Concern about violence, alcohol, drugs, and prostitution prompted the company-controlled *Eddy Argus* in 1889 to lament that what Eddy needed was a religious revival. Commenting that many of the best citizens in town were gradually falling from grace, the newspaper feared that before long Eddy would get the reputation of being a "wicked place."[109] Citizens often heard loud, profane language. One drunk man lay in the street for hours before being removed.[110] C. F. McDonough and H. A. Bennett originally owned saloons in Eddy, but moved to Phenix when the courts upheld Eddy's prohibition clause. Sensing an opportunity, when the judge and most local law enforcement officials were out of town in 1892, the two confiscated a town lot, opened a tent saloon, and did a good business. Arrests soon followed for selling liquor without a license, but the saloon stayed open a week.[111] Local druggists in Eddy attempted to sell liquor from a backroom on the sly, but the Pecos Valley Town Company came down hard. In the 1892 case of the Town Company versus James A. Tomlinson, Tomlinson—a local druggist—was found guilty of selling intoxicating liquors and fined heavily.[112]

Matters became so bad that during the first week of December 1892 local citizens of Eddy gathered in the local Methodist church to "consider some questions relating to the execution of laws relating to public morality."[113] Led by local real estate agent Charles H. McLenathen and

Judge J. O. Cameron, the group decided that law officers had fallen down on the job. Local citizens had obviously desecrated the Sabbath, and saloons sold liquor to habitual drunkards and minors on a regular basis. The group applauded federal judge A. A. Freeman for enforcing morality laws, and the Pecos Valley Town Company in enforcing deed contracts, which stipulated that no one sell intoxicating beverages in Eddy.[114]

In September 1893 city fathers also tried to rid the town of what they considered to be another major vice: gambling. Trustees passed an ordinance prohibiting gambling within the city limits. Charles Eddy and James Hagerman had not prohibited gambling in the deeds to town lots, and several operations had opened in the town, operating under city license. When those licenses expired, city fathers took action to eliminate what they considered a nuisance. Local gamblers grumbled over the new law, since the penalty for breaking it was $100 to $1,000 and imprisonment. Finding Eddy no longer receptive, local gamblers packed their bags and headed for Phenix.[115]

The town of Eddy was having a hard time living up to the standards of respectability portrayed by promoters and their literature. It was one thing to lure prospective customers to the valley. It was another thing to convince them once they got there that the naked prostitution next to them was part of a progressive, civilized, churchgoing community. The more conservative elements of the new town, specifically irrigation company officers, decided to regulate such distractions because they were bad for business and the efficiency of building dams.[116]

Following the meeting at the Methodist church in 1892, McLenathen and his committee secured Deputy U. S. Marshall and Eddy County Sheriff Dee Harkey to clean up activities at the end of town. In fact, outraged that the town of Phenix existed at all, James Hagerman insisted on materially doing away with the place.[117] Company officials complained that "Phenix is filled up with saloons, dance halls, gamblers, prostitutes and those women will come out naked when we drive by there and get in our hacks and buggies. We dare not drive by there with our families or a prospective purchaser for any of our [V]alley land. [We implored Harkey to] keep those women off the streets when they are naked, and out of our hacks when we are passing that way, [and] we will be satisfied."[118]

Harkey subsequently went to Phenix and asked residents to curtail their public displays for the benefit of the nearby Eddy citizens. Apparently, the idea fell on deaf ears, as one of the local leaders threatened Harkey should he attempt to stymie what was obviously a booming business. With the full sanction of Judge Cameron, Harkey commenced what amounted to nightly raids, rousing from shacks adjoining the Phenix dance halls some sixteen partners per night and charging them with adultery under the Edmunds Act, which prohibited polygamy.[119] Harkey then placed the offenders on hacks that were driven, under guard, to Socorro, where they were placed in jail and held for trial in the Fifth Judicial District Court of the New Mexico Territory, a federal court presided over by J. O. Cameron's father-in-law, federal judge A. A. Freeman. Several arrested in Phenix made bail and went back to take revenge on Harkey, but both sides avoided violence.[120]

Politics helped determine Phenix's fate. Business interests in Phenix ran candidates for sheriff, county clerk, and county assessor in November 1894 and lost all three races in an election in which some returns were lost and the results of the sheriff's race were contested. Not to give up completely, the Phenix interests ran two more candidates in January 1895. To ensure their success in the race for constable and justice of the peace, the Phenix group assembled 300 Mexican workers from the various irrigation and dam projects near Eddy and south near Toyah, Texas. The Phenix stalwarts plied workers with food and whiskey in exchange for votes. Provided with music for dancing, the workers, most of whom were alleged to be Mexican nationals, gathered behind a fenced lot in the center of Eddy and celebrated for three days beneath a huge Mexican flag. When it came time for voting, the Eddy contingent challenged almost every Mexican vote and consequently won the election. By the summer of 1895, Ed Lyle, leader of the Phenix group, conceded defeat and agreed to leave New Mexico Territory and never come back if Judge Freeman released the people in Socorro and dismissed their cases. Fifteen days later, Lyle hired wagons and teams to take the remnants of Phenix's illustrious citizenry to Arizona. In the lead wagon, pulled by six horses, a band played a farewell tune as the entourage left Eddy and headed west.[121] Tongue in cheek, a local newspaperman reported, "none who leave . . . are or ever have been tillers of the soil but have followed other occupations."[122]

The population responsible for transforming Eddy from village to town came from a variety of places and backgrounds. A significant Mexican population worked as day laborers or farm laborers. Some herded or sheared goats. The Swiss and Italian colonists who remained in the valley continued to farm. Immigrants from New York, Pennsylvania, and Connecticut tended to be professionals and merchants, although their number included a few farmers, with several more farmers from Illinois and Ohio. Occupations ran the gamut from druggists to clerks, sheep inspectors, lawyers, salesmen, life insurance salesmen, blacksmiths, bookkeepers, photographers, stock raisers, carpenters, telegraph operators, barbers, teachers, nurses, publishers, electricians, ministers, and dairymen. Other occupations included plumbers, painters, postmasters, well drillers, civil engineers, dishwashers, copper miners, gamblers, saloon proprietors, butchers, tailors, shoemakers, teamsters, and a number of women who worked in "sporting houses." The few Chinese in the valley served as launderers, cooks in hotels, and restaurant workers.[123] While the influx of population stabilized the area and formed a solid economy, the impetus for the movement was Charles Eddy's efforts to water the valley.

5

COMPANIES, CANALS, AND CAPITAL

WHILE workers continued to build dams and canals, James Hagerman established a series of companies in the Pecos Valley designed to expand development, to attract capital, and eventually to place control of activities in his hands. Hagerman's goal of expanding canals and reservoirs was based on building a railroad connected to outside markets. As costs mounted, Hagerman restructured and refinanced his operations only to face an economic downturn and devastating flood in 1893.

In 1887, prior to James Hagerman's involvement in the Pecos Valley, Charles Eddy hurried construction of the Halagüeño Ditch because of the New Mexico territorial law of February 24, 1887. According to the law, "it was to be understood that the person who actually began construction work on a canal would obtain the first appropriation right" to a river's water. Eddy was afraid that Long and Jaffa, who wanted to irrigate upstream near Roswell, would beat him to the appropriation.[1] In October 1887 he and Pat Garrett organized the Pecos Valley Land and Ditch Company. The incorporators of the company included Eddy, his brother John Arthur Eddy, Pat Garrett, Colorado polo player Joseph Stevens, Elmer E. Williams, and Arthur A. Mermod, the former neighbor of the Eddy brothers in Colorado. The Pecos Valley Land and Ditch Company was empowered to provide irrigation in return for rental fees and started with $40,000 in capital stock.[2]

On July 18, 1888, to accommodate the expanding vision for irrigation in the valley, Eddy reorganized the Pecos Valley Land and Ditch Company into the Pecos Irrigation and Investment Company. The new company was entitled to issue $600,000 of capital stock (6,000 $100 shares), and was authorized not only to provide irrigation waters but also to provide for "the colonization and improvement of lands."[3] The articles of incorporation of this second company also allowed it to purchase all canals and rights-of-way of Pecos Valley Land and Ditch, including the Halagüeño, and to enlarge the canal to "utilize all of the waters of the Pecos River not [already]... appropriated."[4] Organizers of the new company included Charles Eddy, Charles Greene, and Chicago investors such as Luther P. Bradley, James McKay, and Charles A. Gregory. Charles Eddy moved the records of the former Pecos Valley Land and Ditch Company from La Huerta to the new town of Eddy, where Pecos Irrigation and Investment set up offices. The company established branch offices in Chicago and Colorado Springs to recruit and accommodate new investors.[5] In 1888 the officers of Pecos Irrigation and Investment included the following: president, General L. P. Bradley; treasurer, G. B. Shaw; secretary and manager, Charles W. Greene; vice president, Charles Eddy; and superintendent, Patrick Garrett. Bradley and Shaw, who both lived in Chicago, primarily acted to recruit investors, not to manage the company's land or irrigation development in New Mexico.[6]

Pecos Irrigation and Investment took over the canals and water claims from Pecos Valley Land and Ditch. The new company issued $250,000 of stock and then issued a subsequent $150,000 worth of stock, which sold at fifty cents on the dollar. Newspaperman and promoter Charles Greene served as the company's primary immigration agent, trying to persuade investors in England and Switzerland to buy stock and move to the Pecos Valley.[7] He sold considerable stock in those countries through Basel banker Julius Bernouilli, but his efforts never lived up to Charles Eddy's or later, James Hagerman's, expectations. Eddy believed that Greene spent too much money with too few results. Disenchanted with Greene's management and spendthrift ways, Eddy arranged to have himself appointed manager of the project instead. Bradley resigned as president, and Chicago investor Charles R. McKay succeeded him.[8]

At this point James J. Hagerman entered the Pecos Valley through the introduction of cigar maker Robert Tansill early in 1889. In May Hagerman took a trip to the valley. He then bought, at Tansill's urging, $60,000 worth of par value company stock for between $30,000 and $40,000.[9] Hagerman, like Eddy, believed that Charles Greene's sales and fund-raising efforts had been without merit, but, more important, he lamented that the company had paid too much for the rights of the Pecos Valley Land and Ditch Company, the company started by Eddy, and that few of the planned improvements had been completed. Hagerman noted that the valley would never be a success without building a railroad into the region, a subject with which he was thoroughly familiar.[10] Hagerman presented his findings to the Chicago directors of the company, indicating that unless they came forward with much more substantial capital the valley projects would fail. With Hagerman's logic ringing in their ears, the directors refused to authorize more funding for the project and instead sold their stock to Hagerman. Hagerman proceeded to distribute stocks among his friends and name his own board of directors. Because of his illness and other business concerns, Hagerman directed the company from Colorado and Europe. Charles Eddy served as manager of the New Mexico operations.[11]

Eddy apparently had a profound role in convincing Hagerman that the Pecos Valley would become an agricultural mecca. Anna Osborne Hagerman, Hagerman's wife, expressed concern about Charles Eddy's influence over her husband.[12] Though until 1893 Hagerman had spent little time in the valley, he frequently wrote that the Pecos Valley was the best sort of project he had ever seen.[13]

Hagerman's influence and money buoyed the irrigation company, and by the end of 1889 Pecos Irrigation and Investment had sold 4,000 of 6,000 shares authorized by the directors and had sold $190,000 of the $400,000 8 percent bonds authorized.[14] On December 28, 1889, Hagerman wrote to John R. Holland, an English stockholder who had invested through Greene. Hagerman wanted to sell stock to Holland at par, expecting to keep at least 1,000 shares as his permanent investment. By December 1889, of 4,000 shares issued, Hagerman owned 2,170 shares with a face value of $100 each, the Eddy brothers owned 940, Robert Weems Tansill owned 600, and several different investors owned the remaining 290 shares.[15] At this point Hagerman was extremely confident that he could make hundreds of thousand of acres in the valley productive. He based all of his early estimates in 1889 on the report of E. S. Nettleton, the engineer who first surveyed the valley for Charles Eddy. Hagerman's figures indicate that he thought the Pecos Valley projects could be built for something like three dollars an acre, or less.[16]

Uppermost in Hagerman's mind at the end of 1889 was increasing the settlement of the area to sell land, water "rights," and water to make a profit. Hagerman realized that large companies would not invest in building a railroad through the valley unless there was significant interest in the irrigation company itself.[17] Early in 1890 Hagerman reorganized Pecos Irrigation and Investment into the Pecos Irrigation and Improvement Company to expand the powers of the corporation and to increase available capital.[18] The company was officially incorporated under Colorado law on April 29, 1891. The new Pecos Irrigation and Improvement Company's purposes included (1) acquiring, constructing, and maintaining water ways for irrigation, mining, manufacturing, and domestic and public use; (2) acquiring rights, franchises, and privileges in and title to waters and the use of water in natural streams, springs, and lakes and the right to divert those waters; (3) selling water rights and distributing water, charging and collecting water rentals for receipts and use of water; (4) acquiring lands, by purchase or otherwise, including exchange of water rights, rentals, or stock of company for lands to manage, improve, develop, and sell; (5) acquiring properties, rights, and franchises and lands claimed by Pecos Irrigation and Investment; and (6) acquiring, owning, managing, and selling property, bonds, stocks, and securities of other corporate bodies, including railroads.[19]

Hagerman and Eddy in 1890 began organizing the Pecos Valley Railway Company to build a line from Pecos, Texas, to Eddy, New Mexico, and eventually to Roswell. Hagerman raised $150,000 by par sales of stock in Pecos Irrigation and Investment, which was used to extend canals and construct reservoirs. He hoped construction would bolster support for building and extending the railway.[20] Hagerman also organized the Pecos Construction and Land Company with a capital of $450,000 to construct the ninety-three miles of road from Pecos, Texas, into Eddy. Engineers estimated the cost of construction at $860,000. In mid-1890 Hagerman authorized the sale of railroad stocks and bonds at the rate of $12,000 for each mile. Subscribers put up $860,000 in cash, and in return they got $2,000,000 in par value of Pecos Valley Railway Company and Pecos Irrigation and Improvement Company securities.[21]

Hagerman himself was the single largest purchaser of these stocks and bonds. For example, he personally owned 25 percent of the stock in the Pecos Valley Railway Company. The Pecos Irrigation and Improvement Company, which Hagerman controlled as a major stockholder, owned another 20 percent of railroad company stock. A group of investors recruited by investment bankers Sanford and Kelley from New Bedford, Massachusetts,[22] owned another 20 percent. Charles Otis, the Cleveland steelmaker and Hagerman's close friend,[23] purchased 15 percent; and Sidney Dillon, a director of the Union Pacific, Charles Head, and W. C. Andrews controlled smaller amounts.[24]

By placing various small companies under the umbrella of his new Pecos Irrigation and Improvement Company, Hagerman was able to raise more funds to build reservoirs and eventually to build railroads. Each new company was complete within itself, with managers working closely together. This was more expensive, but was believed to be more efficient and economical in the long run. The concept began with the Pennsylvania Railroad under Jay Gould, when he forced the railroad to build an interterritorial line and convinced other companies to consolidate and expand.

Marcus Hanna, Hagerman's friend and business partner, may have influenced his management of the valley companies, as Hanna relied on a similar system.[25]

Hagerman, like many businessmen of his stature, retained ownership of the company even after he went to the stock markets for capital. But also like many entrepreneurs, Hagerman did not take an active part in day-to-day management of the company. He hired a manager, in this case Charles Eddy. Because the owners and managers had different interests, the owner to make money, the manager to create a stable company, the two groups often clashed.[26] Eddy's role in all of this was magnified, because unlike many managers he owned stock in the valley enterprises, particularly the Pecos Valley Town Company,[27] which influenced his priorities.

Hagerman continued to believe that land sales and irrigation would boost the confidence of investors. In a circular to stockholders, dated May 7, 1890, Pecos Irrigation and Improvement Company officers stated: "It would be gratifying to the management to know that the plans proposed would meet with the approval of every stockholder. We have no hesitation in expressing the opinion that when fully carried out, they will more than double the value of every share of stock and make our enterprise one the best of its kind in the United States."[28] From 1890 through 1893, the company launched an aggressive campaign to attract settlers to buy land and water rights, the right to have water delivered to their property.

The company continued to accumulate land in exchange for water delivery to the land retained by the customer. Land provided the main basis of financing the endeavors in the valley. Investors and "dummy entrymen" filed on sections through the Desert Land Act, then the company bought the land outright for between ten and thirty dollars for the entire section, or the entryman would deed 560 acres to the company in exchange for paid-up water "rights" on his choice of an eighty-acre tract in the section. After securing lands from various individuals in the valley, the company turned around and resold it to new settlers at from thirty to thirty-five dollars an acre.[29]

In making such land sales, the company did not get all the money up front. The company demanded one-tenth down, the balance of payments to be made in nine installments, with interest at 6 percent. The company also sold purchasers of land the right to use water from its canals for $10.00 to $12.50 an acre, payable with one-tenth down and the balance due in nine annual installments. In addition, the company charged owners of "water rights" an annual water rental fee of $1.25 to $1.50 per acre.[30] By the end of 1892 Pecos Irrigation and Improvement Company was owed nearly $1 million by individuals who had purchased land and water rights.[31]

Although Pecos Irrigation and Improvement conducted a very active immigration business, the return on sales did not live up to expectations. While some $1 million in land had been sold, the company realized only $100,000 in the first year, the rest to be paid over a decade. As the cost of building reservoirs and canals increased, sales for land and water-right notes sagged, and Hagerman found it necessary to invest more of his own money to pay for construction and operation of the system.

The stockholder's report for December 1891 indicates a remarkable amount of construction and activity, but Hagerman failed to grasp the enormous complexity of irrigating the valley. The vastly expanding construction of canals and ditches in the valley exemplified Hagerman's and the other promoters' boundless enthusiasm for unlimited growth in the valley. Everyone seemed to accept on faith that the Pecos River would have an adequate water supply to water hundreds of thousands of acres of land.

Hagerman's vision for irrigating the Pecos Valley did not stop at the New Mexico boundary. He purchased from William Walter Phelps some 45,000 acres lying in alternate sections in Texas, just south of the New Mexico line.[32] This land was placed with the Pecos Land and Water Company, another Hagerman company. Plans called for irrigating some 30,000 of these 45,000 Texas acres by building, unlike the massive dams upstream, a simple diversion dam across the Pecos River. Another 40,000 acres of land bought from the state of Texas brought the total of proposed irrigated Texas lands to 70,000 acres.[33]

Optimistically, Hagerman proposed building a third storage reservoir close to the Texas line to include an additional 50,000 acres of irrigable land in southern New Mexico, thus bringing the total proposed acreage in the Territory to 330,000 acres. This did not include the 70,000 acres proposed for irrigation in Texas. Including the 70,000 acres in Texas, in 1891 the grand total Hagerman proposed to irrigate from the Pecos Irrigation and Improvement Company system came to 400,000 acres. The hubris of engineers at the time led them to believe that irrigating 400,000 acres was not only reasonable but a task made easy by the region's level land, and, perhaps more important, by the engineers' and investors' abiding faith in the human use of science to control water and nature.[34]

The Improvement Company's ability to make money from such grandiose plans depended on the sale of perpetual water rights and land at higher prices once it was irrigated.[35] If the dreams of irrigating some 400,000 acres and selling them had come to fruition, it is likely a profit could have been made on such an investment. Hagerman tried to convince investors that the 400,000-acre figure was only the beginning. On November 18, 1891, Hagerman wrote to Lombard, Odier, and Company, a

Swiss investment firm that later invested half a million dollars in the valley enterprises,[36] explaining that California's experience showed that "where irrigation has been carried on for a considerable time . . . after 10 or 12 years when the soil becomes thoroughly saturated, it does not require more than 1/2 as much water as we have estimated. The belief is universal among those competent to judge that in time our supply will be sufficient for more nearly 700,000 acres than for 400,000."[37] In retrospect, it appears that Hagerman and others, including "competent" engineers like Nettleton, had little grasp of the reality of irrigating the Pecos Valley. But during the heyday of irrigation plans in the early 1890s, the sky was the limit.

In the 1891 stockholder's report, Hagerman proposed a further expansion of the corporate structure. Hagerman wanted to increase the Improvement Company's capital to $1,750,000 and authorize $800,000 in bonds at 6 percent interest. The company planned to use $300,000 of the $800,000 to retire the old 8 percent bonds of the Investment Company, the predecessor of Pecos Irrigation and Improvement, and the remaining $500,000 to expand and improve the project. The 1891 stockholder's report also showed that Hagerman had personally advanced some $80,000 in cash to Pecos Irrigation and Improvement and another $33,000 to Pecos Land and Water Company. In addition, the Improvement Company owed Hagerman $100,000 for lands it had purchased from a separate Hagerman company, Hagerman Irrigation and Land Company. Hagerman, Charles Eddy, and other company leaders estimated that over a ten-year period, based on a program of selling water rights for 24,000 acres per year, profits would increase annually until in the tenth year the company would be making $684,120 in water-rights sales and rental fees.[38] Such projections were ambitious and highly optimistic given that irrigated acreage from Roswell to Texas does not exceed 50,000 acres today.

Projecting cash needs for 1892 and 1893, company officials estimated that they would need over $697,000 to complete projected improvements on canal works in New Mexico and Texas. Of this, officials proposed to spend $120,000 to extend the canals in New Mexico, some $155,000 for the proposed storage reservoir in the Seven Rivers area, and $60,000 for construction of a Texas canal to water 30,000 acres in Texas. Another $45,000 would buy the outstanding stock of Pecos Land and Water Company, which owned the Texas land, placing all the land under the control of the Improvement Company.[39]

In 1891 Hagerman and other company officials also decided that they needed more water storage to complete their ever enlarging plans. These plans now included what was to become McMillan Reservoir. The company directors in 1889 had assumed that the flow of the Pecos would be sufficient to water the enormous land developments contemplated in the valley, but Hagerman's plans were more ambitious. So optimistic in the early days, Hagerman had estimated in the fall of 1889 that to complete the upper (Northern) canal to the Felix River and the lower (Southwestern or Southern) canal to the Black River would require issuing $250,000 of bonds in the old Investment Company. He optimistically told potential investors that construction resulting from the issuance of these early company bonds would mean water for about 800,000 acres of land. He estimated it would cost no more than $75,000 to add another 120,000 acres, because as he put it, "the main canal would be extended over nearly level country in easy ground free from work."[40] Besides Hagerman's growing interest in expanding the valley projects, in 1891 he saw a possibility to sell his entire enterprise in the valley and make a profit as he had done with the Chapin Mine and the Colorado Midland. During the early part of 1891, Richard Hutton and Charles Thurlow of Colorado Springs approached Hagerman with an idea of selling the entire Improvement Company to investors in England. During the 1880s and early 1890s, there was tremendous investment by Europeans in American enterprises. For example, large ranching syndicates formed to capitalize on the vast stretches of land available in northwest Texas. Texas, a state that retained its public lands on entering the union, sold millions of acres to English and Scottish investment companies. So many foreigners, especially the British, invested in the United States that the U.S. Senate called for the investigation of a number of British land and cattle companies.[41]

Hagerman realized that his health might prevent him from carrying through his plans for the valley. At the same time, apparently, Charles Eddy's health was not good either. Between the first months of 1891 and the summer of that year, Hagerman attempted to sell all of his companies. Hagerman gave the English investors the option to buy 70 percent of the Improvement Company stock, or 6,300 shares at $220 each, for a total of $1,386,000; all of the Pecos Valley Railway Company stock, 9,160 shares at $40 per share, for $336,400; and 70 percent of Pecos Valley Town Company stock, 1,266 shares at $150 per share, or $189,000. The total value of the option covered amounted to $1,942,000.[42]

Eddy, who owned most of the shares of the Town Company, worried that he would not have enough money to run it and badly wanted the British deal to go through.[43] Thurlow and Hutton spent weeks in the valley examining all facets of the enterprise, gathering as much information as possible to sell it. To confirm their claims, they sent into the Pecos Valley Major George M. Wheeler, who like Clarence King had become synonymous with surveys of the West.[44] Wheeler also made a very favorable report, and Thurlow and Hutton were about to complete the sale to the English investors when fallout from the Baring Bank failure spoiled it.[45] According to Hagerman's son, Percy, Clarence King carried on a considerable correspondence with the British investors trying to pull the deal through, but by December 1891 Hagerman knew that the British were out.[46] Negotiations for selling to the British were still underway when Hagerman left for Europe in July 1891 to meet with Henri Gaullieur and the Swiss investors Lombard, Odier, and Company of Geneva.

By the time Hagerman arrived in Switzerland, Gaullieur had already approached the firm with his impressions of the Pecos Valley. Hagerman's reputation, combined with Gaullieur's favorable opinion of Pecos Valley activities, predisposed the firm to invest in the valley. Lombard, Odier came on board late in 1891 and became major investors in the valley projects. Before a year passed, the Swiss firm bought $250,000 worth of bonds and $250,000 worth of stock in the Improvement Company, most of which were then sold to its clients. Consequently, Swiss investors became one of the most important elements in the investment landscape of the Pecos Valley and in the Improvement Company. Before making such a major purchase of stocks and bonds, Lombard, Odier sent a representative by the name of Sautter to verify Gaullieur and Hagerman's favorable reports. Sautter also became enthusiastic about investment in the valley.[47] The relationship forged between Lombard, Odier, and Company and Hagerman in regard to the many irrigation ventures in the Pecos Valley appeared promising.

But by the end of 1892 Hagerman and the other major stockholders admitted that the establishment of a huge system of irrigation required even more money and was even more difficult than they had originally thought. The 1892 annual report, sent to stockholders on April 20, 1893, stated that funds for the projects were still inadequate. The officers also revealed that the Swiss colonization experiment had been less successful than expected. Many of the immigrants had been inexperienced farmers and had returned to Switzerland within a year or two, forfeiting mortgages and leaving the Improvement Company with land but no income.[48]

Company officers must have realized that their need for more water was imaginary. The Improvement Company could not sell the land or water they had, let alone the thousands of acres they hoped to irrigate. Still, Hagerman and company officers spoke of building not only McMillan Reservoir north of Eddy, but a third reservoir somewhere on the Hondo River in the northern part of the project. In the spring of 1893 the company still publicly projected potential irrigation to water fully some

377,000 acres of land in the valley. This claim rested on the fallacious argument that less water would be needed per acre once the region was irrigated. The Bureau of Immigration promised irrigators that once the ground was thoroughly wet, there would be no difficulty in getting it wet the second time. Yet by the end of 1892 the newly financed Pecos Irrigation and Improvement Company still had not sold $1,236,000 out of the total authorized stock issue of $1,750,000; $300,000 of the 8 percent bonds and $433,000 of the 6 percent bonds were also unsold.[49] The Improvement Company spent more than $1.5 million, including a paper transaction of $840,000, to acquire the property of the older Investment Company, and $733,000 for the construction of canals, laterals, dams, and other improvements. Out of the $733,000 in new construction monies, the company spent over $100,000 for improvements in Texas.[50]

Despite Hagerman's reorganization scheme, he had to provide most of the money himself when a large percentage of the stock and bond issue languished unsold. Bond sales failed to meet the company's expenses, as the monies necessary to expand the project in 1892 were larger than they had been in 1891. Hagerman increasingly found himself paying for construction and improvements on the projects. By the end of 1892 the company owed him a quarter of a million dollars: $126,000 on notes and an equal amount for monies that he had personally advanced. Included among construction costs of the operation were $146,000 for the Seven Rivers area dam, another $67,000 for a dam on the Hondo River, $50,000 in Texas for canal work, and another $50,000 to purchase more lands. The total outlay required in 1892 neared three-quarters of a million dollars. In addition to the cost of the construction work, the company spent $55,000 attracting immigrants to the valley.[51]

From nearly the inception of projects in the valley, company officials planned to expand their ability to supply water to outlying lands. Curiously, only in 1893, after Avalon and much of McMillan Reservoir were completed, did James Hagerman discover that the Eddy-Bissell Livestock Company, not Pecos Irrigation and Improvement, owned much of the McMillan Reservoir site and the lands where Avalon now stood.[52] While Charles Eddy ostensibly worked for James Hagerman and the Improvement Company, it became apparent by 1893 that Charles Eddy, given a long leash, often worked for himself. Hagerman paid $36,000 for the land out of his own personal account following arbitration between the cattle and irrigation companies in New York City. By the fall of 1893 James Hagerman and other investors had spent $2 million on various works in the valley, and, as indicated by the reservoir negotiations, costs were soaring.[53]

Hagerman and company officials, concerned with mounting costs, sent out a circular in April 1893 that discussed the problems and expenses in constructing the irrigation projects. Furthermore, company officials emphasized the need to extend the railroad from Eddy to Roswell and then on to Las Vegas, New Mexico, as soon as practicable. In their minds this was absolutely necessary to provide markets for the region's products and expedite settlement of the valley.[54] By building a line to Las Vegas, the company could connect the valley to markets via the Santa Fe Railway.[55]

By the spring of 1893 Hagerman and other company officials were concerned about the rivalry among the various companies under the project. Although Hagerman owned the Improvement Company itself, Charles Eddy and others had influence in various older companies sheltered under the Improvement Company umbrella. Eddy served as vice president and general manager of both the Improvement Company and the Pecos Valley Town Company. To stop the friction between the Town Company run by Eddy, the Pecos Valley Railway Company run by Jeff Miller, and the irrigation system controlled by Hagerman, Hagerman decided that the management of these interests should be merged. He proposed forming a fourth new corporation, the Pecos Company. Out of $5,000,000 of this new company's stock, $1,458,000 worth would be issued in exchange for all 8,000 shares of Pecos Irrigation and Improvement Company stock, all the stock of the Pecos Valley Railway Company, and all the stock of the Town Company. As part of this latest reorganization plan, Hagerman hoped to raise $1.7 million by selling $1,250,000 worth of railway company bonds and $1,000,000 worth of Pecos Company stock. Hagerman offered the securities in blocks of $2,500 worth of bonds and $2,000 worth of stock for $3,400 in cash. The new Pecos Company, as holding company for all of the other companies under Hagerman's domain in the valley, was organized under New Jersey corporation laws. With this latest reorganization plan, the Pecos Company acquired all of the stocks of the Pecos Irrigation and Improvement Company, the Pecos Valley Railway Company, and the Pecos Valley Town Company.[56]

Pecos Company directors included president James Hagerman, first vice president R. J. Bolles, and second vice president Charles Eddy. Although Charles Eddy was still nominally involved in managing affairs in the valley, the creation of the Pecos Company lessened the roles of Robert Tansill and Eddy and placed more power in the hands of Hagerman and new investors from Philadelphia, New York, St. Louis, and Geneva. The new company's executive committee members included Charles Otis; William McMillan of St. Louis; John Sutterfield of New York, a friend of McMillan; B. F. Ham of Cranfield, New Jersey, a nominal director in New Jersey, where the Pecos Company incorporated; M. L. Kohler of Philadel-

The original wooden flume destroyed by flood, August 1893.
(Courtesy Southeastern New Mexico Historical Society, Carlsbad, New Mexico.)

phia; Henri Gaullieur, representing the Swiss investors; W. A. Hawkins, the company's attorney in Eddy; and T. H. Edsall, Hagerman's attorney in Colorado Springs. Charles Eddy's voice was diminished.[57]

The Pecos Company continued plans for the Roswell railway extension and the expansion of canals, dams, and other waterworks. Anticipating more expenditures for his businesses, in 1892 Hagerman purchased some 22,000 acres of land in the Roswell area for $30,000. Hagerman's apparent purpose was to purchase the land with his own money and at a later time, when Pecos Irrigation and Improvement Company had the funds, sell to the company. Hagerman expected that when the railroad was extended to Roswell, these lands would sell for some three or four times what he purchased them for in 1892. In the meantime, in a move that indicates Hagerman's reluctance to place all his assets in one basket, he established a separate company, the Roswell Land and Water Company, to hold these lands instead of the Pecos Company.[58]

With the establishment of the Pecos Company in the spring of 1893, Hagerman hoped to set the venture on sound financial footing, but two events later that year sabotaged his plans. In August 1893 an unprecedented flood caused serious damage in the area. Heavy rainfall blanketed the Pecos Valley, including its watershed to the northwest.[59] On August 5 waters in the river rose quickly, causing a flood, which damaged irrigation works, including the Avalon and McMillan Dams, railroad structures, and personal property throughout the valley. Floodwaters sent drowning cattle downstream, their carcasses choking the gateways at Avalon, and as Avalon's crest gave way, twenty feet of the dam fell into the river.[60] Some twenty-five gates at the dam's spillway were open to relieve pressure on the dam. Timbers from a bridge at Seven Rivers shot down toward the reservoir cutting out a small footbridge and damaging other structures. Following a warning from Avalon Dam, local townspeople loaded as many belongings as they could into wagons and headed into the nearby hills. Hundreds of residents gathered on the high points outside the city, while others remained in their homes, businesses, or on the streets watching the torrent rush by.[61]

Water in downtown Eddy rose to a level above the knees of horses as they trotted along the main street. According to local newspaper reports, "nothing approaching this flood was ever known, especially on the Peñasco by the oldest inhabitants."[62] The flood washed out James Hagerman's private bridge and dam, which linked Eddy to Hagerman Heights. Local observers saw the bridge rising in the flood and sailing down the stream like a steamboat. The big flume, which carried water from the east side of the Pecos to the Southern Canal on the west side some three miles above the city, washed away just before the torrent reached Eddy. Timbers from the flume washing down the Pecos destroyed structures in their path, including the iron bridge that crossed the river in the center of town. Houses floated down the Pecos like driftwood.[63] The cost of the flume had been some $50,000, and irrigation company officials estimated the damage caused by the flume's components floating down the Pecos at some $10,000. Officials put damage to Avalon at between $10,000 and $20,000. Damage to Hagerman's power dam came to $5,000. Rail service on the Pecos Valley Railway and construction on the line to Roswell was interrupted. The company alone suffered $250,000 worth of damage.[64]

Settlers and investors alike worried about the future of the valley and what action the Improvement Company might take. On August 7, 1893, Charles Eddy issued a statement from the office of Pecos Irrigation and Improvement. Eddy acknowledged that the floodwaters caused enough damage to the company's works to prevent the delivery of water until the damage was repaired. He mistakenly believed that individual landowners would not suffer because their flooded fields would not now need irrigation waters and their property had not been damaged.[65] Relying on a telegram from Hagerman, Eddy announced, "the dam will be rebuilt as quickly as men can do it." Reconstruction crews were to begin work on the dam and

Results of the August 1893 flood. Looking east on Greene Street from
Canal Street in Eddy (Carlsbad).
(Courtesy Southeastern New Mexico Historical Society, Carlsbad, New Mexico.)

the flume as soon as the water receded and cost estimates could be determined. Eddy assured settlers in the valley that the irrigation company
would complete works in plenty of time for the next season. He lamented
the loss to the settlers and to the company, but estimated that damages
would be but a pittance compared to the millions of dollars that the company had invested in the valley thus far.[66]

Even as reconstruction began, the valley began feeling the effects of the
Panic of 1893, a national economic depression caused by overbuilding and
overinvestment that did not end until late 1898. Farmers had no money to
pay for water because prices for farm products in the valley, and across the
nation, fell precipitously. Immigration to the valley stopped almost
entirely. Work on irrigation halted, subscriptions for water dried up, and
the only bank in Eddy shut its doors.[67]

With the economic panic and the flood coming on the heels of the
organization of the Pecos Company, Hagerman and others had difficulty
selling the new company's stock. In addition to these setbacks, the repeal of
the Sherman Silver Purchase Act detrimentally affected silver mining in
Colorado and Hagerman's mining fortune. The Sherman Silver Purchase
Act of 1890 directed the U.S. Treasury to buy 4.5 million ounces of silver

each month at the prevailing market price, then issue legal tender notes, using the silver as backing. Congress repealed the act in October 1893 because the panic caused many to believe the silver standard weakened the gold reserve and that demonetizing silver would pull the nation out of the depression.[68] For Hagerman, the timing of events could not have been worse. The repairs needed because of the flood required cash. Hagerman's wealth was tied up in the stocks of the Pecos Company, the remainder of which could not be sold because of the depression. With the repeal of the Sherman Silver Purchase Act, Hagerman's cash cow, the Mollie Gibson Mine, stopped supplying the revenue he needed.

Despite his personal financial losses, Hagerman felt a responsibility not only toward investors but also toward settlers in the valley. He immediately sought to rebuild the dam, and largely through his own financial resources and those of his friend Charles Otis, he was successful.[69] By October 65 teams of horses and 200 men labored to complete Seven Rivers (McMillan) Dam, which was relatively unscathed compared to its counterpart downstream. At Avalon, 100 teams and 300 men, including "fifty Mexicans . . . from Pecos [Texas]" arrived to rebuild that structure.[70] Workers built the dam 5 feet higher, widened the capacity of the spillway from 200 to 240 feet, deepened it by 3 feet, and built a second spillway.[71]

Despite Hagerman's promises and financial maneuvering, many people fled the valley after the flood, including the Swiss and Italians. The only foreigners who stayed were those too poor to leave, the laborers who worked for the moneyed class of settlers from England, Switzerland, and Italy. Through hard work some of these laborers became respected local townspeople and farmers.[72]

A number of the original investor-settlers also left the valley.[73] Francis Tracy, an early settler and investor, recalled the tough years following the flood: "Everybody who [could] leave the country did, except me, though I guess I could have gone too. There was an almost complete exodus from the community. . . . [T]hings were so bad that I don't think even a single staple was put in a wire fence in the whole Valley in those terrible . . . years." The depression was not over until 1898, and low farm prices discouraged farmers from planting anything.[74]

During the years immediately following the flood of 1893, the Eddy sheriff regularly sold delinquent tax property from the courthouse steps. Unable to produce or sell crops, locals suffered from the lack of income. Included on the nonpayment list were William P. Bonbright, Charles W. Greene, John Arthur Eddy, County Clerk Thomas Fennessey, irrigation company engineer H. S. Church, and Charles Greene, among others.[75] Many of them, such as H. S. Church, had overextended themselves in the valley. In 1889 Church had begun his own ditch on the north bank of the Delaware River, one mile west of where U.S. 285 crosses the Pecos. He developed a farm, which became known as Coade Farm and which passed to Draper Brantley, grandson of the Draper and Brantley families prominent in valley irrigation. Charles Greene lost his land and his fortune. He died not long after in New York.[76]

Shortly thereafter, Hagerman's disenchantment with Charles Eddy's management came to a head because of "numerous irregularities." Eddy and the manager of the railway, Jeff Miller, were constantly squabbling with each other. Each accused the other of being crooked. Eddy asserted that Miller was guilty of graft. Hagerman, from Colorado Springs, ordered an investigation, sending to the valley Arthur S. Goetz from Colorado Springs.[77] Goetz sided with Miller and against Eddy. Since Hagerman's growing inclination was to side against Eddy, Eddy turned in his resignation on April 25, 1894. Hagerman accepted it.[78] Charles Eddy, who along with Pat Garrett had the first vision for irrigating the Pecos Valley, left the valley completely.

Because of the departures of so many investor-settlers between 1893 and 1898, including Charles Eddy, Francis Tracy and his real estate partner, Charles H. McLenathen, assumed increasing importance in the valley, running virtually everything in town, including rental houses, the electric light plant, the Hagerman Hotel, and even the cemetery. For two or three years McLenathen and Tracy tried never to displace a tenant for non-payment of rent, out of fear that the house would be torn down if they did.[79]

In 1894 workers completed the Seven Rivers (McMillan) Dam and the Avalon Dam with funds supplied by Hagerman, who still clung to the hope that the canals and laterals would eventually irrigate some 400,000 acres of land. Hagerman announced in the Pecos Company's first annual report (March 20, 1894) that the "canals and laterals of the company will not be extended any further for the present. While it is expected that about 400,000 acres will ultimately be put under irrigation, the canals as now built cover 200,000 acres which will be sufficient for several years immigration."[80] Regardless of company claims to "cover" 200,000 acres with canal works, the company never actually irrigated that much land. By 1894 Hagerman and his various associates had developed something in the neighborhood of 41 miles of canal ditches and 138 miles of laterals.[81] New Mexico's Bureau of Immigration gave the irrigation companies credit for many more miles of canals and laterals—almost 400 miles.[82] Percy Hagerman in 1934 estimated that the Northern Canal watered 12,000 acres and the Southern watered 25,000; the Eastern and Hagerman Canals were abandoned. Instead of watering 400,000 acres, the system watered only about 37,000.[83] About 25,000 acres are irrigable under the Carlsbad Project today, an irrigation system that serves 700 people on 155 farms.[84]

With the departure of both settlers and investors from the valley, Hagerman's Pecos Company faced the dismal prospect of adding property notes to their assets but receiving no further water rental fees or payments from settlers on those notes. The company was asset rich but cash poor. The widespread flight from the valley combined with operating and building expenses, with maturing interest owed, and with accounts payable, leaving Hagerman's Pecos Company struggling to stay alive.

Besides Hagerman's enormous financial woes, he still had to try and place the valley in a position to attract investors and settlers. Perhaps then he could get out from under his obligations. To this end, Hagerman sought a new manager. Eddy's resignation had gone against the wishes of directors Henri Gaullieur, H. L. Coulter, and others. Eddy became a devout enemy of Hagerman and Charles Otis, and later competed directly with them, securing funds to build a rail line out of El Paso.[85]

With Eddy's resignation, management of the company fell to attorney W. A. Hawkins, a friend of Eddy's who had both intimate knowledge of the company and good management skills, and Charlie Blodgett, an assistant manager and old associate of Hagerman's from the Colorado Midland days.[86] When Eddy resigned, according to Hagerman's son, Percy, Hawkins and Blodgett went to Colorado Springs and likewise offered to resign but indicated that at an increased salary they might stay. At this point, Hagerman cast about for someone to replace Eddy as manager of all the companies. In September 1895 Hagerman sent his son Percy to the valley until a replacement could be found. The younger Hagerman stayed as interim manager until January 1896. Between September of 1895 and January of 1896, Hawkins and Blodgett submitted their resignations.[87] E. O. Faulkner, who came highly recommended by Atchison, Topeka and Santa Fe Railway officials, became the general manager.[88]

During the depression years of the 1890s Hagerman attempted to pull the valley out of its economic doldrums by stimulating crop production. In 1894 he started the Pecos Valley Orchard Company, the conception of Parker Earle, the "foremost palmologist" in the United States, and the Stark brothers, orchardists and nurserymen from Missouri. For years, small orchard operators around Roswell proved that they could grow a number of fruits successfully, especially apples. Hagerman placed 960 acres of the best land on his ranch under control of the Pecos Valley Orchard Company. The Stark brothers accepted orchard company stock for the trees they provided, and Earle agreed to manage the company. Earle planted twice as many

William A. (Ash) Hawkins, about 1886, was Charles Eddy's lifetime personal lawyer. He was involved in Eddy's irrigation and town-building adventures in southeastern New Mexico and in his later machinations in building and selling a railroad to the Phelps Dodge Company and the building of Alamogordo, New Mexico. *(Courtesy Southeastern New Mexico Historical Society, Carlsbad, New Mexico.)*

trees as he should have, and the Stark brothers sold him numerous experimental varieties of apple trees, which they had been promoting but had not planted commercially. The orchard cost Hagerman $95,821.63. He fired Earle and bought the Stark brothers' stock for $11,000. The orchard experiment was less than successful.[89]

In 1895 after the failure of the orchard, Alfred Goetz, who had previously investigated Charles Eddy and Jeff Miller, came up with the idea of raising and selling beet sugar to Milwaukee brewers. Goetz was able to interest the Pabst Brewing Company, the Uhleins of the Schlitz Brewing Company, and others. Hagerman thus established the Pecos Valley Beet Sugar Company, and Goetz secured a French-built beet-processing plant in Canada and moved the entire factory to Eddy. Hagerman meanwhile hired French experts to run the plant, brought in agricultural experts from California, and recruited local farmers to plant beets. The first attempt at raising such crops, however, failed in the fall of 1896. The sugar content of the beets was high, but the purity was only fair, and the tonnage per acre was not close to predictions. Much of the machinery broke down, and new machinery had to be secured and paid for with money that the companies under Hagerman's control no longer had. The Milwaukee breweries were not impressed with the outcome in 1896.[90] Finally the Milwaukee companies agreed to furnish additional monies for machinery and expertise if Hagerman would give up his shares of stock in the beet company to the beer companies. This he agreed to do. The sugar company operated for one more year, with similarly poor results, and then the factory caught fire and burned. Though Hagerman made elaborate plans in 1898 to build another large factory capable of handling 800 tons of beets per day, nothing came of his plan.[91]

Other experimental products, which Hagerman thought might renew the fiscal vitality of the valley, were canaigre, which produces tannic acid, and ramie, a fiber similar to flax. Hagerman personally experimented with growing the crops at Hagerman Heights, across the river from Eddy. A few early experiments were made with long staple cotton, but they were never taken very far.[92]

Financial reorganizations, Eddy's departure, and the depression caused Hagerman to realize the core of the valley's problems: farmers simply could not make enough money from crops to pay their obligations. Reflecting bitterly nearly ten years later, Hagerman recalled that "out of practically $850,000 worth of notes taken for land, which the company had sold on partial payments, not more than the first payment, was ever made. We had about $800,000 in paper that was not worth more than autum [sic] leaves, because the farmers could not make payment."[93]

Despite Hagerman's attempts to stimulate the valley's economy, the flood in August 1893 and subsequent downturns caused company officials to announce the postponement of the Roswell railway extension. The Pecos Company could not collect the needed subscriptions because of the economic depression, and Hagerman's well was running dry. Officials stated that the road would be built sometime in 1894.

James Hagerman's involvement in the Pecos Valley inaugurated a rapidly expanding financial presence in the valley. Relying on the expertise of managers, Hagerman reorganized and created a complex array of companies designed to extend activities in the valley and allow him to retain control. As Hagerman's main goal became the building of a railroad connecting the valley to major markets, his personal investments escalated. With costs far outstripping income, Hagerman hoped to sell all enterprises outright, but instead he refinanced existing companies and lured foreign capital to the valley. A major flood and economic downturn compounded problems in the valley. Many settlers and company officials, including Charles Eddy, fled the valley, never to return.

By 1893 Hagerman had expended $200,000 for 3,492 shares of Pecos Irrigation and Improvement Company stock, $59,625 for 2,973 shares of Pecos Valley Railway stock, $9,500 for 205 shares of Pecos Valley Town Company stock, $151,000 for 150 Pecos Irrigation and Improvement bonds, close to a quarter of a million dollars for 380 Pecos Irrigation and Improvement bonds, some $40,000 of investments in lands in the lower part of the valley, another $71,750 on improvements to his personal residence at Hagerman Heights just outside Carlsbad,[94] some $4,000 worth of Eddy town lots, and well over $200,000 in cash advances to Pecos Irrigation and Improvement Company. By the beginning of 1893 Hagerman had invested $1,236,000 in the Pecos ventures. By the end of 1893, as Hagerman's investment increased to $1,448,823.77,[95] his focus turned to saving the Pecos Valley and his personal investment. As the financial standing of Hagerman's valley enterprises declined, he tried desperately to refinance them and keep them solvent long enough to build a rail extension to Texas.

6

RAILROADS AND RECEIVERSHIP

THE Pecos River flood and the depression of 1893 caused James
Hagerman to sharpen his focus on securing a railroad into the Pecos
Valley. Even as he rebuilt the Avalon Dam and tried to stimulate
the local economy, Hagerman realized that success depended on the extension of a rail line from the valley to points that had connections to major
markets. In the early days of the valley, all materials, crops, products, and so
forth had to be freighted by animal power some ninety miles from Eddy to
Pecos, then from Pecos to various points across the country on the Texas
Pacific Railroad. A railroad through Eddy could cut delivery time by two
days. Early on, Charles Eddy had been optimistic that three great industries—mining, ranching, and farming—would be joined together by a railroad through the Pecos Valley. To the west, Eddy saw in the mountains
huge silver, coal, and iron deposits, and to the east he saw what he considered to be the greatest cattle grazing range in the country—the Panhandle
of Texas. Eddy envisioned in the Pecos Valley the largest irrigated agricultural mecca in the world. With the Pecos Valley railroad, he hoped to unite
all three.[1] But by 1894 Eddy's vision of an agricultural mecca had become
instead Hagerman's headache. As he propped up his failing companies,
Hagerman hoped to fend off bankruptcy long enough to build his railroad.

Charles Eddy's optimism centered around the Pecos Valley Railway
Company, established in 1890, whose investors were largely the same as for
the irrigation company. In March 1890 H. H. Cloud, chief civil engineer of
the line from Pecos to Eddy, noted that construction crews would work rapidly because of a virtually level grade and because few curves existed.[2] In
April 1890 E. G. Shields, who was also selling land and serving as general
land agent for the irrigation company, and a corps of engineers were
locating a route to connect Eddy to the Texas-Pacific line in Ward County,
Texas, near Pecos. The grade from Pecos up to Eddy was insignificant, as it

did not exceed .05 percent per mile, which indicated that at least the line from Pecos to Eddy could be constructed quickly and easily.[3]

On August 30, 1890, after four months on the job, railroad graders had moved within thirty-five miles of Eddy and were pushing forward rapidly,[4] and by the first of November workers had laid fifty-six miles of track and the railway company operated the road for that short distance.[5] Later that month, the advance camp of engineers and graders of the company reached the southern town limits of Eddy.[6] Construction crews completed the rail line from Pecos, Texas, to Eddy, New Mexico Territory, during the first weeks of January 1891. Hagerman heard the news at his home in Colorado Springs on January 13, a Tuesday afternoon. That weekend a formal celebration was held in Eddy with fireworks and hundreds of cheering citizens. The railway chartered a special train carrying Hagerman, Eddy, and many of the directors, stockholders, and executives of the railroad company from Pecos to the town of Eddy. Hagerman's special car, the *Mountaineer*, and railroad manager S. F. Judy's car arrived for fireworks and a big bonfire at the Hotel Hagerman.[7]

In the message sent by the citizens of Eddy to Hagerman on January 10, 1891, the writers were effusive in their praise for the millionaire:

> The people of Eddy and the Pecos Valley greet you. Telegraph line just been completed to our young city and the first telegram to flash over the wires is an "All hail to you and yours"; A recognition of the indomitable energy and perseverance that has joined our valley and town to the markets and trade centers of the world. . . . The labors of your life have been marked with many notable achievements, but none have been greater than this which has changed the wilds . . . and the loneliness of the desert to a peaceful and prosperous agricultural and fruit growing country. You have advanced civilization another step in the unbroken southwest and opened to the commerce of the world another avenue of wealth.[8]

Hagerman responded in kind, thanking the town and encouraging its citizens to work together to develop the valley, noting that the Pecos Valley had "all the natural advantages needed to make it the happy home of thousands of people." Hagerman promised that he and Eddy would develop the valley quickly.[9]

Although built cheaply and easily, the problem with the railway that stretched from Pecos to Eddy was, as one of many critics said in the late nineteenth century, "it began nowhere and ended nowhere."[10] Before the line had outlets to Roswell and then Amarillo, cattle had to be shipped first south to Pecos, then east, and then to northern markets. As a result, the

Pecos Valley Railway did not at first show a profit because it needed a more direct line to northern cities.[11]

But in the spring of 1893 things in the valley seemed to be favorable. Hagerman had just created the Pecos Company to further expand the project, and progress on irrigation works and settlement in the valley were bolstered by the visit of a number of wealthy individuals. In March prominent members of the Pecos Valley Railway Company arrived aboard the special coach *Hesperia*. Although the men had no railroad news for the public, Eddy told the local paper that the company was searching for a northern connecting outlet for their rail line.[12] James Hagerman realized, prior to extending the line from Eddy to Roswell, that the problem with his railroad was that it needed to run north instead of south, and it needed adequate connections with a major railway.[13]

During the spring of 1893 both James Hagerman and Charles Eddy traveled to New York seeking monies to extend the railway into Roswell or Albuquerque or some other city in New Mexico with major connections to northern cities. Hagerman attempted to raise money in New York and London to fund the proposed extension of the line north.[14] An 1893 territorial law that exempted new railways from taxes for six years if they began operation by February 1896 gave Hagerman the boost he needed to secure New York financing for construction north ninety miles into Roswell.[15]

Hagerman announced that construction would begin in early 1894. But during the Eddy to Roswell construction phase of the road, Roswell citizens suffered a series of frustrations. Three and a half years went by after completion of the Pecos to Eddy line before workers reached Roswell.[16] Townspeople from Roswell suspected someone in Eddy was deliberately obstructing progress on the extension;[17] despite rivalry between the two towns, it seems improbable. Holding up work on the project would hurt not only Roswell, but Eddy, which needed outlets later provided by the Atchison, Topeka and Santa Fe Railway. In early 1894 Hagerman had already pledged monies to rebuild flood-damaged Avalon Dam, and most of his remaining wealth was tied up in assets and securities in the valley. Hagerman may not have had the cash on hand to pay workers. When work crews had not started on the extension to Roswell by February 1894, Roswell citizens offered a $50,000 bonus if the rail crews finished the line by September 1.[18]

Once work began in April, delays occurred because of labor difficulties. Hagerman had hired the established railroad builders Hampton and Smith of Kansas City, Kansas, who had built numerous roads in the west and Mexico. However, local workers refused to accept the low pay offered by Hampton and Smith. Consequently, the Kansas company brought in

In 1894 the Pecos Valley Railway extended its line north to Roswell. Dignitaries from Eddy (Carlsbad) arrive at Roswell on the opening day of railroad service. The railway still "started nowhere and ended nowhere" until 1898 when James Hagerman received financing from the Atchison, Topeka and Santa Fe Railway for a third extension of the line to Amarillo, Texas.
(Courtesy Archives and Special Collections Department, New Mexico State University Library.)

workers and equipment from Mexico and other points in the United States. Finally, by the latter part of April 1894, Hampton and Smith secured a large quantity of building materials, and some 40 of the proposed 120 grading teams were on the job.[19] Despite the delays and frustration, local citizens in Roswell warmly greeted the arrival of the first locomotive into that city on October 6, 1894, declaring October 15 as "Railroad Day." During the celebration, a special train moved back and forth between Roswell and Hagerman's Southspring Ranch, near town.[20]

After securing financing for the Eddy to Roswell extension, Hagerman built stock corrals and prepared construction sites at Malaga, Florence, Francis, Otis, Eddy, Lakeview, McMillan, Miller, Hagerman, South Spring, and Roswell.[21] At Roswell, plans were underway to build a two-stall round-house; Eddy to the south had a six-stall roundhouse and car shed by 1894. Company officials built the general offices of the railway company in Eddy. Besides offices, the company also kept repair shops in Eddy. Mechanics hired by the company could rebuild engines on the line or repair the rolling stock, which by 1894 consisted of ten coaches, two sleeping cars, fifteen freight cars, forty-six flat cars, two baggage and mail cars, ten tank cars, and six engines, all equipped with Westinghouse air brakes. The Pecos Valley Railway Company also owned the telegraph system in the valley.[22]

When the first train arrived in Roswell, Hagerman, using typical nineteenth-century language, reminded the valley to carry on work for future generations, because the Pecos Valley would "soon be known as the most prosperous region in the west . . . [with] the most land [and] the best water supply of any spot in the arid region of America."[23] By that time, the valley had shipped out 400,000 pounds of dry cattle hides, 500,000 pounds of high-grade alfalfa, over one million pounds of horses and cattle, 250,000 pounds of sheep, and some 1,100,000 pounds of wool.[24] Another principal commodity carried by the Pecos Valley Railroad was fruit bound for the east. Consequently, the railway advertised itself as the "fruit-belt route," in hopes of reviving immigration to the valley.[25]

Much of the earning power of the valley, however, went to pay shipping charges to the Texas and Pacific line, which the Pecos Valley Railway joined at Pecos, Texas. To make the valley truly successful, Hagerman and Eddy needed a more direct link to large markets in northern cities than that provided by the Texas and Pacific. Hagerman planned to further extend the rail line from Roswell to some city in northern New Mexico along a transcontinental route. Competition between Albuquerque, Las Vegas, and Socorro gave indications, even before the line reached Roswell, that a line would be extended to one of those cities.[26] Albuquerque seemed to have the best chance, since there the Pecos Valley Railway could link up with the Atchison, Topeka and Santa Fe line.[27] In 1892 Pecos Valley Railway Company directors met in Albuquerque, adding fuel to the competitive fire between these northern New Mexico cities. Jay Gould, who owned the Texas and Pacific line, expressed a desire to extend the Pecos Valley Railway from Roswell westward to the gold mining district of White Oaks, providing a northern link with the Texas and Pacific.[28] Gould rode the Pecos line, but "Hagerman gave him to understand that the railroad was not for sale at any price."[29] Having turned Gould away, Hagerman still faced the task of raising money for expansion.

Complicating matters in raising money for a rail extension to Texas was the depression beginning in 1893, the devastating flood, and the shaky financial situation of Hagerman's Pecos Company, the holding company over all enterprises in the valley. The departure of valley farmers and the combined problems of floating debt,[30] bills and accounts payable, and operating expenses after the Panic of 1893 hindered Hagerman's ability to finance both the Pecos Company and his railroad. By January 31, 1895,

Hagerman had sold all but $300,000 of the 1.8 million stock issued for the Pecos Company. But the company needed $411,000 to meet interest obligations on the bonded debt in 1895. The company had to come up with $230,000 to pay its bills. Although he received subscriptions to finish the original and rehabilitative construction work in the valley, Hagerman and the Pecos Company still needed money.[31] The directors attempted to raise the money from existing stockholders, asking them to buy $300,000 worth of new preferred stock. The directors also asked the 6 percent bondholders of the Improvement Company and bondholders of the Pecos Valley Railway Company to take preferred stock in lieu of interest for three years. They refused.[32]

In June 1895 the Pecos Company issued another plan to raise money by offering discounted shares in Pecos Valley Railway Company stock to purchasers of Pecos Valley Railway bonds. Although this plan was intended to raise $600,000 in cash, only some $286,000 was subscribed; prospective investors were frightened away by the large floating debt in the Pecos Company. The whole railway plan failed. By August of 1895 two major investors carried the Pecos Company: James Hagerman and his friend Charles Otis, who now tried to pay the company's outstanding debts of $450,000.[33] Hagerman and Otis could not convince existing stockholders to put up enough money to save the company or to relieve themselves of the burden of having to pay both interest on bonds and running expenses of the company. Failing to sway Pecos Company investors, Hagerman and Otis demanded security for their advances to the floundering company. They placed all company assets in the hands of a trustee in exchange for collateral notes to secure the money they were supplying.[34]

But by October Hagerman and Otis were tired of carrying the company for the benefit of others. Forced into bankruptcy by Hagerman and Otis, the Pecos Company established a reorganization committee in October 1895. The trustees ordered the sale of the Pecos Company's assets. Hagerman and Otis suggested forming a new company, the stock and bonds of which would be distributed proportionately to all previous owners of stock and bonds in the old company who were willing to pay an assessment of twenty dollars per share to help sustain the company. This plan, similar to the previous plan to induce stockholders to provide additional cash by buying preferred stock, also failed, and Hagerman and Otis bought the company's assets at a trustee sale of the International Trust Company in December of 1895. Hagerman and Otis immediately formed a new company called the Pecos Valley Company under their ownership and control.[35] Hagerman and Otis sold some stock in the new company, but generally their interest in the whole enterprise continued to get bigger as they propped up a losing cause. By the end of 1897 Hagerman had spent $2,654,894 in the Pecos Valley.[36]

Spending his own money to prop up the fiscal structure of valley enterprises allowed Hagerman to raise money for a railroad that he believed would redeem his valley investments.[37] Hagerman still owned the Eddy-Roswell line completed in late 1894, which linked the valley but did not link the valley to major markets. His ultimate goal now was to find the capital needed to build a rail line north and east to the Texas Panhandle.

As he sought financial backing, Hagerman realized that it made no sense to financiers to back a rail extension in the middle of nowhere without the rest of the valley's rail system. Lands were only valuable if crops could get to major markets. Hagerman was by 1896 convinced that new construction from Roswell to some point in Texas would be advantageous, with Amarillo the most likely destination. With the reorganization of the Pecos Company into the Pecos Valley Company, Hagerman now offered to sell the entire valley rail system in order to secure financing for the extension to Amarillo where the Atchison, Topeka and Santa Fe Railway Company had a connecting line that could effectively make the Pecos Valley part of the largest railroad system in the country.[38]

The departure of Charles Eddy in 1894 compounded Hagerman's problems in raising cash. Hagerman now found himself competing with Eddy for cash in New York. Eddy had moved to El Paso where, adopting Gould's earlier idea, he envisioned building a rail line north from the border city to the gold fields of White Oaks in the Sacramento Mountains.[39] Eddy, his brother John, and W. A. Hawkins founded a new town, Alamogordo,[40] at the base of the Sacramentos. They incorporated the Alamogordo Improvement Company in April 1898 with $500,000 capital stock and began recruiting settlers. By March 1899 the town had a population of over 1,000. Like Eddy before it, Alamogordo was to be a model city, with wide streets, orderly development, canals, 6,000 trees, and no alcohol. Every deed stipulated that no alcohol could be made, stored, or consumed on the premises, except for one lot in one city block—so the company could control drinking.[41] Alamogordo was designed to rival the town of Eddy. Bypassing the existing town of White Oaks when it refused to pay him a $50,000 subsidy, Eddy built the El Paso and Northeastern Railroad through Alamogordo and Carrizozo and hoped to boost his rail enterprise by connecting with Dawson's coalfield to the northeast.[42]

Eddy's move threatened Hagerman's proposal for a line from Roswell to Amarillo because between 1896 and 1898 Eddy and Hagerman frequently competed for the same money at high interest rates of 12 to 13 percent.

Hagerman often found himself seated in some banker's outer office, hat in hand, staring across at his nemesis, Charles Eddy.[43] Alongside Hagerman for much of this time were his son Percy and Charles Otis. Every time Hagerman convinced the subscribers to finance the new railway from Roswell to Amarillo, some of the subscribers got restless and dropped out. Hagerman then had to devise a totally new plan.[44]

From 1895 to 1898, as Hagerman sought financing in the East, a number of prospective backers[45] examined the merits of a line from Roswell to Amarillo. Potential investors Harvey Fisk and Sons of New York City sent Robert A. Meeker to the valley in 1896 to determine the physical condition of the existing road and the earning potential of the proposed extension for the Atchison, Topeka and Santa Fe. The extension would provide an almost direct route to Kansas City and Chicago and connect with a major railroad stretching from California to Old Mexico.[46]

Meeker thought the Pecos Valley held many advantages, calling the area north of Lake McMillan near Roswell the best-watered area in the arid west.[47] Meeker estimated the equipment needed to support a railroad extension from Roswell to Amarillo, the various costs for improvement of the Pecos Valley Railway, and the profits to be generated for the Santa Fe Railway. Because the surrounding country raised cattle and sheep, he predicted that transporting the animals would significantly increase profits of the rail line. In transporting some 14,000 carloads of cattle, Meeker estimated, the Amarillo rail extension would make about $34.00 profit per car.[48]

After considering the 1895 and 1896 profits on the road from Pecos to Roswell, Meeker estimated the gross earning potential of a new line at somewhere in the neighborhood of $1,054,925 per year. Subtracting $685,701 for operating expenses and taxes left net earnings of $369,223 per year.[49] Meeker's report encouraged the Santa Fe to negotiate with James Hagerman, from whom it had purchased the Colorado Midland a decade earlier. Anticipating the need for a new railroad company to manage the old Pecos Valley Railroad operations and those of the proposed Amarillo extension, Hagerman incorporated the Pecos Valley and Northeastern Railway Company on June 14, 1897. On March 19, 1898, Hagerman created a new company, the Pecos and Northern Texas Railway Company, to operate that portion of the line between the western border of Texas and Amarillo.[50]

During the ensuing months, Hagerman developed a plan so that construction of the line to Amarillo could begin in the spring of 1898. The key to this was a $750,000 loan from the Santa Fe Railway Company. For collateral Hagerman used the stocks and bonds of the Pecos Valley Railway

Company, which he had separated from the Pecos Company. Hagerman got additional financing from the Pullman Car Company, a group of investors led by B. P. Cheney of American Express and Harriman and Company of New York.[51] Before beginning construction, Hagerman and Charles Otis created the Pecos Railway Construction and Land Company. In addition, Hagerman paid the Santa Fe a quarter million dollars in Pecos Valley Railway Company bonds in exchange for Santa Fe transportation and construction materials worth $212,000.[52] The Santa Fe Railway had a vested interest in making sure Hagerman built his railroad: it wanted to keep competitors out of the area. With the Pecos Valley Railway and its extension, the Santa Fe would control a 370-mile-long feeder from Amarillo to Pecos, Texas, which would connect with its Southern Kansas line, allowing the company a dominant route for shipping from Texas–New Mexico cattle country.[53]

Newly capitalized by eastern railroads, the Pecos Railway Construction and Land Company could begin actual construction. On January 31, 1898, Hagerman sent a telegraph from New York City to Eddy, New Mexico, announcing, "Money for extension all raised. Rails and other material already purchased. Road will be completed in October. Make news public."[54] After the initial legal wrangling, on April 15, 1898, the Pecos Railway Construction and Land Company, with Hagerman as president, contracted with Mallory and Company to do the actual construction. Construction on the last phase of the valley's railroad began on May 1, 1898, from the Amarillo end.[55]

Hagerman hoped to complete the railroad in October[56] in order to ship Pecos Valley fruit to the East. This plan was unsuccessful. On January 27, 1899, general manager of the rail line, D. H. Nichols, reported that workers were laying track toward Roswell from Amarillo and were yet forty miles from the city. The gap between work crews closed on February 11, 1899, linking the Pecos Valley to Amarillo, Texas, and major rail lines of the Santa Fe. The final spike was driven near a river crossing north of Roswell. The first full run took place on March 1, 1900.[57] Hagerman and Otis still owned the Pecos Valley and Northeastern Railway, with Hagerman owning 985 shares of preferred and 19,643 shares of common stock. Hagerman and Otis operated the company independently "under a friendly and favorable contract with the AT&SF."[58]

Late in 1900 Hagerman again headed east. He learned that Charles Eddy had closed a deal with the Rock Island Railroad. Eddy convinced the company to extend a line west to near Santa Rosa, New Mexico, to connect with his El Paso and Northeastern terminus near Carrizozo. After the Rock Island complied, Eddy extended the El Paso and Northeastern to Santa

Rosa and built a 132-mile spur from Tucumcari to Dawson, in the process crossing the Atchison, Topeka and Santa Fe Railway line and the St. Louis and Rocky Mountain line.[59]

In Hagerman's mind, Eddy's deal represented a threat to his Pecos Valley and Northeastern Railway. Eddy's line would parallel the Pecos Valley rails for 200 miles, and would cut into the business of transporting cattle, sheep, and other products, "seriously reduc[ing] the rates on the cattle business left to us."[60] Understandably, Hagerman anxiously wanted to sell his stock in the Pecos Valley and Northeastern Railway to the Santa Fe. In this, he was successful.

In December 1900 the Santa Fe Railway Company purchased the Pecos Valley and Northeastern Railway for $2,675,902, at $17.50 a share for preferred and common stock, or 96 percent of the stock and more than two-thirds of the bonds of the railway.[61] Santa Fe estimated that it spent less than $10,000 a mile to obtain control of eastern New Mexico rail traffic. According to Percy Hagerman, his father recovered some of his ten-year investment in the valley, receiving $557,000 for his railroad stock, $162,000 from the Pecos Railway Construction and Land Company for surveys and expenses, and $208,000 for land sales connected with the extension.[62]

Separated from the railroad during reorganization, Hagerman's other enterprises in the Pecos Valley floundered. The Pecos Valley Company and its subsidiary companies—Pecos Valley Irrigation and Improvement Company, Pecos Valley Town Company, and the Roswell Land and Water Company, plus several land companies—were defaulting on interest payments and moving toward receivership.[63]

To make matters worse, in 1896 disagreements arose between bondholders possessing 8 percent bonds from the old Investment Company and those holding 6 percent bonds from the subsequent Improvement Company. At the heart of the disagreement between bondholders lay conflicting claims about the assets each group owned in the event of a foreclosure. The 8 percent bondholders insisted they had a lien on all property of the Investment Company, while the 6 percent bondholders asserted they had a lien on all property of the Improvement Company. These claims put the two groups in conflict, but none of the assets could be easily sold. The 8 percent bondholders argued they had a prior right to property since they had invested with the earlier Investment Company. The 6 percent bondholders, however, charged that, unlike the Improvement Company, which could actually own and sell land, the Investment Company had only obtained, through its articles of incorporation, the right to "colonize and improve lands."[64] According to the 6 percent bondholders, in a court of law such language did not imply the right to own property. Therefore, the 6

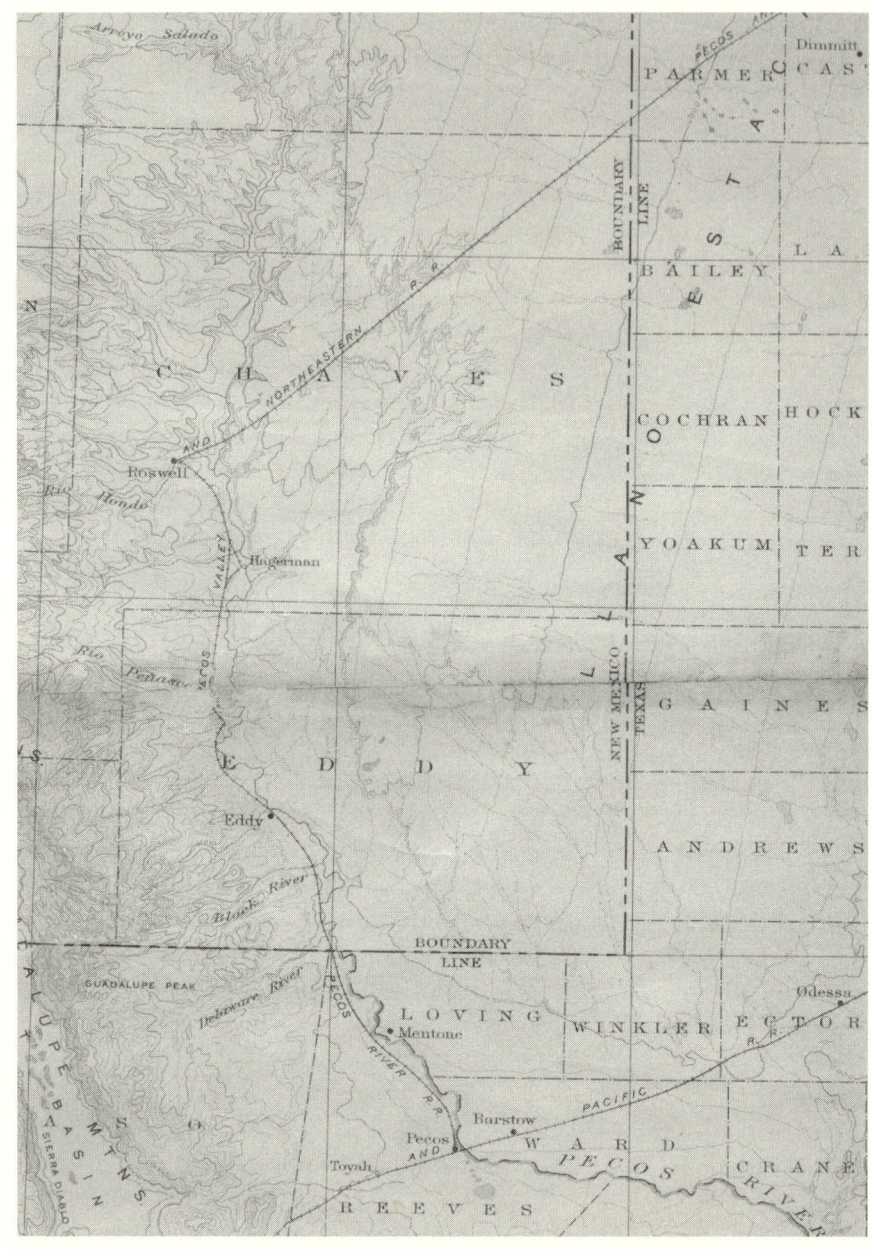

The Pecos Valley and Northeastern Railway.
*(From Map of Texas and Parts of Adjoining Territories [Washington, D.C.: Department of the Interior,
U.S. Geological Survey, Charles D. Walcott, Director. Sheet XI Topographic Map. Compiled by and under
the direction of Robert T. Hill. Drawn by Henry S. Selden and Willard D. Johnson, 1899].)*

percent bondholders declared a superior right to property. Adding to these declarations, James Hagerman insisted that in the event of foreclosure, Pecos Irrigation and Improvement, as a quasi-public corporation, like a railroad, was obligated to pay back monies first to those who had kept the company "a going concern,"[65] in other words, himself. All three parties realized that in the course of litigation they might get nothing.

The 6 percent bondholders had to resolve difficulties of their own before they could negotiate with Hagerman and the 8 percent bondholders. Lombard, Odier owned or represented $250,000 worth of the 6 percent bonds—all of the issue, except for $76,000 held by the Pecos Valley Company itself. The Swiss company's representative in the valley, Henri Gaullieur, inspected the valley in 1896, finding depression, destruction, and panic. Those connected with the Pecos Valley Company believed that Gaullieur would recommend litigation on the part of the Swiss syndicate. Litigation, Hagerman believed, would hold up work getting a railroad to Amarillo.[66]

Francis Tracy[67] realized that once Gaullieur's report was presented to the syndicate, "we would never be able to favorably present our contrasting viewpoint—the viewpoint of those who knew the valley and all its idiosyncrasies so intimately."[68] Tracy and real estate partner Charles McLenathen wrote separate letters to Frederick Dominice, who had lived in Eddy in the early days of the community and represented Lombard, Odier; Dominice had returned to Geneva in 1893 but was still familiar with valley affairs. Tracy and McLenathen thought he would be sympathetic to their plight. The Tracy and McLenathen letters resulted in Gaullieur becoming dismayed with everyone, including the syndicate in Geneva.[69] Gaullieur removed himself from the fray, and Lombard, Odier appointed a new representative: Francis Tracy.[70] The company also hired Charles Eddy's attorney, W. A. Hawkins.

Hagerman successfully slowed down the receivership process until he had raised the money for the railroad. The Illinois Trust and Savings Bank of Chicago and the Central Trust Company of New York, which held the trust deeds securing all of the bond notes, sued for payment. The Pecos Valley Company could not raise the money and was forced to sell its property to pay off debts, including the investment bonds.[71] Charles S. Kelley and Tansill, representing the 8 percent bondholders, and Francis Tracy, representing the 6 percent bondholders, met with Hagerman in New York and agreed to a reorganization plan in May 1898, four months after the railroad financing deal had closed.[72]

According to the reorganization plan, Hagerman would pay all outstanding bills and accounts of the Pecos Irrigation and Improvement

Company, including workers' salaries[73] and taxes up to January 1, 1898, plus an additional payment of $15,000.[74] Hagerman also agreed to deed to the receiver or a new corporation, if one was created, all lands in the lower valley; in turn the receiver would deed to Hagerman all property in the upper valley near Roswell, in Chaves County, including the Northern Canal, the Hondo Reservoir lands, and water-right notes in Chaves County.[75]

Shortly after the New York meeting, Hagerman told the reorganization committee that he could not go along with the original agreement. Hagerman had difficulty providing the cash required to pay back salaries. Consequently another meeting was called, this time in Colorado Springs, in an effort to negotiate some new understanding for the receivership. During this heated meeting Hagerman declared that he could not raise the cash needed to pay off back salaries of $25,000 to $30,000. Tansill, who had become receiver in July, intervened and told Hagerman that if he would give him a note for $50,000, Tansill would cash it and finance the deal. Hagerman and the committee signed the new agreement for receivership on August 27, 1898.[76]

Hagerman carried out his part of the bargain and confined his operations to the northern part of the irrigation area around Roswell. All of his business in the lower part of the Pecos Valley ended in 1899. The Pecos Irrigation and Improvement Company, which still owned the lower valley's irrigation works and appreciable lands, would be managed by Robert Tansill as receiver, and there was nothing left of the Pecos Valley Company.

Hagerman gave up on the lower valley. The soils were poor, the canals leaky, and the land had been irrigated too much, which ruined it. After selling his Colorado Springs property, he retired to the Roswell area and lived on a 6,000-acre ranch property formerly owned by John Chisum, where he built a three-story house. He purchased another ranch called Whetstone. The ranches cost Hagerman $75,000. Hagerman later sold some of his Roswell property because its "management [was] too great" for a man of his age and health.[77] With Hagerman's withdrawal from the lower part of the Pecos Valley, all of his financial machinations and attempts to keep the various irrigation companies afloat ended.

In Eddy, events continued as usual. The night before Tansill officially became the receiver of the Pecos Company, the earthen approach at the northern end of the flume carrying water to the Main Canal washed out. Officials in Eddy wired A. C. Campbell, Hagerman's attorney at Socorro, New Mexico, where Hagerman and Tansill were attending court to finalize the receivership of the company. Campbell pocketed the telegram detailing this latest catastrophe on the river and watched the proceedings. After Tansill was appointed receiver, Campbell produced the telegram.[78]

With Hagerman gone and the threat of entangling complications out of the way, the irrigation company limped into the twentieth century. As receiver, Tansill reported that 1900 "can fairly be considered a successful year; in spite of an unusually dry season our water supply has been ample and its distribution has been most satisfactory." According to Tansill, crops had been good, and despite a drought elsewhere in the country, good prices had been realized for local farmers. It appeared, in large part because of the Amarillo extension, that by the turn of the century the Pecos Valley was beginning to recover from the previous six or seven years of questionable economic viability.[79] In Eddy and vicinity settlers filled every residence, and a shortage of accommodations ensued. The recently reopened First National Bank doubled its capacity to keep pace with the improving economy. In 1900 the lower valley's total water contract acreage in good standing was 13,372 acres, although area farmers only irrigated 7,900 acres. In 1901 the irrigated acreage increased by 2,000 acres. But according to Tansill's report, land values continued to stay very low, in part due to the removal of railway offices and shops from Roswell by the Santa Fe Railway Company.[80]

In May 1899 Tansill encouraged changing the name of the town of Eddy through a city election. Tansill chose the name Carlsbad, trying to capitalize on the popularity of the city of Karlsbad in Czechoslovakia, which was famous for the medicinal qualities of its mineral water. He hoped to advertise the mineral springs in the area and to assuage the feelings of Hagerman, to whom the name Eddy had become almost unbearable. Voters approved the name change, and Eddy became Carlsbad, New Mexico, although the county retained its name.[81]

By 1900, two years after Pecos Irrigation and Improvement Company passed into receivership, Tansill, supervised by a committee of creditors, reestablished the beet sugar factory. He hired Arthur S. Goetz, Hagerman's former secretary, as manager. Tansill also replaced Francis Tracy with Goetz as manager of the reorganized irrigation company, prompting Lombard, Odier to call for reinstating Tracy, their representative in the valley. Tansill complied.[82]

As Tansill and Tracy tried to place the company on firm financial ground, Percy Hagerman, reflecting with bitterness on his father's investments in the Pecos Valley, thought that the entire enterprise of irrigation and agriculture had been a failure: "Every other thing that has been tried down here, with the one exception of apple growing in Roswell has proven a failure. I have made up my mind that this country has got to go back to stock raising and depend on that for whatever prosperity it has. That means that it will never be a decent country to live in; no cattle country never is [*sic*]. All the expensive efforts that have been put forth to make this a white man's country have been futile and always will be."[83] Tansill and Tracy, however, continued to try to irrigate the Pecos Valley. The company remained on shaky ground, and in 1901 its financial standing was in jeopardy. Financial records showed a narrow margin of earnings on the balance sheets for that year. The receiver collected $6,000 on notes, accounts, and water rents; $9,468.00 had been expected.[84]

Because of the financial woes of Pecos Irrigation and Improvement Company, Tansill, Tracy, and others organized a new company to assume ownership of Pecos Irrigation and Improvement assets. They created the Pecos Irrigation Company in September 1900 with $325,000 of stock (3,250 shares at $100 each), which was distributed proportionately among the original 6 percent and 8 percent bondholders.[85] For the first three years, interest would be payable in script at the option of the company. Under this organization, former bondholders were now stockholders, there being one share for each bond. There could be no conflict of interest between bondholders and stockholders, no matter what disaster might happen in the valley. R. S. Benson, J. O. Cameron, and A. M. Pratt agreed to act with Tansill and Tracy, the manager of the company, as incorporators and directors for the first year's operation.[86] According to its charter, the company was to "purchase or acquire the reservoirs, canals, contracts, notes, real estate, franchises and properties of the Pecos Irrigation and Improvement Company, to hold, enjoy, use, collect, enforce, and dispose of the same." After some eleven desperate years of operation, Pecos Irrigation and Improvement Company was absorbed by Pecos Irrigation Company in 1901.[87]

As a consequence of the agreement between Tansill and Tracy, the new company was incorporated for a period of fifty years and now owned the irrigation plant; canals, including the Hagerman Canal southeast of Carlsbad; all water rights on the Pecos River for 200 miles; and more than 30,000 acres of land. Avalon and McMillan Dams appeared to be safe and in good working order. Despite complaints of water losses in the canal system, Tracy and others believed that the whole system was in fair condition, with very few areas of serious leakage or danger. The one glaring exception to this was the old wooden flume, which was fast outliving its usefulness.[88] In 1900, two years before the Newlands Act, Tansill suggested that an effort be made to sell the irrigation plant to the federal government.[89]

7

PRIVATE IRRIGATION DROWNS IN DITCHES OF ITS OWN MAKING

DESPITE the best intentions of James Hagerman, Charles and John Eddy, Robert Tansill, and Francis Tracy, the Pecos Valley of New Mexico would not be developed through private initiative. Private irrigation companies failed because their projects were planned, engineered, and promoted by speculators and those not thoroughly familiar with the demands of irrigation. More important, private companies never had enough money to carry through their plans. Although irrigation companies expanded rapidly between 1870 and 1890, slowing only after the Panic of 1893, irrigation systems were shoddy—built too quickly and too cheaply. So-called experts misjudged water resources. Few systems had adequate drainage, and engineers failed to use water supplies fully or prevent alkali buildup. Settlers were slow in coming and those who did come knew little about irrigation farming. Consequently, 90 percent of all privately owned irrigation companies were near or in bankruptcy by 1902.[1]

Try as they might, westerners could not solve this problem with private capital, initiative, and hard work. As more and more companies fell into receivership, the West began seeking assistance from the federal government. Reclamation came to be seen as necessary—a way for the United States to utilize western resources. An outgrowth of Progressivism, reclamation helped save the West and many eastern capitalists from financial disaster, while at the same time encouraging a dependency on the federal government. In 1894, the U.S. Congress attempted to help troubled westerners by passing the Carey Act, which granted land to each state to finance irrigation. Few took advantage of the program, and irrigationists began calling for a new plan.[2] By 1900 both the Democratic and Republican Parties favored reclamation, but President William McKinley expressed only disinterest.[3]

Change came in 1901 with Theodore Roosevelt's administration. Roosevelt had seen failed irrigation schemes and understood the role of irrigation to the West. As an admirer of John Wesley Powell, Roosevelt believed in organized federal reclamation. Thus, in his inaugural speech of 1901, Roosevelt laid the basis for watering the West: "Great storage works are necessary to equalize the flow of streams and to save floodwaters. This construction has been conclusively shown to be an undertaking too vast for private efforts. . . . [I]t is properly a national function. . . . [T]he Government should construct and maintain these reservoirs."[4]

Many groups and individuals supported federal reclamation, including the United Mine Workers, the Chicago Federation of Labor, other labor unions, the National Board of Trade, national irrigation associations, and trade organizations. Newspapers also supported reclamation, but federal reclamation met opposition from states' rights supporters; older, eastern states; and from farmers fearful of competition, who tended to oppose federal involvement in agriculture.[5]

Nonetheless, with the approval of the president and both major parties, plus grassroots support, Congress began working on a bill to allow federal reclamation. Francis G. Newlands from Nevada introduced the bill in the House. Newlands had previously lost $500,000 in irrigation and believed that the federal government had a duty to help irrigators.[6] Senator Henry C. Hansbrough of North Dakota introduced the bill in the Senate. The bill passed Congress in 1902—by 146 to 55 in the House and with no division in the Senate—largely because Roosevelt supported it.[7]

The Newlands Reclamation Act had several provisions, including a stipulation that property owners actually live on the property and possess no more than 160 acres. Another provision required property owners to form a water users association as a corporate entity, which could legally allow them to negotiate with the federal government.[8]

The Newlands Act also provided that 95 percent of monies from federal land sales be spent for reclamation, through the Reclamation Fund, and that income from rents on public lands fall into the fund as well. The remaining 5 percent of the fund would go to education.[9] The Reclamation Fund in 1900–1901 had $3,144,821 available from sale of land. By June 1902 the fund had accumulated $4,585,520 more.[10] No funds were made available for clearing or equipment, guidance for farmers in growing crops, or marketing procedures.[11]

Leadership of the new Reclamation Service fell to Chief Engineer Frederick H. Newell,[12] the former chief hydrographer of the U.S. Geological Survey.[13] Reclamation tried to employ the best engineers that Newell could find.[14] He wanted projects in every western state, to avoid serving the needs of one section over another. Newell also wanted to dispel antipathy toward the Reclamation Service—of which there was already plenty. According to one western writer, "the engineers who staffed the Reclamation Service tended to view themselves as a godlike class performing hydrologic miracles for gratified simpletons who were content to sit in the desert and raise fruit."[15]

A national Board of Consulting Engineers, consisting of A. P. Davis, Joseph Barlow Lippincott, George Wisner, H. N. Savage, J. H. Quinton, W. H. Sanders, and B. M. Hall, planned all reclamation work.[16] The board immediately ordered preliminary work across the West, including mapping, drilling, water measurements, underground water storage, and engineering studies.[17] Water users in the Carlsbad area, including the Pecos Irrigation Company, desperately tried to convince Reclamation to consider the Pecos Irrigation Company's works, which were rapidly deteriorating.

Federal investigations into the Carlsbad area had begun with Elwood Mead, Wyoming state engineer from 1888 to 1899. Already an authority on irrigation engineering and law, Mead included the Carlsbad irrigation area in a series of reports submitted to the U. S. Department of Agriculture in 1899.[18] In the reports, entitled *The Use of Water in Irrigation*, Mead addressed the general issue of irrigation in the West. Specifically addressing Carlsbad's situation, he indicated the need to line the canals to eliminate water losses.[19] The Improvement Company, however, was bankrupt in 1898 and had no money to implement Mead's recommendations. Instead, Tansill, as receiver representing the creditors, and Tracy, who managed the company's physical operation, began to lobby to have the irrigation system purchased and/or renovated by the federal government. In October 1902, at the request of Robert Tansill, the chief engineer of the Reclamation Service, Frederick Newell, visited the Pecos Valley as workers rehabilitated the flume north of Carlsbad. Newell investigated the project from McMillan Reservoir south. He considered the possibility of purchase by the government and also looked to developing a flood storage system along the Pecos River. During this same trip Newell also investigated the area on the Hondo River, now owned by James Hagerman, which had been proposed as a reservoir site years before. Newell was careful not to make commitments to either project, not allowing himself to be pinned down by declaring publicly that the government would purchase lands for either site.[20]

Following Newell's visit, Reclamation hired William M. Reed to evaluate building a reservoir on the Hondo. Reed was no stranger to the valley, having been connected with irrigation companies throughout the valley for ten years. His first job had been as an instrument man for Pecos Irrigation and Investment Company at Eddy; he later served as engineer and water superintendent in the Roswell area and chief engineer in the area around Carlsbad from late August 1898 to 1900. Reed then worked as an engineer through his private firm in Roswell. In 1902 Reed took charge of hydrographic work in the area for Reclamation, placing measuring stations on the Pecos and Hondo Rivers. In January 1903 he supervised Reclamation's preliminary surveys on the Hondo Reservoir project, measuring the amount of silt deposited in the Hondo Reservoir site area and gathering information regarding gypsum formation, leakage, and so forth. After completing his investigations, Reed recommended a Hondo Project.[21]

Rivalry between Carlsbad and Roswell grew intense when the federal government began showing an interest in the Hondo site just a few miles outside the city of Roswell. What was at stake was not only government money and an irrigation system but the very flow of the waters that promised to make the Pecos Valley bloom.[22] When Reclamation began feasibility surveys, local landowners formed a corporation called the Rio-Hondo Reservoir Water Users Association in accordance with the provisions of the 1902 Reclamation Act.[23]

In the middle of June 1904 Secretary of the Interior Ethan Allen Hitchcock,[24] on Reed's and then Newell's recommendations, reached his decision concerning the Hondo Project. Following his boss's decision, A. P. Davis, acting chief engineer for Reclamation, notified Reed that Reclamation's Board of Consulting Engineers recommended building the Hondo Project and directed Reed to arrange for the purchase of the Hondo Reservoir site. The consulting engineers recommended including 10,000 acres of land within the project for irrigation purposes. They estimated the total cost to be $275,000, or $27.50 an acre.[25] The *Roswell Record* ecstatically published Davis's letter, notifying the community of Hitchcock's decision. Agreements were secured that provided for the surrender of the water rights on lands in the vicinity of the reservoir and for repaying the government for the cost of the works. During his lower Pecos days James Hagerman had established companies to the north, near Roswell, as well. Reclamation purchased from his Felix Irrigation Company the rights to the floodwaters of the Hondo River and the land of the Hondo Reservoir project for $20,000.[26]

Those downstream in Carlsbad asked themselves whether or not the establishment of the Hondo Reservoir meant the retention of seasonal floodwaters and normal winter flow of that stream. And if so, would it not "be feasible for other projects to retain the same flow on the Felix and other [Pecos] tributaries which are unquestionably the rightful property of the primary corporation [Pecos Irrigation Company]?"[27] Francis Tracy, man-

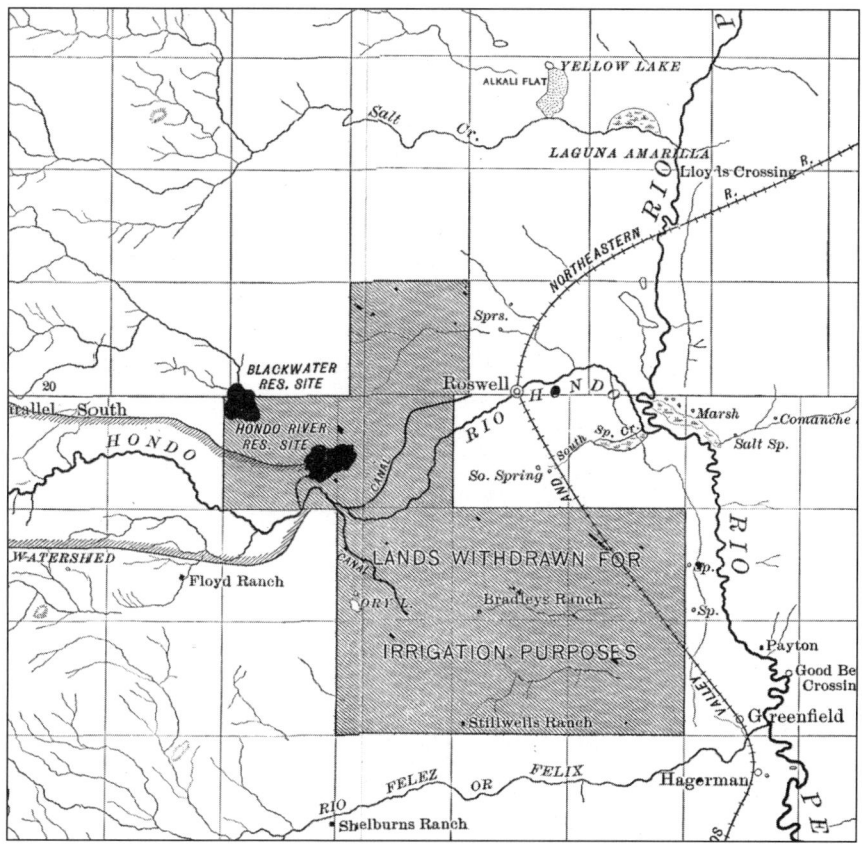

The Hondo Reservoir site and land purchased for irrigation.
(From F. H. Newell, Second Annual Report of the Reclamation Service, 1902–3,
58th Cong., 2d sess., 1904, H. Doc. 44).

ager of the Pecos Irrigation Company, suspected that Reclamation contemplated using not only the winter and flood flows of the Hondo, but also the floodwater of all the streams and tributaries emptying into the Pecos at and below the town of Roswell. Tracy and others argued that all such waters, including winter volume and flood flows, were the property of Pecos Irrigation Company, as a result of its predecessor companies' prior appropriation.[28]

Fearing that Reclamation might not buy the Pecos Irrigation system since it was planning an upstream project on the Hondo, Pecos Irrigation officials protested the building of the Hondo Reservoir. Tracy telegraphed the protest to newspapers across the country, making complaints against the secretary of the interior public. Pecos Irrigation and the Pecos Water Users Association were concerned that if the government could take the floodwaters and the winter flow of the Hondo River and impound it for use in the reclamation of new lands, it might in turn impound and use waters of the Felix, the Peñasco, the Seven Rivers, and various draws and tributaries up and down the length of the Pecos River.[29]

The Carlsbad protest resulted in a hearing held at Roswell on September 6–9, 1904, styled "Hearing in the Matter of the Protest of the Pecos Irrigation Company to the Honorable Secretary of the Interior against the Construction of the Hondo Reservoir by the Reclamation Service." A. P. Davis, assistant chief engineer; W. H. Sanders, consulting engineer; and B. M. Hall, consulting engineer for the Hondo Project, represented the Reclamation Service, which held the proceedings. Also present were Judges A. A. Freeman and J. O. Cameron, who represented the protesters from Carlsbad; W. C. Reid, who represented James Hagerman, chief spokesman for building the Hondo Reservoir; and William M. Reed, district engineer, who represented the U.S. government.[30] Although the hearing in the Hondo controversy ostensibly had to do with the validity of building a reservoir in the upper part of the valley, Hondo proponents and reclamationists placed just as much focus on the inadequacies of the irrigation works in the lower part of the valley. In many ways, the hearing on Hondo turned into an inquisition directed at the Carlsbad protestors over whether or not their claims of prior appropriation were justified given the allegedly excessive waste on the project. The key question that concerned the Carlsbad contingent was whether or not what they considered to be perfectly legitimate water rights would be respected.

As the key witness for Pecos Irrigation Company interests, Francis Tracy claimed that the irrigation system around Carlsbad embraced some 100,000 acres and had water rights for more than 50,000 acres of land. According to Tracy and others, the irrigation system had actually supplied water to over 50,000 acres of land. The title to much of this land had been acquired under the Desert Land Act, which, according to the Carlsbad contingent, validated their water rights.[31] Tracy, using language from the original appropriation, claimed that Pecos Irrigation Company, through its predecessors Pecos Irrigation and Improvement, Pecos Irrigation and Investment, and Pecos Land and Ditch Company, had a right "to all the normal flow of the Pecos River and its tributaries and without its reservoirs to so much of each flood flow as can be diverted into its canal and put to beneficial use."[32] As a consequence of these rights, the predecessor companies had built two reservoirs, and, Tracy argued, Pecos Irrigation had a right to all waters necessary to fill these reservoirs before any subsequent appropriators, including Reclamation, could have one drop of water.[33]

Tracy and others admitted that the Pecos Irrigation Company had ignored earlier violations of its water rights by Hondo irrigators. Too little water had been taken by private ditches to justify the cost of litigation. But when the company realized that the government was serious about using irrigation waters at Hondo, officials decided that to do nothing in response would "reflect a huge neglect of its interest" especially the interests of settlers who held water contracts that the company was bound to honor.[34] Tracy and other protestors anticipated that the parties in the upper part of the valley would argue that building the Hondo Reservoir would not significantly affect the water supply of the Pecos Irrigation Company. Thus, in the protest, Tracy and others argued that after the expense of $275,000 of public money—the estimated cost of the Hondo Project—and after settlers had taken charge of the lands under the project and paid the government twenty-five to thirty dollars per acre, it would be too late to try to prevent injury to settlers in the lower part of the valley. Tracy undoubtedly also knew that to retain appropriative rights, the possessor had a limited time in which to challenge violators.

When asked how much land had been irrigated by water from the Pecos Irrigation Company system, Tracy replied that the company and its successors had sold water contracts for 54,795 acres served by the Southern, East Side, Black River, and Hagerman Irrigation and Land Company canals. Subtracting Hagerman Irrigation and Land Company canals, this left some 52,515 acres of water contracts under the East Side and Southern canals. But of the 52,000 acres, Tracy admitted that only 14,403 acres were actually in the hands of water consumers; Pecos Irrigation Company itself owned 30,140 acres of the total. Avoiding pointed questions from A. P. Davis concerning what part of the land was good land, Tracy told the Reclamation official that all the land had been proved up under the Desert Land Act, and that water, at one time or another, had been placed on all of the land. But Tracy finally admitted that he could not "stand up here and say just how many acres is [sic] good arable land; I should say 20,000 acres."[35] When asked by Davis if the company could supply water to 30,000–35,000 acres of land through its irrigation works and canals, Tracy responded that the company could deliver water with various repairs to "a great deal more land than that." W. C. Reid, lawyer for Hagerman's interests, asked Tracy to estimate the value of the land per acre with water. Tracy stated that the company was prepared to sell land with water for twenty-five dollars per acre.[36]

Hagerman and attorneys representing the Roswell interests responded with two major points. First, much of the water in the Hondo River had been appropriated by others before the lower Pecos Valley companies appropriated any water. In July of 1888, when Pecos Irrigation and Investment Company adopted articles of incorporation and appropriated all unappropriated waters of the Pecos River, the First New Mexico Reservoir Company had already used waters from the Hondo, and the Investment Company recognized First New Mexico's right to do so. Hagerman and the government argued that this understanding was made implicit when the lower valley company(s) offered to buy First New Mexico's water rights, and then when it sold these same rights to James Hagerman's Felix Irrigation in 1898.[37]

Secondly, Hagerman argued that the appropriations from the Hondo River by the successive lower valley irrigation companies had always been intended for use in the Northern Canal. Never had the Pecos companies considered using it for their lower valley operations. Essentially, Hagerman argued that any water Pecos Irrigation and Investment had appropriated from the Hondo River was for its Northern Canal, not for the canals and reservoirs located in Eddy County. The Northern Canal had been transferred to Hagerman when the property split in 1898. Thus, Tracy and company officials in 1904 could make no claim to the waters of the Hondo. Hagerman came close to accusing Tracy of deliberate deception since "many of the alleged facts state[d] . . . suppress[ed] and ignore[d] vital facts and acts which are fatal to his claims, and to which he himself was principal party."[38]

James Hagerman testified about the history of irrigation along the Pecos River and his involvement with it. When asked about the history of Lake Avalon and Lake McMillan, Hagerman replied that Avalon was built in 1890, McMillan in 1893, and that in December 1888, six months after Pecos Irrigation and Investment was formed, First New Mexico Reservoir and Irrigation Company took out a charter for the building of the Hondo Reservoir. The owners of First New Mexico Reservoir filed maps in Washington, D.C., for the reservoir site and completed a small amount of work, but sold out the entire enterprise to Charles Eddy, who thought it would add considerable water to the lower valley interests. According to Hagerman, he and Eddy initially questioned First New Mexico Reservoir and Irrigation Company's right to use the waters of the Hondo. After investigation, they decided that they could not stop First New Mexico from using the water, so the partners bought them out.[39] Hagerman believed the Hondo land was the best of the valley, and he and Eddy felt they could store irrigation water very close to the river with little waste, which made purchasing the First New Mexico Reservoir lands a good proposition. Hagerman stressed that the Hondo was never intended and never thought of or talked about as a feeder to the irrigation works in the lower valley.[40]

Pecos Irrigation and Improvement began working on the Hondo

scheme in 1892, but when the dam at Avalon washed out in August of 1893, financial problems forced the company to temporarily halt work on the Hondo. Hagerman insisted that the reason work stopped was a lack of money, not lack of faith in the project.[41] When asked why he had purchased the Northern Canal and property in Chaves County, Hagerman, reflecting bitterness over affairs in the lower valley, replied, "I purchased it to get out of carrying a big elephant that I was tired of. The facts are these: as early as 1896 I had made up my mind that the lower Valley had no future, and that the upper Valley had a great future, but to make it available and to save the situation, I knew we had got [sic] to do just what we originally intended: that is to extend the railroad from Roswell to Amarillo, so we would have a northern and eastern outlet."[42]

During the bankruptcy reorganization of the Improvement Company, Tracy and Tansill raised $18,000 to $20,000, but needed $40,000 to $50,000 for reorganization. Eager to sell the property, primarily because they thought it had little value, Tracy and Tansill sold it to Hagerman for $50,000. According to Hagerman, "they bet on that horse, and I bet on this one,"[43] not mentioning that he had already bet over $2 million on the southern "horse." Hagerman testified that Tracy's claim to floodwaters had no validity whatsoever, and that the Hondo waters they demanded were just an afterthought.

The hearing then shifted its focus toward whether Pecos Irrigation and the lower valley made enough use of its waters to be entitled to them. Hagerman argued that there had been no coherent method for laying canals or selling land, that money had gone into shoddy canals and unproductive land, and that sizeable water losses occurred. In a pointed stab at Charles Eddy and the process of selling water contracts in the lower part of the valley, Hagerman commented that during 1893, 1894, and 1895, before he came into possession of the northern properties, many settlers who had entered into water-rights contracts under the system canals left the country. He pointed out that the irrigation company[s] sold water contracts on land scattered throughout the countryside—depending "very largely on where the general manager [Charles Eddy] . . . individually owned land [and] water to sell. If he owned forty acres a long distance off, the company would build a long lateral to it, and if his sister [Mary Fox], who owned a good deal of land, wanted to sell it, they built laterals to it."[44] Pecos Irrigation Company and its predecessors sold land in the lower valley forty acres in one place and forty acres half a mile distant without reference to the concentration of landholders.[45]

Hagerman's testimony pointed out that no rational, efficient development plan existed for locating canals or distributing water, leading to tremendous waste. Hagerman also downgraded the land values of the lower valley. His attorney, W. C. Reid, pointed out that there was some question regarding the value of the lands owned but never irrigated by Pecos Irrigation Company.[46]

Hagerman described other lands of little or no value, including 10,000 acres south of Black River, which had been purchased by Swiss settlers in 1892 to establish a colony near Malaga. Irrigation company officials had never put any water on the land except for "proving up" purposes. Following Tansill's appointment as receiver in 1898, the company tried to spend money only on projects that were a sure thing, abandoning the Southern Canal south of Black River.[47] Hagerman testified that farmers had never cultivated land five or six miles south of Eddy on the East Side Canal because the canal, though fairly large, was part of "an old scheme" of Charles Eddy's to sell land to Hagerman and the company. Subsequently the company spent a large amount of money enlarging the canal, and Hagerman estimated that while farmers might ultimately irrigate 1,000 acres of good farm land there, the scheme to irrigate large acreages nearby was a total failure. He did not realize the true value of the East Side Canal until late in the history of the company.[48]

While Hagerman described what actually existed, the Carlsbad contingent used against him an old prospectus from 1891 designed to entice immigrants. The prospectus, with Hagerman's signature, showed the company's purported ability to irrigate thousands of acres. Hagerman could only plead ignorance, noting that if he had properly understood the facts, he would not have invested millions of dollars in the lower valley. He accused the Carlsbad contingent of "trot[ting] out old prospectuses and things of that sort" and claimed that he had been duped—that when he first went to the valley, Eddy and others pointed out where the surveys showed canals should be located. Eddy had on hand an engineer—either Nettleton or Blauvelt—who explained the whole scheme to Hagerman. Hagerman believed it and placed a large amount of money into the venture. Hagerman called Eddy a very plausible man in whom he truly believed and continued to believe until 1894.[49]

Revealing his total confidence in Charles Eddy and other company officials, Hagerman recalled that in August 1893, when the Avalon Dam washed out, instead of quitting he made the mistake of trying to save local farmers by raising money and rebuilding the dam.[50] In the latter part of 1893, Hagerman said, he began to realize the true physical conditions of the lower valley: "Leaky canals and the impossibility of making them tight, the poor soil, the soil underlain with gypsum, and the fact that on such lands you could not raise good crops, no matter how much water you put on them."[51] Hagerman admitted that it was not until 1897 that he finally realized completely that reports made to him, including those of engineers Nettleton and Blauvelt and manager Charles Eddy, had been false.[52] Hagerman's charges against the engineers and his bitterness suggest that Nettleton, Blauvelt, and other engineers worked for Eddy, not Hagerman.

Cameron took great pains to discredit Hagerman's claim of innocent trust in Charles Eddy. He wondered why, if Hagerman made trips to the valley, built a residence of some value near Eddy, and expended a great deal of money, he had not taken the trouble to examine the various dams and canals in the irrigation system. Hagerman responded that he had mistakenly trusted the engineers and placed too much confidence in them.[53]

Cameron badgered Hagerman, trying to show that his confidence in the project at least up until 1894 indicated a fervent belief in the lower valley. The lawyer suggested that his investment and approval of engineering advice was a tacit statement of his belief in the southern irrigation enterprises. Frustrated with the examination, Hagerman replied, "I won't answer any more of your questions about those old things," and "I think one of the purposes of your questions is to try to embarrass me and show that I made a mistake in statements. I think that is your main purpose in asking these questions."[54] Cameron asserted that was exactly what they did not want to do, because it was to their advantage to show under what faith Pecos Irrigation Company had acted.[55]

Hagerman's testimony revealed the bitterness of a man who thought he should have known better. James Cameron asked him if there was not considerable good land along the Southern Canal near Black River and whether it was not "a very much better proposition than you found it." Hagerman recalled a particular trip to the valley when he was traveling down the Southern Canal with Charles Eddy and engineer Charles Blodgett in 1894 and came upon a bend in the river known as Gyp Bend. Hagerman noticed how much water disappeared from the area, and brought this to the attention of Eddy and the engineer, who simply dismissed it. Hagerman claimed that leaks through the gypsum formation at that point of the valley wasted a tremendous amount of water that could have been used for irrigation, a point emphasizing the Hondo group's claim that Carlsbad wasted water.[56]

Hagerman's anger intensified as he recalled that during 1892 several hundred thousand of his dollars went into the irrigation enterprises under the direction of Charles Eddy. During the early 1890s, Eddy bought lands, built the East Side Canal, and, according to Hagerman, did "a lot of other things . . . not worth a cent."[57]

Hagerman's health and other business had allowed him only infrequent trips to the valley prior to 1900, the main reason for Charles Eddy's de facto management there.[58] To Hagerman's detriment, from 1890 to 1894 he never stayed in the valley more than two or three times, and then for only a day or two. The second time he was there, railroad crews had just finished laying track from Pecos, Texas, to Eddy in January 1891. Hagerman became very ill soon after, went to Europe, and spent most of the year there. When he came back to Colorado, he faced legal complications involving the Molly Gibson Mine, and consequently he saw the valley only once in 1892.[59]

Although Hagerman's assessment of irrigation works was damaging to Pecos Irrigation claims to all the water in the river, W. M. Reed's testimony was even more damaging. Reed had long been associated with irrigation enterprises in the lower and upper parts of the valley and in 1904 was district engineer in charge of the Pecos River District under the U.S. Geological Survey. As such, he had charge of investigations pertaining to the Hondo Project, a silt survey of the Pecos Company's McMillan Dam and Reservoir, and other projects in the district. After investigating the Hondo project in 1903, he recommended that the project go forward.[60]

Reed contended that he never thought diverting part or all of the Hondo River would lessen the flow of the Pecos River flow or injure Eddy County irrigators. Reed claimed that if irrigators in the lower valley lacked water, the reason had nothing to do with dispersion of water upstream, but rather wasteful water usage and leakage in Avalon and McMillan Reservoirs. According to Reed, "it was . . . simply the reservoirs could not hold sufficient water that came to them; a great deal of water was wasted by . . . farms and . . . land was almost ruined."[61] Based on Reed's figures, reservoirs in the lower part of the valley were losing at least 6,000 acre-feet of water, or a flow of 100 cubic feet per second, each year. Reed thought the lost water might irrigate 10,000 acres of land. Most important, he thought leakage in the lower Pecos Valley equaled or surpassed the amount to be withdrawn by the proposed Hondo Project. This was the key point. Reed's assertion about waste implied that Pecos Irrigation must fix its irrigation system. Legally, if the company was not using all of its appropriated water productively, others could use it. Reed effectively minimized the effects of the Carlsbad group's argument against the Hondo Project.[62]

Reed identified another problem facing the reservoirs of the lower valley, particularly Lake McMillan. Silt carried down the Pecos during floods built up in the reservoir, reducing the amount of water it could hold. Engineers, including Reed, thought the silt still piling up in McMillan might plug leaks evident at the base of the reservoir and along its laterals, although McMillan's storage capacity steadily decreased.[63]

The hearing at Roswell pointed out not the possible injury to water users and the irrigation company in the lower valley, but rather the inadequacies of the company's reservoirs and irrigation systems. The Hondo proponents made the Carlsbad protestors appear greedy for waters to which they might not have a moral right. In any case, Carlsbad appeared vulnerable on the issue of wasted water and ineffectual irrigation works.

Afraid that competition between Roswell and Carlsbad might "envenom and embitter both towns if allowed to run unchecked,"[64] the *Carlsbad Argus*, in true partisan form, claimed that "most of his [Hagerman's] testimony was worthless to the [Hondo] Water Users' Association." The *Argus* believed that Hagerman actually helped the cause of the Pecos Irrigation Company in its attempt to sell its works to the government.[65]

On September 10, 1904, the Board of Consulting Engineers that had held the hearings submitted a report endorsing both of Hagerman's arguments, that Charles Eddy intended Hondo water for the Northern Canal and that Carlsbad wasted a great deal of water. According to Davis, Sanders, and Hall, all of the Hondo's flow at Roswell had been taken into the Northern Canal for irrigating upstream, and the amount of Hondo water reaching the Pecos River was insignificant. The board reported that for ten or twelve years waters had been customarily used in the Northern Canal and never reached the Pecos or Lake McMillan. The engineers offered evidence showing that the Pecos Irrigation Company was unable to conserve any Hondo waters that might reach Carlsbad and that such additional waters would be "a useless addition to an excess supply."[66]

Turning their investigation toward Carlsbad, the engineers concluded that southern storage facilities, though defective, furnished adequate water storage for 13,300 acres, more lands than Pecos Irrigation Company irrigated at the time. The right of prior appropriation allowed the company to water continuously the 13,300 acres. The men estimated that adequate storage capacities existed in the Avalon and McMillan Reservoirs, if the company made them more efficient, to water more than twice the 13,300 acres even during the driest years in the valley. The board criticized Pecos Irrigation Company and irrigators under their system for wasting four or five times as much water as they beneficially applied to the land: "A claim to water right, based upon such use, involves the claim that at least three acres in another region remain desert in order that one acre continues this waste. . . . [T]he claim made in this company's protest that they have the right to extend such wasteful use and preference to extensions where reasonable economy will be practiced under federal supervision is clearly untenable."[67]

The Board of Consulting Engineers concluded that the government had consent for construction of the reservoir from all parties having valid claims to the Hondo, and it recommended immediate construction.[68] The protestors from the lower end of the valley were obviously dismayed at the decision. Tracy and others leading the protest feared that the new Reclamation Service, in selecting project sites across the West, would bypass Carlsbad since the Hondo site lay just one hundred miles away.[69]

What happened the following month, in October 1904, added insult to injury in the lower valley, but may have resurrected that part of New Mexico. On Sunday, October 2, 1904, floodwaters raced down the banks of the Pecos River and partially destroyed Avalon Dam. Observers estimated the flood to be twice as great as the one that had caused so much destruction in 1893. The flood surge that followed the main washout at Avalon kept Tracy and other officers of the company vigilant in watching the waters rise in McMillan Reservoir. At one point Tracy sent a telephone message advising residents in the valley to seek higher ground.[70] Although Carlsbad residents had some idea about the dangers of the flood from reports in Roswell, such reports ended when floodwaters washed out telegraph wires and railroad bridges.[71]

Francis Tracy, on October 12, 1904, duly reported the damage to stockholders of Pecos Irrigation Company. The flood had partially destroyed Avalon Dam, parts of the Southern Canal, and the earthen approaches near the concrete aqueduct. McMillan Dam had withstood the floodwaters, although 2,000 feet of earth embankment near the spillway had washed away and the headgates were damaged. Tracy estimated that Pecos Irrigation Company needed $50,000 to repair the irrigation works in order to deliver water the following spring. Avalon was the keystone of the irrigation system, and its destruction left farmers without water. Tracy contacted Lombard, Odier in Geneva, as well as Sanford and Kelley in New Bedford, Massachusetts, the two major investment groups in the valley.[72]

The investment bankers would not provide the money necessary to rebuild the dam, but Sanford and Kelley called on Massachusetts's congressman William S. Greene to lobby for the Reclamation Service to buy the property. Sanford and Kelley told Greene that Reclamation could "put it in proper condition so that the settlers may be provided with water for irrigation in spring, otherwise they will be in pretty bad shape and stand a chance of having their property made valueless."[73] The valley was too valuable in their estimation to be allowed to go to waste, and the investment firm fully anticipated government takeover. Pointing out the urgency of the situation, Sanford and Kelley requested immediate action. Pleading with Greene, they asked his advice about sending papers to Senator Henry Cabot Lodge imploring him to apprise Reclamation of the necessity of direct and immediate action. Sanford and Kelley were no doubt primarily interested in the welfare of their stockholders, but they carefully pointed out the importance of settlers' interests in the valley as well.[74]

W. M. Reed, who one month earlier had served as main witness for proponents of the Hondo Reservoir, quickly responded to needs in the lower valley. Following the flood, Reed toured the valley to assess damages. He found the concrete power dam standing without apparent injury, but the river had cut a new channel around the west end of the dam. Flooding destroyed an electric powerhouse, cotton gin, two highway bridges, and the railroad bridge, but Carlsbad itself was not significantly damaged. At McMillan, Reed noted damage to an auxiliary earthen dam west of the main structure, and downstream he found the Pecos River racing through the middle of Avalon Dam. Out of thirty-two gates in the dam's spillway, only eight still stood, and they were badly damaged, while the river had broken the Main Canal directly below the dam and caused minor damage to earthen approaches at the flume.[75]

While Reed was inspecting conditions in the lower valley, the Carlsbad Commercial Club[76] began lobbying him, explaining what they considered to be major problems in the valley. The group did not believe Pecos Irrigation Company could repair the many structures damaged in the lower valley. They informed Reed about the company's financial straits and expressed fears that roughly 14,000 acres of cultivated and improved land might revert into desert.[77]

Sending a telegram from the Pecos Valley to Chief Engineer Frederick Newell, Reed reported that flooding partially destroyed the distribution dam at Avalon, and the company did not have the resources to replace it. Reed wanted Newell or a consulting board to visit Carlsbad and take immediate action to provide irrigation the following season. Newell found it impractical for a consulting board to visit Carlsbad because other projects across the West demanded attention. According to Newell, "immediate action in Carlsbad [was] impossible."[78] Reed doggedly tried to convince Newell of the impact of flood damage to the valley. Acknowledging that locals mounting the lobby effort might themselves exaggerate the emergency, Reed still thought the situation serious enough to inform his boss about the headworks at Avalon and the spillway at McMillan. Floodwaters had cut the spillway down to the riverbed, and, according to Reed, McMillan needed a new dam to make the reservoir of any value.[79]

The Carlsbad Commercial Club repeated the need for help. Francis Tracy and locals anxious to have the government pay attention to their plight pointed out the destruction caused by floodwaters. Matching Tracy's figure of $50,000, the *Carlsbad Argus* pointed out that McMillan Dam sustained $2,500 in damages; Avalon, $15,000; the Southern Canal, $2,500; the La Huerta bridge, $3,000; Hagerman Heights, $42,000; and a privately owned power dam, cotton gin, and farms damage in the neighborhood of $7,000 to $10,000.[80]

Following the 1904 flood, the Reclamation Service began taking more interest in the possible purchase of the Pecos Irrigation Company's facilities.[81] Reed recommended that the water users in the valley form an association in compliance with the Reclamation Act of 1902.[82] Encouraged by Reed, in October 1904, less than a month after the Hondo hearing and a week after the flood, fifty valley residents met at the Eddy County courthouse. Members of the new association received one vote for each forty acres covered by a water-rights contract. The goal of the group was to convince the federal government to purchase the irrigation works of the Pecos Irrigation Company. The Pecos Water Users Association incorporated under the laws of New Mexico Territory; among the charter members were R. S. Benson, Charles H. McLenathen, William A. Finlay, George H. Webster Jr., J. C. Keogh, G. W. Witt, and Richard J. Bolles, most of whom were early investors in the valley. Officers of the association took prompt action in establishing contacts with local and national Reclamation officials.[83]

Reclamation policy made it clear to the valley water users that it would not deal with individuals concerning the development of a reclamation project near Carlsbad. It was imperative that the Pecos Water Users Association serve as the instrument through which farmers could voice their concerns and reach agreements with the federal government. Charles McLenathen, secretary of the association, and his real estate partner Francis Tracy presented a formal request to the Reclamation Service asking that engineers be authorized and instructed to design and supervise repairs on the "Carlsbad Project."[84]

Officers of the Pecos Water Users Association recommended that the group formally call on the Pecos Irrigation Company to obtain a definite answer as to whether the company would repair the project and irrigate lands during the 1905 season.[85] The company had no money for repairs.

Reclamation officials directed W. M. Reed, W. H. Sanders, and B. M. Hall to formulate a project board for repairs in the valley. In December 1904 Reclamation began a seventeen-month-long investigation and preliminary surveys in the lower valley to repair and rehabilitate the system.[86] Because of significant damage caused by the flood, Reclamation's main priority was to repair facilities sufficiently to extend waters to farmers the following spring. The government paid for labor, the company for materials. But financing substantial repairs of the Pecos Irrigation Company's works by the water users alone was an impossibility. Beyond temporary repairs, water users simply were not able to raise the money on their own, so the federal government began an inseparable partnership with private interests in the Pecos Valley to rehabilitate and improve the nineteenth-century irrigation works of Pecos Irrigation Company.

8

MR. WRECKLAMATION MAN

WHAT began as the U.S. Reclamation Service's attempt to supply water to farmers in the spring of 1905 developed into a full-scale investigation, purchase, and rehabilitation of the entire Pecos Irrigation system. Although scientific evidence and history suggested that Reclamation had no business in the valley, the Service came because of political pressure from the White House. Reclamation engineers brought to the valley a wariness and distrust generated by information revealed at the Hondo hearing and by political pressure to accept a project with which they had serious reservations. This initial wariness and distrust was compounded by the fact that Reclamation was taking on a project largely in the private hands of absentee landowners. Because of the political considerations and special nature of the project, Reclamation moved reluctantly into the valley.

B. M. Hall served as Reclamation's chief engineer for New Mexico Territory. When Reclamation started to examine the Carlsbad area for more than just temporary repairs, A. P. Davis, acting chief engineer, warned Hall not to underestimate costs for the valley. Davis advised Hall in December 1904 that "great care should be taken that you are not thus embarrassed as it would be greatly to our discredit to enter upon such an enterprise and fail to carry through as promised."[1]

As Davis counseled Hall about cost overruns on permanent repairs, Mexican work crews and equipment began to arrive in January 1905 for temporary work on the project. Reclamation's E. W. Myers served as project engineer under B. M. Hall's supervision. Workers first built a road paralleling the east fill on the upstream side of Avalon Dam, so they could access the river and haul timbers to a framing yard. They built several turnouts from the road to hold the teams and wagons dumping earth into the fill.[2] Mexican workers and Anglo farmers worked together, the Mexicans loading the earth, and the farmers, driving their own teams, moving it.

Neither group worked steadily, as numbers of each frequently took off from work for a day or a week.[3]

Reclamation's situation in the valley during these early days was not ideal. Pecos Irrigation and Reclamation engineers worked together to supervise the temporary patching of dams and canals, while local water users provided funds and labor. G. W. "Boston" Witt, builder of the flume and longtime resident, served as carpenter foreman. Reclamation's E. W. Myers considered him incompetent, and Pecos Irrigation's engineer, Vernon Sullivan, replaced Witt. Oscar Weaver, a company ditch rider, served as earthwork foreman. Myers and Hall considered him inexperienced and therefore incompetent as well. Pecos Irrigation used seven drag scrapers and five wheel scrapers to move the dirt, utilizing what was available, although Reclamation officials considered the wagons and horses inferior.[4]

If wagons, horses, and foremen were shoddy, Myers and Hall saved much of their criticism for the labor force and technology employed. "[A]ll concrete was mixed by hand by Mexican labor, working with shovels and was wheeled to the work in barrows," complicating repairs.[5] Hall and Myer's comments reflected prevalent racial biases. Myers called the Mexican laborers "very inefficient."[6] Hall stated that "without constant supervision his [a Mexican's] output for a day will fall far short of that obtained from eastern labor, probably in many instances the output per Mexican being less than half that obtainable with other labor elsewhere, while the difference in the rate of pay: 12–1/2 cents per hour as against fifteen cents per hour for eastern labor does not equalize the difference in efficiency."[7]

Weather also hindered repair efforts, particularly in February when three snowstorms and cold weather kept crews from working two-thirds of the month. Rains and melting snow kept water levels in the Pecos River high, forcing crews to raise the gates of McMillan to full height on March 11, when a work bridge at Avalon's diversion dam washed out. Hall was disgusted that, despite working in an arid region, his work crews waded knee-deep in mud every time they went to work.[8] With these problems, Reclamation went beyond the $35,000 raised by water users for temporary repairs. The Service justified cost overruns by repeating their complaints about labor, equipment, and weather.[9]

Despite repair problems in the valley, Reclamation considered appeals from the Pecos Irrigation Company and the Pecos Water Users Association to purchase and rehabilitate the entire project. To remove any obstacles to government purchase of his company, Francis Tracy wrote to Territorial Governor Miguel A. Otero and asked him to consider an amendment to section fourteen of "An Act to Authorize the Formation of Companies for the

Workers riprapping the front face of Avalon Dam on the east side of the Pecos River.
Bradbury and subsequent contractors relied heavily on Mexican and
Mexican American labor.
*(Record Group 115, Records of the Bureau of Reclamation, Public Relations Photograph Collection,
Series 3—Historic, Photographs by Project Sites, Box 004, National Archives and Records
Administration, Rocky Mountain Region, Denver, Colorado.)*

Purpose of Constructing Canals for Irrigation and Other Purposes, etc.,"
approved by the state legislative assembly February 14, 1887. The legislation allowed the creation of the earliest irrigation companies in the lower
Pecos Valley. The amendment would authorize Tracy, with only a two-
thirds vote of all stockholders, to negotiate government purchase of Pecos
Irrigation's plant. Tracy feared the territorial attorney might insist on a
unanimous agreement from all stockholders, because some 14 shares out of
3,029 belonged to unlocated foreign investors.[10] While Tracy was successful
in this effort, Reclamation rebuffed his first proposal, feeling that the price
he wanted for the company canals and dams, $300,000, was "absurdly
high."[11]

Newell agreed, however, to Hall's recommendation for a topographic
survey of the main canals and a proposed third dam and reservoir site
between McMillan and Avalon. Pecos Company officials had always entertained the notion of building what became known over the years as Reser-

voir Number Three as a means of bolstering water storage. Newell also approved the creation of a general map of the region, and in early 1905 an investigating board went to the valley to determine the feasibility of a Carlsbad Project. The investigating board consisted of W. M. Reed, W. H. Sanders, and B. M. Hall.[12] In June, before the investigating board arrived, the newly repaired dam washed out again. Pecos Irrigation had no money for repairs. Neither did the water users. Board members, in the company of Carl E. Grunsky, an engineer appointed by President Roosevelt as a special advisor to the secretary of the interior and the director of the U.S. Geological Survey, immediately went to the valley. Grunsky's visit was important since all plans made by engineers for projects went before him. Grunsky stayed in the valley for two days investigating the irrigation works.[13] The investigating board delivered its report on August 31, 1905, calling for Reclamation to take the project under its jurisdiction and rehabilitate it. Grunsky approved the project.[14]

The secretary of interior and the Reclamation Service director still had to approve the project. In October 1905 Newell, wishing to avoid any public relations pitfalls, advised B. M. Hall that the secretary of the interior was alarmed at the large expenditures of the new Reclamation Service and recently refused to approve any new projects, fearing there would not be enough money in the budget to complete those already in hand. Reclamation started work on the Roosevelt Reservoir on Arizona's Salt River and was trying to finish the Truckee-Carson Ditch in Nevada. Newell advised Hall that Reclamation and its engineers had to "be extremely cautious and diplomatic not to embarrass or anticipate the action of the secretary or cause the people to feel that we are not keeping faith with them. They must at all times be assured that we are not free agents that our business is simply to present facts to the department, and that the reclamation fund is not sufficient to build all of the needed works in New Mexico, much less those in the whole country . . . while developments in New Mexico have proceeded rapidly, they have gone ahead far more rapidly elsewhere, and similar and more pressing cases are before the department."[15]

To complicate the issue, Newell received a letter from George Wisner, one of Reclamation's consulting engineers who had participated in the 1905 investigation of feasibility on the Carlsbad Project. Wisner reported that relations between Pecos Irrigation Company, the people in the valley, and Reclamation officials were strained. Reclamation engineers, having spent $35,000 of the water users' money on repairs, did not endear themselves to Francis Tracy or the water users when flooding damaged many of those structures again in late 1905. Wisner noted that such failures engendered mistrust and advised Reclamation to take up the entire project at an early date and to complete stopgap repairs quickly so that farmers in the valley could be supplied with water the following year.[16]

Newell, apprised of the new damage, asked W. H. Sanders, one of the three engineers on the investigating board, to write him a confidential letter concerning conditions in the vicinity of Carlsbad. Newell told Sanders: "I am disappointed at the failure of the work under Mr. Hall both at the dam where the washout occurred and at the earthwork at both ends of the aqueduct. I should like to know as to whether good engineering ability was displayed. In particular, I fear that Mr. Hall had placed too much confidence in Mr. [E. W.] Myers who does not have engineering experience or ability."[17] Newell, also concerned about finances, wrote Hall for an estimate of future expenditures for the years 1906 through 1908 if Reclamation was to make permanent repairs on the project. Newell assembled these figures to present to the Interior Department.[18]

In late 1905 W. H. Sanders responded to Newell's requests for a confidential report of the conditions at Carlsbad. Sanders visited Carlsbad in February 1905 and noted that work had commenced on the temporary diversion dam at Avalon. He advised E. W. Myers to fill the front part of the dam structure completely with rock to give sufficient weight for stability. He indicated that sheetpiling should be used in the dam and that the waters of Lake McMillan should be drawn down several feet. Later, when an unusual flood flow occurred, McMillan Reservoir was not lowered as Sanders had recommended and the flood endangered work already completed. During the 1905 repairs Sanders discovered that instead of using the pile driver as he suggested, workers drove the sheetpiling by hand, and some of the sheetpiling might not have been placed at all. Furthermore, he noted that construction of timber work showed poor workmanship.[19]

Blame for the engineering mishaps fell on the shoulders of local project engineer E. W. Myers, who had assumed much of the responsibility of the project in the absence of territorial engineer B. M. Hall. In defense of Myers, Sanders noted that even had the dam closure been completed, the storm that passed through would have washed the entire structure away. It was impossible to construct a spillway of sufficient capacity with the amount of money provided by water users for a temporary diversion dam. In light of the engineering mishaps, W. H. Sanders and George Wisner, another member of Reclamation's investigating board, advised placing W. M. Reed in charge of the work. Because Reed had worked previously with the lower valley irrigation companies, they thought the people of Carlsbad "ha[d] every confidence in him."[20]

While Reclamation had made no long-term commitment to a lower

Pecos project, local newspapers, the Pecos Water Users Association, and the commercial clubs of Roswell and Carlsbad used the problematic repair work to convince the public that a full-fledged Carlsbad Project was in the works. The groups lauded Reclamation's prompt attention to rebuild the irrigation works and to save the Pecos Valley by rehabilitating the entire system. The pressure Davis and Newell had earlier feared was now being used to force Reclamation's hand, even as Newell continued to be concerned about the expense and quality of the temporary repairs and about the attitude of Carlsbad residents. Although Reclamation was making more repairs than anticipated in the valley, at this point both Davis and Newell wished to avoid Carlsbad, its leaky canals and dams, and the Pandora's box that might be opened if they made a permanent commitment to the valley. Besides their strong reservations about full involvement at Carlsbad, Newell and Reclamation faced pressure from other sources. Newell reported that Secretary of the Interior Ethan Hitchcock "is becoming alarmed at the large expenses involved elsewhere and apparently is not willing to consider further construction in New Mexico. I fear that the people of the Pecos Valley have assumed that we are free agents in the matter, and that having taken up and examined the Carlsbad Project we are committed to it. While there may not be funds for New Mexico, there is a large surplus which must be invested, if possible, in Oklahoma, and I hope you will use every possible effort to assist Mr. Hall in bringing the Oklahoma work to the point where early construction can begin."[21]

All the while Newell and Reclamation were considering where to spend money on various reclamation projects, representatives from Pecos Irrigation Company and investment brokers Sanford and Kelley tried to influence the secretary of the interior, senators, and congressmen to sway the doubters at Interior. Representatives of the valley hit Newell and the Reclamation Service in Washington and at the local level. Alpheus A. King, secretary of the Irrigation Commission of the New Mexico Territory, and G. A. Richardson, representing the Roswell Commercial Club, sent letters to the Reclamation Service and notified the media in a cooperative effort to make it appear that Reclamation had already accepted responsibility for totally rehabilitating the project. Richardson and King endorsed resolutions to the Reclamation Service, many of which Newell feared the public and administrators in Washington might misunderstand. One resolution was especially galling to Newell: "We heartily endorse and approve the action of the reclamation service in hastening to the relief of the people of Carlsbad." According to Newell, such language was "superfluous and somewhat exaggerated." Assured that Reclamation had systematically examined the possibilities of developing the Pecos River valley, Newell carefully responded that "great care has been taken, however, not to appear to favor any locality or render special assistance."[22]

Not only had the Reclamation Service not accepted the project, its engineers were not at all sure that the government should ever consider rehabilitating the project. In light of revelations about wasted water at the Hondo protest the year before, Reclamation was reluctant to take on Carlsbad beyond temporary repairs. In fact, the weaknesses of the lower valley—leaky canals and dams, gypsum, and poor soil—caused Reclamation to look at other sites across the West. A. P. Davis, who often acted in Newell's absence, argued that the government should remain committed to restoring the project only in terms of temporary repairs. Davis noted that McMillan Reservoir leaked worse in 1905 than it had in 1904 and that its storage capacity was precarious, making the hope of irrigating new lands preposterous. Davis explained: "There is no reason why the government should touch this project except to save improvements already there. . . . Developments on reservoir three [are] unfavorable, the proper repair and difficult maintenance of the project will cost all the land will stand, without any payment for the present system. The estimates sent in did not include maintenance which will be very heavy."[23]

While Davis thought landowners would have a hard time paying for extensive repairs, Reclamation's on-site investigating engineers proceeded with their survey of the area.[24] During their investigations between Avalon and McMillan Dams, engineers realized that they were about to find oil or asphalt. Hall sent Newell a sample of a drill core and asked him to verify that the lands had been properly withdrawn from settlement. He argued that the public should not be informed about the discovery to prevent a rush to the region for oil claims.[25] Newell verified that withdrawal of the lands on April 14, 1903, prevented all entry selections by the public. While he did not believe any monies brought about by the discovery of oil in the Carlsbad area could be spent on the Carlsbad Project itself, Newell told Hall to investigate the discovery thoroughly so Reclamation might benefit from it later.[26]

Undeterred by Reclamation's lack of enthusiasm, the Roswell Commercial Club, the Carlsbad Commercial Club, and the Irrigation Commission of New Mexico Territory increased their efforts to convince Reclamation to take on the entire Carlsbad Project.[27] The groups attempted to ensure that the report made by Reclamation's board of engineers would induce Reclamation to purchase the Pecos irrigation system and not only repair but rehabilitate its reservoirs and canals. A constant presence was Francis Tracy, the manager of Pecos Irrigation Company lands and the most vocal of all the water users in the valley. He began his lobbying in 1904 when he

Charles L. Ballard of Roswell, about 1904. Ballard had served with Theodore Roosevelt in the Spanish-American War. Following the inception of the Reclamation Service, Ballard and a committee representing the irrigation interests of Carlsbad went to Washington to convince Secretary of the Interior Ethan Allen Hitchcock to place Carlsbad on the priority list of projects. The secretary turned the committee down cold until Ballard called on his old friend Roosevelt, who was now president. Roosevelt told Hitchcock to undertake the Carlsbad Project despite his reservations.
(Courtesy Archives and Special Collections Department, New Mexico State University Library.)

met with Frederick Newell. During 1905 Tracy made three subsequent trips to Washington to see President Roosevelt, Secretary of the Interior Hitchcock, Newell, and other engineers.[28]

Ultimately Reclamation's hand was forced when a committee of interested farmers and irrigators traveled to Washington and met with President Theodore Roosevelt. One of the group, Charles L. Ballard, had special ties to Roosevelt, having served in his Rough Riders outfit during the Spanish-American War. Ballard's influence helped induce Roosevelt to pressure the Interior Department and Reclamation to adopt Carlsbad as one of its initial projects.[29]

Roosevelt had strong regard for Charles Ballard and his abilities. On May 7, 1898, Ballard had mustered in at Santa Fe as a second lieutenant in the First Regiment of the U.S. Volunteer Cavalry (the Rough Riders), assigned to duty with Troop H. From Santa Fe, Ballard and the others traveled to San Antonio, where they joined Colonel Leonard Wood and Lieutenant Colonel Theodore Roosevelt, along with Arizona and Oklahoma squadrons. Only part of Ballard's regiment proceeded to Cuba after traveling to Tampa, Florida. Ballard and the rest went to Montauk, Long Island, where they stayed until the rest of the regiment returned from Cuba. Ballard's regiment mustered out on September 15, 1898. Despite not having charged up San Juan Hill, Ballard received high praise from fellow New Mexican Captain George Curry. Theodore Roosevelt, who signed

Ballard's discharge papers, gave him glowing praise, saying, "This officer I regard as one of the very best in the regiment, I know none in whom I grew to place more trust, and none whom I would be more anxious to have with me if again called into the field." Roosevelt invited a number of officers, including Ballard, to Oyster Bay before returning home.[30]

Later, Ballard responded to a request from the War Department to join the Volunteer Cavalry in the Philippines. Ballard accepted a commission as second lieutenant in Troop L on August 26, 1899, and recruited thirty or more cowboys and George Curry to go with him. In the Philippines, Ballard acquired the jacket of rebel leader General Emilio Aguinaldo.[31] In 1905 Ballard was elected to the upper house of the New Mexico legislature as a Democrat. Ballard succeeded Senator Albert B. Fall, a Republican, who had defeated Granville Richardson two years earlier.[32] When Roosevelt was inaugurated, Ballard and thirty-nine other Rough Riders were appointed a Guard of Honor, accompanying Roosevelt's carriage from the White House to the Capitol for the oath of office and back to the White House. As president, Roosevelt was willing and able to reward his Rough Riders. They showed him loyalty; he gave them high political positions.[33]

After his appointment by Roosevelt to the governorship of New Mexico, James Hagerman's son, Herbert Hagerman, appointed Ballard to his staff in 1906, the same year Ballard was elected sheriff of Chaves County.[34]

Following the washout of the irrigation system in 1904, Judge A. A. Freeman of Carlsbad asked Ballard to go to Washington as part of a committee to request that the dam be restored. Six committee members left for Washington to see Interior Secretary Hitchcock. Freeman, as chairman of the committee, presented Carlsbad's plea to Hitchcock, who "turned [them] down cold." Ballard then volunteered to see Roosevelt personally and ask for his assistance. Roosevelt agreed to have lunch with Ballard the next day. Following lunch, the two men sat down to talk. Reminiscing about their meeting, Ballard remembered Roosevelt saying, "Hello, Charlie, I suppose you want the job of U.S. Marshall of the Territory of New Mexico? Well you may have it. What else do you want?"[35] Ballard refused the position and then laid out the situation along the Pecos, making a strong appeal for his support. Roosevelt appeared interested and asked Ballard to bring the committee to his office the following day. The next day, the committee met with Roosevelt and Hitchcock. According to Ballard, Roosevelt, after hearing the committee's pleas, turned to Hitchcock and said, "Mr. Secretary, this matter appeals to me very strongly, and I would like to see this dam built." Hitchcock replied, "Very well, Mr. President, the matter is on my desk now." Roosevelt told Hitchcock to "put it in as fast as men and money can do it."[36]

Although the Ballard connection to Roosevelt seems to have played a major role in getting the president's ear, other political and personal connections may have played a part as well. Joseph Stevens, intimately connected with irrigation projects in the lower Pecos Valley during their infancy and whose property there was managed by his cousin Francis Tracy, also served as one of Roosevelt's Rough Riders. In fact, Stevens may have purchased some important machine guns for the group.[37] Stevens's name and that of his father, Frederick, long associated with the Chemical Bank of New York, could not have hurt in Roosevelt's decision to favor the Carlsbad Project.

Political connections and Roosevelt's strong hand, along with strong local boosterism, combined to trump the engineering concerns and expertise of the new Reclamation Service. Despite Newell's and Davis's reservations about the project, by the end of 1905 Reclamation had formalized an agreement to take the Carlsbad Project under its wing. Frederick H. Newell, Reclamation chief engineer, informed territorial engineer B. M. Hall in November 1905 that the secretary of the interior had approved the Carlsbad Project subject to certain conditions. Reclamation agreed to pay Pecos Irrigation $150,000 for its irrigation works (the company retained its land), and estimated the cost of reconstructing the dams and canals at $450,000.[38]

Reclamation drew up a contract with the local Pecos Water Users Association in which water users committed to repay the cost of maintenance and operation on the project plus the cost of purchasing and repairing the works. Expecting to provide water to 20,000 acres, Reclamation initially estimated that landowners would have to repay Reclamation at the rate of thirty dollars per acre served by the project.[39]

The quantity and quality of acreage selected for projects had to be sufficient to guarantee repayment to Reclamation for its service. Reclamation agreed to purchase the dams, canals, and irrigation works of Pecos Irrigation Company if the company could show good title to the properties. Newell authorized Hall to begin working on the larger project, which was placed under the immediate direction of local district engineer W. M. Reed, only after the conditions were met.[40]

Thus, the Reclamation Service began a careful examination of soils in the area to help it decide which lands could bear the costs of construction and maintenance under the project. Soil Engineer Thomas H. Means,[41] from the Bureau of Soils, a newly created bureau in the Department of Agriculture, arrived in the valley in early 1905 to conduct the investigation. Hall emphasized the importance of Means's latest mission when he assured A. P. Davis, acting chief engineer in Washington, that "we will keep the

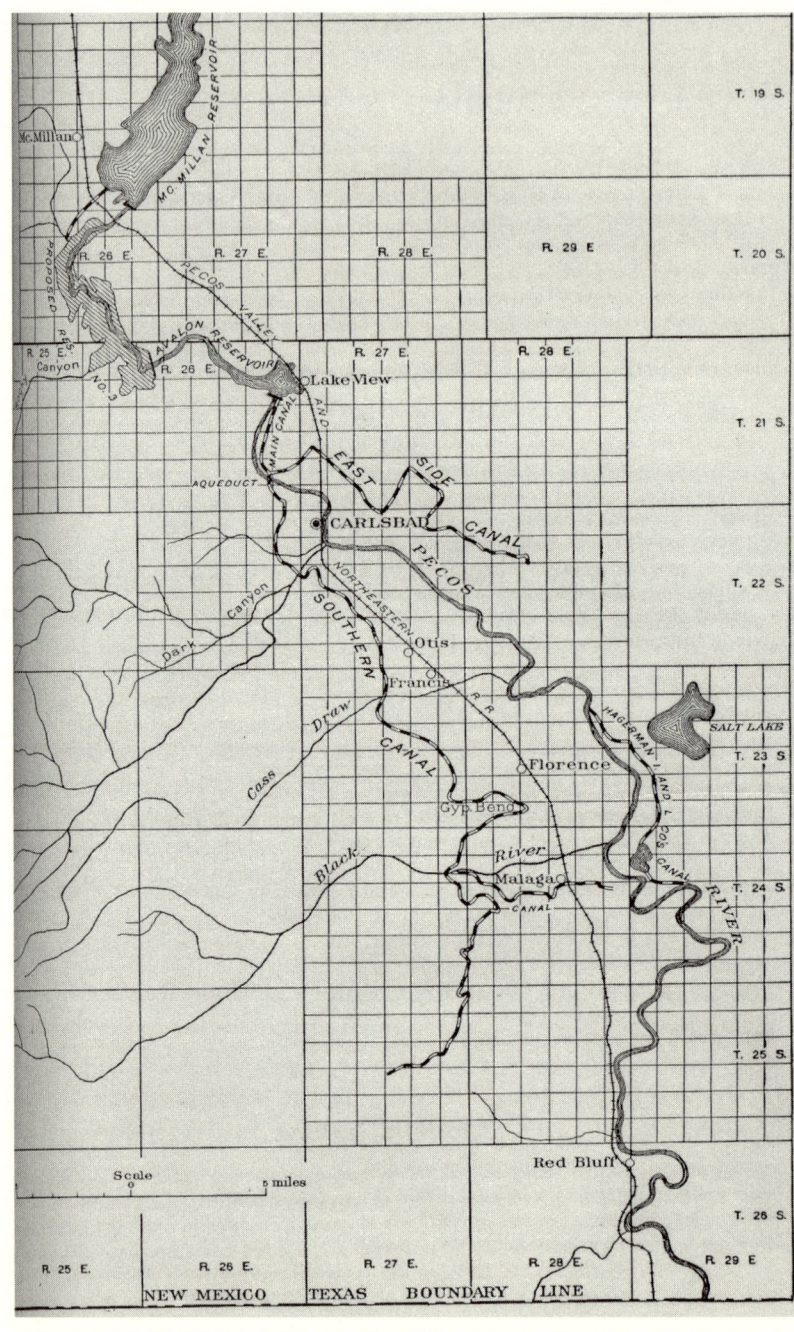

The Carlsbad Project, 1905.
(From F. H. Newell, Fourth Annual Report of the Reclamation Service, 1904–5, *59th Cong., 1st sess., 1906, H. Doc. 86).*

people well in hand and see to it Means is not lynched,"[42] referring to the delicate issue of selecting lands for the project.

As Means's team examined the soil, crops, water, and climate, they found a small number of farmers growing profitable crops and showing the potential of the valley.[43] But they found many more farmers failing to make a living. While the valley attracted a large number of settlers, most of them were inexperienced in irrigation or had little knowledge of agriculture. Failures occurred regularly, and abandoned farms littered the area. The situation, according to Means, resembled other failed projects in the West: "Such a depression in the prosperity of every irrigation scheme fostered by a corporation seems to be inevitable and when other schemes of similar magnitude in the United States are compared, the Pecos Valley does not seem to have suffered more than many others."[44]

Means noted that in the Pecos Valley, as in other parts of the western United States, engineering and technology did not match the needs of the projects, and financing dried up. Means noted other problems at Carlsbad. Inexperienced settlers, the lack of agricultural and irrigation knowledge, misleading information provided by irrigation company "agriculturists," and the focus on unsuitable crops all contributed to agriculture failure in the valley.[45]

In addition, river water used to irrigate crops was charged with salines. The river carried an average of 310 parts of soluble matter per 100,000 parts water. Out of that amount, some 152 parts were made up of salts harmful to the soil that might accumulate and interrupt plant growth. Canals and laterals seeped, leaving adjoining lands waterlogged, and when the water evaporated, fields were impregnated with alkali. Means concluded that irrigation water was so highly constituted of gypsum and alkali that it was gradually destroying the soils of the project area, and eventually would prevent growing all except the heartiest of alkali-resistant crops.[46] Although fruit, asparagus, onions, cantaloupe, celery, and sweet potatoes showed positive results on well-managed farms, the valley soils lacked the nutrients required for many crops.

Overall, Means, like R. B. Marcy fifty years earlier, thought the Pecos Valley was best suited as grazing country. Its mild climate made the valley well suited for cattle production, yet cattle ranges in the area failed to produce enough feed to fatten the stock for shipment. Consequently ranchers shipped their cattle to grain-producing districts in the Midwest for fattening and finishing before shipping them to packing houses in Chicago. Means thought the valley should stress the production of alfalfa and grain to serve the valley's cattle industry. But alfalfa in the valley suffered a root disease in the second year that weakened the plant so much that harvests suffered. Means suggested raising feed grains, especially kaffir corn, which produced thirty-six bushels per acre on average and up to 4,000 pounds, or seventy-one bushels, on good lands. He argued that barley or oats during the winter months, with alfalfa and kaffir corn grown during the summer, would be especially profitable.[47]

Farmers in the Carlsbad area grew cotton in sufficient quantities to warrant the construction and operation of a gin. Means noted that cotton suffered from the lack of fertilizer, but that yields as high as one and a half bales per acre grew in the region. Farmers experimented with Egyptian cotton, and Means thought the crop had a bright future.[48]

Means noted other problems in the valley related to farming techniques and management. Although tenant farming in the valley was gradually giving way to farming by actual owners of the property, he still noted a great deal of "shiftless" farming in the valley, where farmers poorly cultivated, never manured or fertilized, and overwatered their lands, thinking that irrigation allowed them to dispense with cultivation. The only tools utilized by farmers appeared to be a disc harrow and plow, while cultivators, grain drills, or deep plows, which were of great value on irrigated land, were nowhere to be seen.[49]

Although Means cited a number of problems, from salty water to farming techniques and crop selection in the valley, he thought lands to be served by the proposed project could easily handle a charge of twenty-five dollars per acre payable in ten years as part of the government's required repayment plan. The key, according to Means, was management of the land. Poorly managed lands would not pay; so in his estimation it became largely a matter of individual effort as to whether a farmer was successful or not: "any intelligent man willing to work should successfully manage any first class land under the Carlsbad Project and be able to pay for his water right and adequately support his family."[50]

Parts of the Means report upset Francis Tracy, who realized that adverse publicity would discourage settlement in the valley, especially on the poorer quality lands his company owned. Following resistance from Tracy and other community leaders, the Bureau of Soils retracted its statements regarding alkali and poor water, issuing a statement indicating that the real trouble on the project was poor drainage, not bad water.[51] Although Means never suggested it, his report pointed to any number of reasons why Reclamation should not have taken the project. Inexperienced farmers, poor soils, poor water, corporate ownership of lands, tenantry, and lack of equipment indicated that Carlsbad would be problematic. Despite the scientific evidence against the project, the political decision to take the project had already been made.

To satisfy Reclamation, in late December 1905 the Pecos Water Users Association called a meeting, requesting all members of the association to be present. The most pressing issue was to establish clear title to any land that might be included under the project's 20,000 acres. Since at least 40,000 acres lay under ditch, it was imperative that water users perfect title to all their lands so they might be included under the project. B. M. Hall reminded the association that section eight of the December 5, 1905, agreement between Pecos Irrigation Company and the government required the Pecos Irrigation Company, as the largest landowner, to join other water users in guaranteeing to protect the U.S. government against claims not subscribed to the project.[52]

As representative of the largest land holdings in the valley, with one vote for every forty acres, Francis Tracy could dictate the stance taken by the Pecos Water Users Association. Although he personally owned only one share in the Pecos Irrigation Company, he controlled by proxy all shares in the company. A. M. Hove, secretary of the association, traveled to Washington to advocate the cause of the Pecos Valley and to urge the government to proceed with construction as soon as possible. Water users sent telegrams to Interior Secretary Hitchcock encouraging him to begin work. The water users advised Hitchcock that the association would make any necessary title corrections as soon as possible, and that valley citizens were willing to put up bonds insuring the government against any loss due to defects of title in the valley. Water users pleaded with Hitchcock to begin work immediately and accept a bond in whatever amount needed to move forward.[53]

Although many studies were incomplete and eligible lands had not been selected, authorization for Reclamation to begin long-term construction work came on February 24, 1906, and bids were opened for Avalon Dam's reconstruction. No bids were received, and Reclamation took charge of the project, from design to hiring subcontractors and labor to construction. Preparatory work began May 1, 1906, with actual construction beginning June 1.[54] Workers incorporated part of the old Avalon Dam into the new structure. Avalon was rebuilt, much as it had been before, using the earth and rockfill design, but with new floodgates and three spillways, each considerably larger than their predecessors, for better flood control. Construction of the dam progressed smoothly and was completed in November 1907.[55]

Elsewhere on the Carlsbad Project, workers completed the rehabilitation of the Pecos River flume. Reclamation spent $18,000 fixing broken concrete, widening and strengthening the footings, and lengthening the flume. A new siphon across Dark Canyon two miles south of Carlsbad was designed to transfer the Main Canal's water beneath the floor of the canyon to avoid recurring washouts. The concrete siphon was 400 feet long and was built during the winter of 1906–7. Reclamation began lining some canals with concrete to eliminate seepage.[56]

Reclamation also began studies to determine which lands should be included in the project. To help assign lands, Means had earlier classified lands by quality. He concluded that 30,000 acres could be classified as first-class land, with good drainage, nutrients, and the ability to produce profitable harvest. Thus, with 30,000 acres of first-class land, and Reclamation only accepting 20,000 acres under the project, the stage was set for competition between landowners. Local landowners, especially Francis Tracy, president of Pecos Irrigation Company, began a long crusade for inclusion of the company's good lands and expansion of the project to include lands not designated as first class.[57]

The government used a number of criteria for selecting eligible lands. L. E. Foster, author of the report, gave first priority for inclusion to lands with first- or second-class soil (based on the system set up by the U.S. Department of Agriculture) that were privately owned, cultivated, and possessed Pecos Irrigation Company water rights. Second priority was given to land owned by Pecos Irrigation Company with first- and second-class soil. Third priority went to private lands that had water rights from Pecos Irrigation Company but which were not then cultivated. Fourth in priority were those lands in private ownership, which at some point had been in cultivation and had water rights from Pecos Irrigation.[58]

Farmers near Carlsbad anxiously awaited the Foster report, which would recommend lands for the project.[59] The assignment and subsequent reassignment of lands under the Carlsbad Project developed into a tedious and involved process. Once Foster made his recommendations for land inclusion, B. M. Hall reviewed the selections with members of the Pecos Water Users Association, who had to approve them before he added his endorsement.[60] Notified by mail of the government's decision, some water users came into Reclamation's Carlsbad office to discuss the matter. In some instances the government made slight changes. Engineer W. M. Reed decided to eliminate lands subject to failure and substitute quality lands based on the Means and Foster reports. Exclusion of poor land benefited the owners of such lands since they would not have to pay Reclamation's per acre fees. While the changes satisfied most the valley's population, the

Reclamation District included numerous acres of poor land that Reclamation hoped to reject. Owners forced inclusion because they "held bona fide water rights from the old irrigation company and would not surrender them, and had not land of good quality at that time to which the rights could be transferred."[61]

Foster's plan included the holdings of many prominent members of the Pecos Water Users Association, including R. S. Benson, an officer; Richard J. Bolles, absentee landowner, early investor, and friend of James Hagerman; William J. Fox, Charles Eddy's brother-in-law; A. M. Hove, secretary of the association; and the La Huerta Company, which was owned by Tracy. Tracy personally owned close to 1,000 acres in various locations across the valley. Others whose land fell under the project were C. H. McLenathen, Tracy's partner, who also owned a number of town lots and land in the valley; I. S. Osborne, connected with the Pecos Water Users Association; Vernon L. Sullivan, a consulting engineer in the Eddy area who later became New Mexico's state engineer and served on irrigation projects in both New Mexico and Texas; and Joseph S. Stevens, much of whose property in La Huerta was managed by his cousin Francis Tracy. Other landholders included Mary Tansill, the widow of cigar manufacturer Robert Tansill.[62]

Government inclusion of land owned by Pecos Irrigation Company involved a long drawn-out process because of the number of times the land had been sold. Individual landowners also had to perfect title to their property in anticipation of government involvement in the valley. This included land owned by James Hagerman and his wife, a number of foreign investors, Credit Lyonnaise of Paris, Holt Livestock Company, Amos Bissell of Eddy-Bissell Livestock Company, and Elmer Williams.[63] Reclamation wanted to ensure that those included under the project had clear title to the land. Members of the Pecos Water Users Association met in Carlsbad and agreed to send letters to members not living in the valley, notifying them of the government process of perfecting title and the selection of project lands.[64] The government also required a legal notice about how project landowners could get irrigation water. The local newspaper carried a notice stating: "All persons owning land under the Carlsbad Project, or claiming a right to water from the irrigation canals of the Pecos Irrigation Company, and expecting to have water therefore after the United States has taken charge of the said Carlsbad Project are hereby notified that they can only procure water by joining said Pecos Water Users Association."[65] By joining the association, landowners gave up the right to bargain individually with the government.

W. M. Reed had accurately predicted that objections to selected lands would come from nonresidents of the valley who were "really more in the nature of speculators than water-users." In December of 1905, J. O. Cameron wrote to Richard Bolles in Colorado Springs, apprising him of a meeting of the Pecos Water Users Association on the eighteenth. Cameron urged Bolles to agree to inclusion of his lands under the Carlsbad Project, as he was the only landowner who had previously received water from the irrigation company who had not joined the association or signed a contract agreeing to sell his land to a member of the Pecos Water Users Association. The secretary of interior required a contract with the association stating that all project lands would be brought in under the 160-acre rule and all excess lands sold, or that no claims would be brought against the government "on account of outstanding water rights."[66] Cameron encouraged Bolles to join the association and place his excess lands with the association as trustee for later sale.[67]

Bolles joined the Pecos Water Users Association and transferred his property to the association in trust, but he brought suit in local district court to obtain waters promised his land under previous agreements with the Pecos Irrigation Company.[68] Bolles owned water contracts for 1,440 acres within the Carlsbad area. In the suit, Bolles wanted Pecos Irrigation to pay him $13,200 in damages. He claimed that Pecos Irrigation had broken a contract to provide water from the Southern Canal to his property following the October 1904 flood, which washed out Avalon Dam. The local judge sided with Bolles against Pecos Irrigation, giving hope to Bolles and others that Reclamation might have to furnish water to more than 160 acres. But on appeal, the Supreme Court of New Mexico reversed the decision. The court stated that Pecos Irrigation did not try to abandon its obligations between the washout of Avalon Dam in 1904 and the sale of irrigation works to the federal government. And, the court believed, there was not breach of contract when the irrigation company sold its works to the government, it being understood that, as a corporation, Pecos Irrigation was of limited duration. The company had no obligation to honor its water contracts after government purchase of its dams and canals.[69]

Bolles's law suit complicated Reclamation's process of land inclusion. Since Bolles and others claimed vested rights to water under private contracts, should Reclamation consider this when mapping out the projects, regardless of how many acres a landowner held? The Bolles suit reflected a period of adjustment and appeal for land inclusion. Reclamation decided to give water-right holders until the end of July 1906 to appeal the results of land inclusion.[70]

In August 1906 Tracy objected to Reclamation's inclusion of only 5,527 acres of Pecos Irrigation Company's 55,000 acres. Tracy wanted another 6,000 acres included, but Reclamation officials refused.[71] As Tracy continued his crusade for more land to be watered under the project, Reclamation still had questions about the company's holdings. The heart of the problem was the 160-acre rule stipulated by Reclamation. Pecos Irrigation Company owned thirty-four times that amount under the project. Local water users, anxious to have the government move forward on the project, wrote the secretary of the interior indicating the title to land and rights to water of Pecos Irrigation Company, the major landholder in the area, were "substantially good, that if work is delayed . . . irreparable ruin will come to 3,000 people in this valley."[72]

Reclamation was caught between its own rules restricting use of water for each owner to 160 acres and the expediency of providing water to bona fide settlers on the project. The government finally accepted the titles to Pecos Irrigation's superfluous acreage, admonishing the company to sell it quickly. Because the dam was not far enough advanced to turn water into the canals until late March 1907, and then only partially, farmers were again without water in 1906. For two years in a row farmers under the project had not had water, and much of the land had relapsed into desert conditions. Reclamation provided water in April 1907, so late in the season that perhaps only 25 percent of lands under the project could be tilled. The lack of water directly affected the cattle and sheep industry in the area, since stockmen were heavy consumers of products from local farms. Stockmen, faced with an economic downturn, could not afford to purchase feed from the farms, and there was little produce available.[73]

The Pecos Valley irrigators were not the only ones unhappy with Reclamation. Many people across the West expressed displeasure.[74] In response to irrigation problems, William Ellsworth Smythe of Nebraska formed the Irrigation Congress in 1905 to promote solutions for irrigated agriculture in the West. The Irrigation Congress became more powerful in 1911 with the creation of the National Irrigation Association, which lobbied for improvements on existing projects. The two groups aroused public sentiment, which led to better conditions, including financial arrangements.[75]

According to agreements made between Reclamation and the Pecos Water Users Association, association members had to make their first payment toward construction and operation of the project on March 1, 1908. This requirement resulted in more conflict. Water users believed that they would be allowed a full crop before having to pay the first installment. But conditions under the project had not allowed water users to grow a decent crop for 1906–1909 while Reclamation repaired the system. The association appointed a committee that included Francis Tracy, president of Pecos Irrigation Company, A. M. Hove, secretary of the Pecos Water Users Association, and William J. Fox to draft a statement of protest to the new secretary of the interior, James R. Garfield, against paying the first installment.[76] The protesters noted that the project was unique in that, unlike the "usual" projects contemplated by the Reclamation Act, the Carlsbad Project dealt with lands that had been entirely in private ownership and that had carried vested water rights. Carlsbad was not a construction project in an undeveloped region, offering public lands to new settlers. It was a project that called for purchase, reconstruction, repair, and development of an irrigation system already in existence. Fully some 13,000 acres of land had been irrigated under the private irrigation system of the Pecos Irrigation Company and would be returned to cultivation. Another 7,000 acres of raw land would be added to the project. Thus, Reclamation should consider the special circumstances when determining a repayment schedule.[77]

Tracy and the protest committee pointed out that about one-half of the land under the project was owned by persons "not qualified to perfect water rights"—that is, owners exceeding the 160-acre limit. Tracy referred to absentee landholders, many of whom owned shares in his company. Although Tracy and others had tried to induce settlers to locate in the valley, their efforts had been somewhat unsuccessful. Tracy and the committee noted that while immigration had begun and many sales had been made, the financial situation resulting from a lack of water in the valley effectively stopped immigration and land sales. Home seekers were cautious because they had to pay the first installment for construction and operation of the water plant in advance; the committee saw this as a reason why would-be settlers decided not to locate in the Pecos Valley. Furthermore, the association was of the understanding that payments to the federal government would occur at the end of the crop season, not the beginning, which in their minds made much more sense because farmers would have reaped profits from their harvests.[78]

Perhaps encouraged by the Bolles suit, others pushed for water on lands exceeding the 160-acre limit. The issue of absentee ownership of excess lands was an acute one, especially for Francis Tracy, who represented thousands of acres within the project. According to stipulations in the Reclamation Act, landowners under Reclamation projects could not perfect water rights for more than 160 acres and had to live on the land. Many absentee landowners had acquired a number of acres in excess of the 160-acre rule before the Newlands Act came about.[79] F. E. Bryant, one absentee owner, wrote a letter to *Field and Farm* in Denver:

Everybody was pretty hard up and they thought it was a good thing for the government to have it [the irrigation works]. Personally, we did not but signed up reluctantly *pro bono publico,* you might say, with a pistol at our heads figuratively speaking. The argument was, "if you don't come in you will get no water" well, we came in, and now on March 1st Mr. Wrecklamation Man says he is ready to deliver water but only to bona fide residents who own 160 acres or less. Unfortunately, we happen to own 760 acres we have paid our assessment . . . and tried to sell the land at a fair valuation but there have been no buyers.[80]

Field and Farm's response indicated that the magazine thought Bryant and the absentee landholders should not lose their water rights simply because of government intercession. In fact, *Field and Farm* encouraged Bryant to go before the federal district court with mandamus proceedings to require Reclamation to furnish water based on vested rights held by Bryant and others. Although *Field and Farm* thought the Reclamation Act did not provide for government purchase of established irrigation systems, it indicated that Reclamation had no power to deprive a water consumer of his rights obtained under previous legal arrangements, in this case through rights vested through Pecos Irrigation Company or predecessor companies.[81]

Bryant was not the only large owner. Pecos Irrigation Company alone had over 5,000 acres under the project. Joseph Stevens, Tracy, the La Huerta Company, and others owned thousands of acres.[82] Together, these large landowners represented a huge voice in the Pecos Water Users Association. Tracy and partner C. H. McLenathen were anxious to provide waters to the lands owned by Pecos Irrigation, Stevens, and others, and requested "assistance" from the association.[83] The partners claimed that property owners made every effort to comply with the Reclamation Act, which required them to live in the valley and sell off their excess property. Since the land would be seriously damaged without water and since water rights existed for their clients, Tracy and McLenathen strongly urged the Pecos Water Users Association and Reclamation to provide water to their lands Because their clients were acting in good faith to sell the land, the partners reasoned that they should continue to get water. The two businessmen noted that in comparison to others in the valley, their absentee clients sold a proportionately larger share of all lands over the previous year, 1906–7, than had any other landholders in the valley.[84]

Aware that absentee landowners controlled a large percentage of lands under the project, the Pecos Water Users Association contacted them and explained the fine points of the Reclamation Act of 1902, specifically the fact that water "shall not be sold except to bona fide residents on the land or occupants residing in the neighborhood." Reclamation had ruled that "neighborhood" meant twenty miles in a direct line of the project. The act stipulated that water would never be delivered to land owned by nonresidents who had never claimed the Pecos Valley as their home or had no intention of making their homes there in the future.[85] When in 1908 the Interior Department became concerned that fifteen to eighteen large owners still held half the land in the project in tracts of more than 160 acres, officials asked the association what it planned to do about the situation. The association pointed out that even among those with tracts of less than 160 acres, there were more than 130 absentee landowners. Interior had to enforce the absentee provision of the Reclamation Act or work out a compromise.[86]

Apparently the Pecos Water Users Association, with encouragement from Tracy, made its case.[87] In March 1908, after giving careful considerations to the conditions in the Carlsbad Project, and "in order that the holders of excess lands and those who find it impracticable to reside upon the land or in the neighborhood may not suffer serious injury by deprivation of water" acting Interior Secretary Franklin Pierce agreed to temporarily suspend the rules. Water rental contracts would be extended to absentee landholders during the irrigation season of 1908, with each property holder's case to be considered separately on its own merits.[88] Pierce added that he did not want to continue the policy past the 1908 irrigating season and expected landowners to make all possible efforts to comply with the Reclamation Act.[89]

McLenathen and Tracy were two of the more prominent voices clamoring for water rental contracts, receiving contracts in 1907 and 1908. Interior was well aware of Tracy's position as agent for several absentee landholders and made it clear to the Pecos Water Users Association that rental agreements were not advantageous to the project and the temporary contracts should be terminated as soon as possible.[90] A number of landowners requested permission from the association to shift their stock and water from one piece of property to another more suited to farming.[91] Interior granted this as well.

Starting with temporary repairs on the irrigation works, Reclamation, in subsequent investigations, pointed to significant problems at Carlsbad. Carlsbad had many of the features for which Reclamation would later be criticized. Largely exposed during the 1904 Hondo hearing as having leaky canals, leaky dams, and wasted water, Carlsbad, under further examination was also revealed to have salty water, poor soil, and uninformed farmers. Despite significant reservations by A. P. Davis, Frederick Newell, and others, Reclamation and the Interior Department bowed to political pressure locally and from the White House to rehabilitate the irrigation works in the lower Pecos Valley. To complicate matters, Reclamation was saddled with a project made up of absentee and corporate landowners who owned far more acreage than the watered acreage allowed under provisions of the Reclamation Act.

9

SUCH OF THE LANDS ECONOMICALLY PRACTICABLE TO IRRIGATE

WHILE some absentee landholders resisted Reclamation's insistence that they move to the Pecos Valley and occupy 160 acres of land, others energetically tried to dispose of excess real estate. In April 1908 McLenathen and Tracy issued a brochure on the history of the Carlsbad Project and ballyhooed the numerous agricultural products that thrived in the district. Here, in public, Tracy and McLenathen had only praise for government intervention, suppressing any private complaints because of the need to attract buyers. As they had in the heady days of the 1890s, Tracy and McLenathen proclaimed the benefits of irrigation and listed the virtues of plums, grapes, cherries, nectarines, apricots, and persimmons, alfalfa, kaffir corn, cotton, sugar beets, turnips, poultry, beekeeping, and dairy products, asparagus, tomatoes, and cauliflower. Within the pamphlet, McLenathen and Tracy included an insert listing irrigated land and soil conditions under the project, including numerous orchards. For example in La Huerta, a suburb of Carlsbad, they were selling a ten-acre peach orchard for $4,000 and five-acre raw lands for $500. The real estate firm had listings from across the valley, from Black River to near Otis to several areas in and around Carlsbad.[1] Tracy sent the brochure to Frederick Newell to assure him that they were busy trying to dispose of excess property.

McLenathen and Tracy were part of a much larger effort to sell lands in the valley. In 1906 the Pecos Water Users Association voted to have the board president appoint a committee to devise plans and raise money to advertise members' lands and resources.[2] Association board members contacted the firm of Ward and Thomas, a newspaper, magazine, and outdoor advertising agency in Chicago that touted itself as the largest in America. In August the company told the board, "we believe we can prepare a plan for advertising the district in which you are interested which will bring a great many settlers to your country."[3]

Harvesting wheat on the McClellan Ranch, Carlsbad Project, New Mexico.
(Record Group 115, Records of the Bureau of Reclamation, Public Relations Photograph Collection, Series 3—Historic, Photographs by Project Sites, Box 004, 8NS-115-95-112, National Archives and Records Administration, Rocky Mountain Region, Denver, Colorado.)

Reclamation was anxious to settle the valley, encouraging the Pecos Water Users Association to do whatever it took to get people there. Not only was Reclamation interested in bringing in settlers, but so were railway officials, whose lines ran through the region. In late December 1907 E. M. Meyers, the general freight and passenger agent for Santa Fe Railway's New Mexico rail system, asked the secretary of the association if he might secure some of the association's advertising funds. Meyers argued that not only did the government want a return on the monies spent at Carlsbad, so did the Santa Fe Railway. He believed that a cooperative advertising venture would help all parties concerned.[4]

Apparently the joint advertising by Reclamation, the railway companies, the Pecos Water Users Association, and McLenathen and Tracy reached many prospective settlers. In 1907 and 1908 the association received hundreds of requests for maps and land prices, most coming from Chicago, but also from Ohio, Iowa, Colorado, California, Minnesota, North Dakota, and Wisconsin. The association responded that settlers could obtain raw land with water in the valley for $30 per acre. Improved land sold for between $40 and $75 an acre, and orchard tracts in La Huerta sold

Growing corn on the Carlsbad Project.
(Courtesy Sacramento Mountains Historical Society, Cloudcroft, New Mexico.)

for a minimum of $100 an acre. Payment was one-third cash with a balance of one to two years and possibly longer, negotiable by the purchaser.[5] To answer the avalanche of correspondence, the Reclamation Service created its own circular about the district and sent it across the country.[6]

Those trying to sell lands included individuals and corporations connected with Pecos Irrigation Company and its predecessor companies dating back to before the turn of the century. Leading stockholders included Mrs. Robert Weems Tansill; G. B. Shaw, vice president of American Trust and Saving Bank of Chicago; Joseph Stevens, who now resided at the Knickerbocker Club, New York; Lombard, Odier, and Company, investment bankers from Switzerland; George H. Webster, president of Chicago Savings Bank; August Uihlein, secretary of Schlitz Brewing Company, who had built a sugar refinery in the valley; and the Pecos Valley Northeastern Railway Company.[7]

In October 1907 Pecos Irrigation Company, under manager Francis Tracy, hired emigration agents to sell company land as quickly as possible. This had always been the company's goal, but Tracy anticipated better sales with the free publicity generated by the upcoming opening of the Carlsbad Project in early 1908.[8] Tracy entered into an agreement with Malaga Land and Improvement Company, managed by E. E. Hartshorn, to promote land in the Malaga area, south of Carlsbad. Hartshorn and Malaga Land and Improvement used questionable tactics and claims to promote the land, not necessarily in keeping with reality. Malaga promoters convinced clients that their lands would soon come under jurisdiction of the Carlsbad Project. Some of the lands offered for sale by Malaga had been sold to that company by McLenathen and Tracy. Malaga did not have enough money to pay for them at once, so the deeds were placed in escrow. The company soon expected to sell enough land, however, to pay McLenathen and his partner. To do so, Malaga Land and Improvement promoted lands that Reclamation would probably never water.[9]

In 1908 W. M. Reed, now the supervising engineer of projects in Arizona and New Mexico; Louis Hill, local supervising engineer; and by extension, the director of Reclamation, A. P. Davis, discovered the connection between McLenathen, Tracy, and the Malaga Land and Improvement Company. Reed realized that Malaga was nothing more than a front for Tracy trying to sell dry land south of Carlsbad withdrawn by the government in anticipation of the Carlsbad Project but not actually included within the final project boundaries.[10] Louis Hill alerted Reed that McLenathen was representing himself to prospective buyers as secretary of the water users association, when in fact A. M. Hove was secretary of the association.[11]

Reed was more than happy, in light of comments made by Tracy and McLenathen regarding his management of the Carlsbad Project, to send the director a whole batch of advertising matter, blank contracts, and other papers concerning the sale of land through Malaga Land and Improvement Company. Reed noted that the company's advertising matter was conspicuously unavailable in any areas near Carlsbad where people would see through the facade, but that Malaga's principal point of operation was in Kansas. Reed obtained advertisements from the company's local agent in El Dorado, Kansas, and figured out what the company was doing. Referring to Malaga Land and Improvement Company, Reed informed the director: "A close inspection shows that their work is very artistic. They have used great ingenuity in avoiding, or trying to avoid, placing themselves criminally liable, yet they are undoubtedly deceiving people and know that they are doing so. . . . [T]hey are trying to sell land that has not been admitted to the project, but are so wording their advertisements that those who are not familiar with the circumstances would be led to believe that they are getting lands that has a water right or is subject to a water right."[12] Reed also enclosed a deposition by Tracy used in a lawsuit filed by Malaga Land and Improvement Company. Ironically, the company brought charges against a disgruntled customer who made statements reflecting on the "good intentions of the company." Reed, who did not trust Tracy, thought the deposition shed light on Tracy's character.[13]

In the deposition, Tracy admitted that over 25,000 acres of land owned by Pecos Irrigation Company had been conveyed by a deed of trust to the water users association to negotiate water rights with Reclamation. The judge asked Tracy how many acres had been sold by Malaga Land and Improvement Company; Tracy answered that the company sold some 5,500 acres. When asked whether any of the 25,000 acres had been excluded from the agreement with Reclamation and released from the trust to the Pecos Irrigation Company, Tracy said no.[14] Tracy, McLenathen, and Malaga were selling lands deeded to the association prior to knowing which lands would be irrigated. Given their quality, the Malaga lands were far down the list for possible inclusion under the project.

Defending his activities in trying to sell land, Tracy recalled his negotiations on behalf of Pecos Irrigation with the federal government and the water users association. He argued that the government's failure to build further reservoirs to water Pecos Irrigation's lands justified negotiating with Malaga Land and Improvement to dispose of excess property. Tracy estimated that Reclamation could irrigate 50,000 acres if the project were expanded as he saw fit. He testified that the greater portion of the company's lands were "good irrigable lands," a statement contradicting the government's findings.[15]

Tracy and McLenathen told the Reclamation director that Malaga Land and Improvement Company had sold 80 percent of all lands sold in the valley. McLenathen felt "very grateful to them [Malaga Irrigation and Improvement]. . . . [T]hey have greatly improved the country around Malaga."[16] Whether or not that was true, Malaga sold land at high prices—five-acre unimproved tracts for $150 each. Although the company "guaranteed that each purchaser will receive at least five acres of good irrigable land," none of the Malaga land sold included an irrigation right.[17] A number of customers duped into buying such lands complained to the government. Malaga's president, E. E. Hartshorn, claimed that through brochures and promotional literature company officials tried to convey that the land was of average fertility and situated so that it could be easily irrigated from the government irrigation system, but this was wishful thinking on his part. Hartshorn claimed that since the land was "within the limits of the irrigation belt, [it was] in line for water when future assignments [came from] the government," and McLenathen told Reclamation's acting director, Morris Bien,[18] that he thought it was generally believed by valley residents that the government would extend the project in the near future and Malaga was simply reflecting popular opinion.[19]

McLenathen claimed that when company officers realized their literature needed to be revised, they called for a conference in Carlsbad among themselves, the local commercial club, and members of the Pecos Water Users Association.[20]

The Malaga people, Hartshorn, and his partner, Gardner T. Sanford,[21] stated their case and submitted their literature for examination so that the commercial club might revise it to reflect true Malaga conditions. This is almost laughable considering that Tracy, McLenathen, and other supporters made up the membership of the club. McLenathen admitted that literature published by the company was "undoubtedly misleading," but he tended to think that the omissions of information were unintentional. Typically, in McLenathen's mind the only problem with the literature was that those who prepared it were not familiar with the Reclamation Act and rules established by the Department of the Interior. McLenathen excused the oversight by the Kansas advertising company, with its main office in far-off Wichita where the circulars were printed.[22]

According to McLenathen, Malaga compiled circulars from previous valley irrigation publications and did not discriminate between watered land and unwatered land. Whether due to incompetence, which is highly doubtful, or a manipulation of the facts, McLenathen and Tracy's friends at Malaga sold a great deal of land. McLenathen admitted that 5-acre to 160-acre tracts in Malaga sold very effectively, "there being just enough speculative about the proposition to be received favorably by the average American, with $10.00 to his credit."[23]

There is little doubt that Malaga Land and Improvement Company tried to dispense of unwatered lands owned by Pecos Irrigation Company. Reclamation's acting director, Morris Bien, pointed to a misleading statement on page three of the brochure: "The project will without doubt be extended to include an additional 10,000 acres next year in the vicinity of Malaga [and also that] the country is the very best part of what is known as the Carlsbad Project being at the junction of the Black and Pecos Rivers. . . . [A]ll of these lands are situated within the limits of the land withdrawn for the purposes of the Carlsbad Irrigation Project under government control. All the lands are within the limits of the irrigation system as conducted by the Pecos Valley Irrigation and Improvement Company."[24] Bien told McLenathen, "these statements, together with others of like character have more or less foundation, but evidently will deceive those not familiar with the conditions on the project."[25]

The Pecos Water Users Association was placed in the unenviable position of having to explain to various purchasers of Malaga land just what was going on. Generally, the message sent to them noted that soil engineers had examined all of the land withdrawn for the project and had selected 20,000 acres specifically to be irrigated from the system. The association kept a record of all owners allotted water on the project, and Malaga Land and Improvement Company owned no land with water. The association informed numerous purchasers that they had bought land that had no water whatsoever, and "no one could say when this land could get water for irrigation, if ever."[26]

In 1908, even as Tracy touted the merits of the valley to attract settlers and the federal government completed rebuilding projects in the valley, Tracy condemned Reclamation's management, in terms of construction, technique, assessment of maintenance and operation charges, and other features. W. M. Reed, who had been a nemesis of Tracy's since 1904 when the upstream Hondo Project was approved, wrote the director of Reclamation to prepare him for a barrage of Tracy correspondence criticizing management at Carlsbad and making charges of a personal nature against Reed. Reed wanted copies of any letters criticizing him so he could answer the charges. Tracy and C. H. McLenathen were, in Reed's words, "making a great hullabaloo in order to further their interests." Although Reed denounced the Tracy-McLenathen charges, Reed did think that Project Engineer L. E. Foster allowed area farmers too much water and suggested that farmers had improperly prepared their lands.[27]

While Reed saw few other problems on the project, McLenathen and Tracy did. They first launched an attack on the 1908 failure of headgates installed at Avalon Dam. Tracy estimated that a 1908 flood, which destroyed some of Reclamation's work in progress, might not have been so destructive had Reclamation installed a different style of gates.[28] McLenathen, for his part, criticized the notorious leakage problems at Avalon and McMillan. He demanded that Reclamation repair holes in the bottom of the McMillan storage reservoir, which in June 1908 held no water. Supervising Engineer Louis Hill saw no great advantage in spending large sums of money on repairs that would last only a short time and would not prevent water from escaping through thousands of other leaks in the area.[29] Reed also discounted McLenathen's claims about losing water. He thought Avalon had plenty of water to supply all the acres under irrigation.[30] Nonetheless, the Service did renovate McMillan Dam in 1908. Rotting headgates were replaced, and workers repaired damaged spillways. Reclamation also constructed a 4,000-foot-long dike on the eastern shore of the reservoir to block off an extensive gypsum deposit.[31]

Tracy also complained that his company was hampered by errors in water-right allotments. Tracy may have been right. Water-right allotments were finally corrected by issue of new maps approved by Secretary of the

Interior Richard Ballinger on November 1, 1909.[32] The Interior Department issued the map only fifteen days before Reclamation issued an order to sell excess lands by February 15, 1910.[33] By 1910 Pecos Irrigation had 2,000 of their original 5,600 project acres left and a limited time to sell.[34] Tracy and Pecos Irrigation focused much of their attention on selling the land, although they obtained an extension to August 24, 1910. Nonetheless, the public announcement by Reclamation to sell all excess property caused the bottom to fall out of the real estate business.[35]

Not only was Tracy unhappy about having to sell the land quickly at deflated prices, but he and other water users again protested the government's first construction charge, due on March 1, 1909.[36] The protest centered on the lack of water delivery until so late in 1907 that crops failed. Water users also complained that the irrigation works were not complete and repayment was an "utter impossibility."[37] They succeeded in changing the due date from March to December 1909 and again to March 31, 1910. Pecos Irrigation and the Pecos Water Users Association also objected to maintenance charges that they understood would be fixed at seventy-five cents a year per acre for ten years. In 1909 Reclamation raised the charges to $1.35 a year per acre. Pecos Irrigation and the Pecos Water Users Association filed another protest with Secretary of the Interior Richard Ballinger against the raise in maintenance charges and leveled charges of incompetency against Chief Engineer Arthur P. Davis.[38] These protests and others across the West convinced Congress to establish graduated construction charges on projects, including Carlsbad.

As Reclamation amassed more projects and expenses, Tracy and water users at Carlsbad wanted more water and contended that water storage on the Pecos was inadequate. During 1910 they repeatedly requested investigations to establish a third reservoir site—an idea in existence since the beginning of large-scale irrigation in the valley and one that would be repeatedly raised over the next forty years. Frederick Newell did not favor continuing investigations in the region. Nonetheless, he wrote to Louis Hill about hiring a geologist to examine a third reservoir site near Carlsbad. In early June the director of the U.S. Geological Survey suggested that an assignment could be made, but on June 18, 1910, Chief Engineer A. P. Davis argued that it would not be wise to spend more money on investigations in the Pecos Valley. Newell agreed and figured that the main objective at Carlsbad should be to get back the money spent on acreage already watered.[39]

Nevertheless, Tracy and water users persisted. By December 1910 members of the Pecos Water Users Association reminded Washington officials of the historic background of the project and how it was originally designed to water a great deal more land than that currently supplied by Reclamation. Based on the government's prior appropriation of 300,000 acre-feet a year for the project, association members argued that Reclamation could irrigate 50,000 acres of land, even allowing for evaporation, seepage, and other losses. At the urging of Tracy, the association reminded officials that the agreement for purchase of Pecos Irrigation Company properties was incomplete. Tracy long considered the government's cash consideration of $150,000 below the physical value of the property. More important, he argued that the federal responsibilities included reconstructing and developing irrigation for all project lands "economically practical to irrigate." Tracy considered many more valley lands irrigable.[40]

According to the water users, Pecos Irrigation Company had brought 75,000 acres of irrigable land within the limits of the irrigation system. Of that amount, early entrymen proved up between 40,000 and 50,000 acres under the Desert Land Act, and the company "reclaimed" the same through water supplied by canals and laterals from the Pecos River and its tributaries. Considering that laterals and canals had already been built to such lands, some water users thought Reclamation had an obligation to water at least 50,000 acres.[41]

Despite misgivings, Reclamation conducted investigations into construction of the third reservoir and the feasibility of expanding water storage at Lake McMillan. Investigators at McMillan noted sinkholes large enough for a man to walk through leading downward into gypsum beds. So numerous were cavities in the reservoir floor that the ordinary flow of the Pecos was not sufficient to fill the reservoir even when all the gates were closed. Numerous springs at McMillan did not compensate for surface waters lost in an area covered with sinkholes. While sinkholes were numerous at McMillan, they also appeared conspicuously along the western margin of the proposed location of Reservoir Number Three. Investigators saw several holes from a few feet to fifty feet or more in diameter, and in places large blocks of the surface sank into the earth.[42] Investigators discovered that although river water disappeared through the porous bedrock at McMillan, it returned to the Pecos between McMillan and Avalon, the diversion dam downstream. Yet it was obvious that not all water returned above Avalon Dam. Underground passages existed that took at least part of the lost water from McMillan to points beyond the control of Avalon. Reclamation engineers surmised that such passages might "render the proposed reservoir as useless as McMillan is at the present time." After such damning information, it was puzzling that investigators ended their report on a positive note: "Our examinations show that on the whole the site of the proposed reservoir number three is unusually good for the Pecos Valley, where gypsum is so widespread. While conditions are not ideal for a reservoir it is believed, nevertheless, that the site will prove much better than that of Lake McMillan, and that if results of the suggested test pits and drilling are favorable, the site will prove reasonably satisfactory."[43]

A third reservoir would make it possible to add more lands to the Carlsbad Project, a notion that caused a schism in 1911 within the ranks of the Pecos Water Users Association and between "actual farmers" on the project, those with small acreage who lived there, and absentee landholders, many of whom owned large holdings. At the heart of this split was the issue of expanding the project. As Tracy sold lands under the project to satisfy Reclamation and his own stockholders, he also lessened his voting power within the association. When Reclamation took over the project, he held almost 150 votes. But, in a project with 20,000 acres and one vote for every 40 acres, Tracy, who now controlled only 10 percent of the votes, still wanted to expand the project. Adding new lands to the project meant a higher selling price, since much land had already been sold, but a new voice from small farmers arose to challenge him.

Dryland owners, especially Pecos Irrigation Company and interested parties connected with the company, wanted to include their lands under project jurisdiction. Although Tracy was neither an officer nor a director of the Pecos Water Users Association in 1911, he still had considerable influence within the body. And, with control of 2,000 acres under the project in 1911, he held at least fifty votes.

Scott Etter, who was closely allied with Tracy, played an increasingly prominent role in the valley's politics. Born in Illinois in 1875, Etter had attended Blackburn College and practiced law for a time in Illinois and Missouri. He came to Carlsbad in 1909, and from 1909 to 1910 he farmed in the area, taking advantage of the climate to improve his health. In 1910 the Pecos Water Users Association elected him secretary and treasurer.[44] During the spring of 1911 Etter moved to the forefront of expanding the Carlsbad Project, trying to add 1,500 dryland acres to the project.[45]

In 1910 and early 1911, dryland farmers urged the Pecos Water Users Association Board of Directors to try to expand the project, but the board delayed pursuing the idea, preferring first to develop the project under its original acreage. However, by May 1911 Etter and the board decided to support increasing the watered acreage. On May 2 the Pecos Water Users Association passed a resolution advocating an expansion and sent contracts to area dryland farmers. Etter was adamant that farmers make no changes in the contract, suggesting such changes would hold up the process: "The terms of the contract speak for themselves, and if you desire your land to be entered in the request which will be made to the government immediately after May 31st, sign the enclosed contract, and mail it together with forty cents per acre, which will go towards defraying the expenses of the United States engineers in making the examinations of your lands."[46]

Of the nine board members of the Pecos Water Users Association, the majority appear to have sided with dryland property owners represented by Etter and the Tracy interests. P. J. McShane, president of the association, and J. W. Fanning, however, opposed expansion of the project and increasing the control of large dryland interests. At a board meeting on October 28, seven members voted to send Etter to Washington, D.C., ostensibly to ask for an extension of payments on the project. Both Fanning and McShane immediately opposed this.

According to correspondence sent to W. M. Reed, Reclamation's district engineer, the specific interests of Etter traveling to Washington revolved around the desires of Francis Tracy and board member E. E. Hartshorn to expand the project in the attempt to sell dry lands near Malaga, southeast of Carlsbad. Tracy, Hartshorn, and Etter advocated a Carlsbad project that would encompass at least 40,000 acres.[47]

Tracy, Hartshorn, Etter, and others also attacked the competency of Reclamation engineers and officials on the Carlsbad Project in order to win support for a third storage reservoir and increased project acreage to cover the Malaga acreage and more. Even as the trio pursued more watered acreage under the project, others criticized their actions as nothing more than promoting the interests of large speculative landholders over small farmers. In October 1911 the writer of an anonymous letter disapproved of the board's water-use decision, reached, in his mind, because of pressure by the Tracy interests. The author tried to impress Reed from the standpoint of what he considered real farmers in the region. According to the letter, farmers had complete confidence in the ability and integrity of Reclamation, "were opposed to knocking by any clique in the past, present or future, and desire[d] to deal directly through a representative committee, and not as in the past through a few individuals who are either the paid agents or financially obligated to the promoters whose sole aims are to sacrifice farmers' interest for the individual benefit. The farmers are well organized and only need another water users election to forever eliminate the old-time knockers."[48]

An avowed enemy of the Tracy faction, P. J. McShane warned Reed that the Tracy-Etter group was not representative of local farmers. If they succeeded, Reclamation's goal of "a real farmers' project" would be postponed indefinitely. According to McShane, prior to Etter leaving for Washington in November to push for expansion of the project, Tracy's real estate partner, and now mayor of Carlsbad, Charles H. McLenathen, held a meeting in his home to instruct Etter on his responsibilities in the nation's capital. McShane believed the Tracy interests wanted to test the legality of Reclamation's control over the project in order "to wrest the project's control from its present management." McShane allied himself with farmers intent on maintaining the project as it was. He claimed that various persons tried to buy him off, but that he could not be bought or sold by anyone. McShane mistrusted Etter when he first arrived in the valley in 1910, and he had opposed the original officers of the Pecos Water Users Association, many of whom had ties to Pecos Irrigation. He had tried to defeat them for several years. McShane was concerned that Etter and Tracy were now "going at things in a different manner and should they be able to show apparent results for this trip [to Washington] it will be difficult to defeat them."[49]

The split among the water users did not go away. Farmers in the Loving area of the project, south and slightly east of Carlsbad, believed that the Tracy interests controlled the Pecos Water Users Association. L. E. Foster and W. M. Reed, well acquainted with the rancor in the association, reported to A. P. Davis and other officials on a regular basis. Davis, in turn, had direct communication with Frederick Newell, and passed on information given him by Foster and Reed. Foster reported to Newell that farmers in the Loving area held a mass meeting endorsing work on the project and expressing full confidence in Reclamation officials.[50]

In the meantime, Tracy, who had attended the National Irrigation Congress meetings held in Denver, then went East. Although Foster did not know the purpose of Tracy's trip, he suspected it had something to do with enlargement and extension of the Carlsbad Project. According to Reed, Tracy and Etter planned to join up in Washington, visit the secretary of the interior and request further extension of the Carlsbad Project, while protesting the quality of certain repairs made by Reclamation in 1911. Reed was of the opinion that several farmers under the project were perfectly happy with Reclamation's progress so far and did not want Etter or Tracy to represent their interests.[51]

P. J. McShane and farmers from the Loving area contacted A. P. Davis during the first weeks of November 1911. McShane telegraphed Davis on November 3. W. R. Hamill, chairman of a farmer's mass meeting near Loving, also telegraphed Davis. Hamill, like McShane, thought Etter's entire purpose in going to Washington was to further Tracy's interests. Etter "ha[d] gone in entire ignorance and absolute defiance of the desires of the actual farmers and we request that he be given no hearing and no representative and further that the matter of our payment and all other matters be taken up by local Superintendent Foster or other agent designated by you and committee of real farmers of the project."[52] When the Tracy faction discovered that McShane had been sending telegrams to various Reclamation officials, they tried to depose him as president of the Pecos Water Users Association on November 18, 1911.[53] When the association confronted McShane with evidence that he had indeed sent telegrams to Reclamation requesting that they not listen to either Tracy or Etter in Washington, McShane gleefully produced such a telegram and read it to the board.[54]

At one point during the meeting, Charles McLenathen tried to defuse some of the bitterness between the McShane and Tracy camps. He handed a paper to Scott Etter, who had returned from Washington, and on motion had Etter read it aloud. In effect, the letter showed that members of the standing committee of dryland owners, of which the Tracy interests were a part, had collected $6,000, paid into the treasury of the association. Then, at the beginning of November 1911, the committee had sent Tracy from Peaksville, New York, to Washington to represent them in a conference between Interior Secretary Walter Fisher[55] and Etter, secretary of the association. McLenathen and the dryland owners claimed that the group had sent Tracy and Etter to Washington to act in their behalf, not on behalf of the Pecos Water Users Association, and therefore their action was free and voluntary.[56]

Those like McShane who closely watched Tracy and McLenathen thought Tracy's fight in 1911 was just a continuation of one carried on for several years, but according to Reed this was his supreme effort. The only difference between Tracy's fight to expand the project in 1911 and that of earlier years was that he had adopted a new method—trying to remain in the background and directing the fight from "undercover."[57] According to Reed, "he was somewhat disappointed at the last end, as his hand was plainly seen and there was a reaction to such an extent that his plans were thwarted to a considerable extent and it became necessary for them to rush to the Washington office with the hope of carrying their points before the reaction could have its effect."[58]

Reclamation leaders in Washington as well as in Carlsbad considered Tracy a major thorn in their sides. This would be the case for another forty years as Tracy continued his crusade for expanding the Carlsbad Project. According to Reed, "until some decisive step is taken by which Mr. Tracy will be eliminated from the Carlsbad Project, we will have trouble, as everybody else has who had anything to do with the project since Mr. Tracy became preeminently identified with Carlsbad."[59] He was right. Tracy continued emphasizing a part of the contract between Pecos Irrigation and Reclamation, referring to a specific paragraph about expanding the project: "The Reclamation Service shall include in said project such of the lands of the party of the first part in the opinion of the Secretary of the Interior, it is economically practicable to irrigate."[60]

Part of Tracy's campaign in advocating expansion was to lay blame on Reclamation officials working the project. In the fall of 1911 Tracy launched a series of letters to government officials in Washington, including President Taft, condemning Reclamation's workmanship in Carlsbad. Specifically, Tracy criticized the Service for what he considered shoddy repairs. As an extension of his claims directed toward Reclamation, Tracy believed that had Reclamation done its job, the Service could easily have expanded the project and watered twice the acreage. In a 1911 telegram to Taft, Tracy, who had tried unsuccessfully to discuss problems at Carlsbad with Secretary of the Interior Fisher, asked the president if there was "redress or relief or adequate appeal for settlers from idiotic incompetence of [the] Reclamation Service. [The] Carlsbad Project is now suffering from stupid obstinance of Arthur P. Davis and W. M. Reed, in retaining for three years in spite of protest, spillway gates at Avalon dam, which could not be kept open in times of flood."[61]

Tracy went on to lambaste Reclamation engineers, discussing how on July 24, 1911, with both McMillan and Avalon Reservoirs full and spillway gates closed, both reservoirs, according to Tracy, were "needlessly and seriously damaged." Tracy thought Reed and Davis neglected emergency measures for the people under the Carlsbad Project and concealed the results of their incompetence. Tracy recommended that the president send someone into the valley who was not connected with Reclamation, someone whom he could trust and in whom the president had confidence to determine what had happened in the valley. To back up his claims, Tracy asked Taft to refer to Senators Carter, Jones, and Wood, as well as the Army Board of Engineers, who Tracy claimed had publicly condemned the spillway gates in 1910.[62]

While Tracy argued that still unfulfilled water provisions for Pecos Irrigation lands were the chief consideration in his accepting the government offer eight years before, Newell believed that Reclamation had completed all of the practical work on the project and Interior Secretary Fisher concurred. In Newell's mind the project had been doomed to fail until Reclamation stepped in to resurrect lands and orchards imperiled by the failure of Tracy's company to furnish water in the first place.[63] The reason Reclamation accepted the project was due to a delegation of farmers from

Carlsbad who, portraying the situation as desperate, moved President Roosevelt and former secretary of the interior Ethan Hitchcock to go against the personal views of some top officials at Reclamation.[64]

Newell informed engineers W. M. Reed and A. P. Davis of his decision not to expand the project. Tracy was undeterred and continued to hammer on Reclamation officials using Pecos Irrigation Company's water rights as a basis for extending waters to lands not included under the original boundaries of the project. When Newell questioned Reed about the legitimacy of such water rights, Reed responded that the alleged water rights at issue expressed nothing more than an "inchoate right." While some of this land had been briefly watered to establish Desert Land claims, it had not been cultivated, nor had the water been used in a beneficial way as the law required. Establishing a water right was questionable at best. According to Reed, title to most of the Desert Land entries had eventually fallen to the Pecos Irrigation Company. Reed insisted that any hardship experienced after the sale of the company's works to Reclamation fell where it properly belonged—on the company itself. In some cases where settlers made entries in good faith, all rights had been lost by abandonment of the lands long before the government purchased the system. Reed reiterated that generally Reclamation had included all lands that had "a shadow of a bona fide right to water, therefore including considerable second class lands which otherwise would not have been included under such a project."[65]

Francis Tracy continued his campaign; he was obsessive and annoying. Not one to mince words, he often reviled those involved in the Carlsbad Project, from local officials all the way to the upper echelons of government. According to Reed, Tracy himself was the storm center of problems on the project. He believed that Tracy took any disagreement with his views as a personal affront.[66]

Tracy's criticism, however, did have an effect. In October 1911 Frederick Newell wrote to the acting director of Reclamation A. P. Davis in Washington, D.C. Tracy had been sending letters to everyone, including acting secretary of the interior Samuel Adams. Adams, apprised by Tracy of the "incompetence" of Reclamation officials, put pressure on officials in Reclamation and Interior to rectify the situation. Because of correspondence from Tracy and others, Adams found it necessary to travel to Carlsbad himself and asked Newell to fit a meeting into his schedule. At the end of September, Newell wired Adams that he would meet him on the way to Carlsbad. He wanted Adams to send him a statement of complaints received by President Taft and Interior. Newell also wanted a copy of Senator Carter's report from Senate hearings held in Carlsbad in 1909 so he could refresh his memory on the details of the project.[67] In response to

Newell's request, Davis sent him the telegram Tracy had wired the president and memoranda on the subject based on field reports.[68] Davis found Tracy's claims ridiculous and believed that responsibility for any problems on the project should be laid on Tracy himself: "The Reclamation Service has been hampered by a lack of funds to complete the Carlsbad Project and some cheap work had been done, due mainly to the attitude of certain parties under the leadership of Mr. Tracy, to avoid the re-payment of expenses on this project. Under these conditions it was necessary to keep the expenses down as low as possible."[69]

On October 17–18, 1911, Newell and Adams visited Carlsbad and met with a subcommittee of members of the local Pecos Water Users Association. The subcommittee included President P. J. McShane, Secretary Scott Etter, and Directors E. E. Hartshorn, D. F. Doepp, and W. B. Wilson. The group met with L. E. Foster and W. M. Reed, and then visited McMillan Reservoir where it found a large force of men and teams at work repairing washed-out earthen dams. The group then traveled to Avalon and observed a portion of the west end of the dam washed out during recent flooding. Following visits to the dam sites, the Washington visitors traveled to the far southern end of the project where they saw a portion of the Main Canal near "Gyp Bend." This was an area of the river infiltrated with gypsum and plagued by large losses of water.[70]

After assessing various communications between local Reclamation officials, between Tracy and Reclamation, and others involved in criticism of the project, Samuel Adams responded directly to Tracy's contentions that Reclamation had botched the project. Adams wrote that the Interior Department had given full consideration to all the matters Tracy presented in connection to the United States from Pecos Irrigation Company. Adams found nothing to cause Interior to recognize any of Tracy's claims. He told Tracy that investigations in the valley did not warrant a belief that Reservoir Number Three could be safely used to store any amount of water. Neither would Reclamation consider the possibility of irrigating additional lands under the project.[71]

By 1912 Tracy was seen as a mere irritant, and Newell advised officials in the Interior Department to ignore Tracy's requests for reports and other paperwork, which would only serve to further his arguments against Interior and Reclamation.[72] Accusing Reclamation of deliberately mislaying his letters,[73] Tracy launched a barrage of articles in favor of expanding the project under the general title of the "Greater Carlsbad Project."[74]

Tracy also recruited lawyers and politicians to further his cause. In February 1914 he consulted with Armstrong and Botts, attorneys from Carlsbad, who in turn contacted H. B. Ferguson in Washington, D.C.

Tracy's stable of attorneys tried to make the case, now well known by 1914, that plenty of good land existed within the vicinity of Carlsbad to utilize surplus waters that the government had acquired through appropriation. Ferguson represented the new state of New Mexico, and Tracy hoped he would cooperate with Armstrong and Botts in securing a third reservoir.[75] Ferguson responded quickly. A week after the Carlsbad attorneys talked to him in Washington, Ferguson contacted A. A. Jones, a New Mexican recently appointed assistant secretary of the interior.[76] Ferguson sent Jones the copy of a letter and pamphlet from Tracy along with information from the Carlsbad Commercial Club, all making mention of water wasted on the Carlsbad Project. Ferguson thought such water could be put to beneficial use in New Mexico and asked Jones to look into the matter, feeling secure that here was an authority within the Interior Department who could place pressure on Reclamation or others and "bring relief" to Carlsbad and the surrounding area.[77]

Because Tracy no longer dominated the Pecos Water Users Association, it did not hurt his cause in 1914 that he was president of the Carlsbad Commercial Club and that Etter served as director of the club.[78] Although Tracy, Etter, local attorneys, and others continued to lambaste Reclamation and demand the expansion of the project, that did not stop them or the commercial club from promoting Carlsbad and the Pecos Valley.

During the many years that Tracy criticized Reclamation and demanded expansion of the project, he also tried to sell land. In 1912 the Carlsbad Commercial Club, the Realty Board, and the water users association inaugurated a strenuous campaign. To expedite the sale of excess lands under the project, the association encouraged local real estate brokers to sign an agreement governing the manner of sales and commissions. Francis Tracy, who controlled much of the area's real estate, was elected president of the real estate board. Under the agreement, brokers listed all saleable lands under the project exclusively with the secretary of the water users association. The association created one stipulated price and one listing, copies of which would be placed in the hands of all real estate men operating under the project. The idea was to have individual agents publicize project lands in Oklahoma, Kansas, Nebraska, Iowa, Missouri, and Illinois.[79] Local real estate men including Tracy and others raised $800 and turned it over to the association directors, who used the fund to prepare and print massive amounts of advertising material, which they sent to the Midwest.[80] While brokers made deals between 1912 and 1914, most rank-and-file members of the association had no knowledge of it. Real estate men had free access to listings and the backing of the association's board of directors but not from the general members.

In the summer of 1912 a representative from Civic Development Company of Kansas City came to the valley and made arrangements to sell lands and promote development by advancing money to farmers planting alfalfa, peaches, and other fruits. This arrangement fell through and was never put in force. In the fall of 1912 another company representative made arrangements for listing lands directly with Civic Development Company, with the water association board of directors serving as guardians to protect the rights of members.[81]

The local realty board, which had raised funds to promote valley lands, was also drawn into the scheme with Civic Development Company. The company scheduled an initial excursion by potential investors in November, but no excursion took place. A second trip failed to materialize as well. Finally, another Kansas company called Farmers Irrigated Land Company brought potential buyers into the area and made a few land sales, some 300 acres in 1913. Prices set by the company were high, and apparently even the realty board did not have a list of selling prices. Because prices were so high, sales began to drag. At the end of six months the company had failed to sell the stipulated 2,000 acres, and the listing contracts lapsed. The manager of the company, a Mr. Rudd, resigned, and Scott Etter took his place. By February 1 Francis Tracy withdrew all of his lands and started to sell them on his own. This move signaled a split between Tracy and Etter and was the beginning of the end of the development and emigration scheme. Eventually Farmers Irrigated Land Company terminated its agreement with the directors of the Pecos Water Users Association, Scott Etter retired from the company, the local realty board cut itself loose from the water users, and things were much as they began in 1912.[82]

In 1914 Francis Tracy, having gone his own way during the land promotion, now felt free to criticize Etter's representation of the association in Washington and in valley land sales. Tracy attacked Etter in the local paper, asking the public what happened to the $2,800 advertising fund and what monies Etter may have received from land companies in Kansas. Responding to accusations against him, Etter claimed that Tracy, while also in Washington, requested the return of Pecos Irrigation Company buildings now used by Reclamation and the water users. Etter then accused Tracy of selfishly placing the interests of Pecos Irrigation above the water users while in Washington, trying to get an allotment of water for his lands first. Tracy, of course, denied this and claimed no conflict of interest in representing both the Pecos Water Users Association and Pecos Irrigation Company. Tracy's long history of self- and company promotion indicated that using the water users association for his own interest was to be taken for granted. He argued that if not for his efforts and those of Pecos

Irrigation, drylands in the Malaga area, where his company owned considerable property, would revert to cattle range or desert.[83]

At the same time Tracy and others were criticizing the work of Reclamation, they were visibly involved in printing brochures and other propaganda indicating that Carlsbad and vicinity was an ideal place to live. Such propaganda allowed them to sell lands. In fact, in brochures and letters the organization noted that the U.S. Reclamation Service had "transformed 20,000 acres of arid land into rich, fertile, and productive soil, producing annually: six tons of alfalfa per acre, 1–1/2 bales of cotton per acre, $300.00 in sweet potatoes per acre, $300.00 in cantaloupes per acre, $400.00 in peaches per acre, $500.00 in pears, and $200.00 in apples per acre."[84] In 1914 the commercial club boasted that Carlsbad had one of the best climates in the United States, a population of over 3,000, two national banks, two lumberyards, two hardware stores, three drugstores, five hotels, two livery stables, three barbershops, three jewelry stores, a telephone system, an electric light system, an up-to-date fire department, an ice plant, a public library, a free reading room for men, two newspapers, ten lodges of secret and fraternal societies, power dams, two furniture stores, two automobile garages, two bakeries, a cotton gin, a tuberculosis sanatorium, a courthouse, and a jail.[85]

Amid such apparent progress in the valley, the water users association board of directors requested that Reclamation turn over management of the project to the water users in 1914. The directors sent the request without the knowledge of water user members. Small farmers in the Loving area, finally apprised of the plan, rejected it. They believed the plan favored excessively large landholders since they still controlled the association. M. L. Muggeridge, president of Local 86, Farmers Education and Cooperative Union of America, wrote Interior's Franklin Lane on April 13, stating that small farmers and regular members of the association opposed turning management over to the association. They feared that the directors, who were the largest landholders in the area, would, as before, try to ramrod decisions down the throats of the other members. Average farmers in the Carlsbad area, according to Muggeridge, resented large landowners and managers like Francis Tracy. Small farmers argued that they moved to the valley to stay, while Tracy and others who held proxy votes in the association attempted to avoid construction, maintenance, and operation costs placed on the project by Reclamation. The small farmer had no choice but to pay what was expected of him.[86] Once again, Tracy had apparently been caught trying to increase control in the valley at the expense of small farmers.

And although Tracy, the water users association, the commercial club, and others tried to convince Reclamation officials of the need for expansion, in April 1916 Assistant Interior Secretary A. A. Jones informed Frank H. Richards, secretary of the club, that extension of the Carlsbad Project by building a third reservoir or a proposed high-line canal to irrigate lands west of such a reservoir would not be possible. Congress had appropriated funds only for completing projects currently under construction.[87]

Despite protestations to the contrary by McLenathen, Tracy, and others, Reclamation had largely resuscitated the irrigation works of the lower Pecos Valley of New Mexico. By repairing the dam at McMillan, reconstructing the dam and diversion works at Avalon, and repairing the East and West Canals near Black River, Reclamation eventually provided irrigation water to 20,000 acres of land. As early as 1910 Reclamation delivered water to 13,500 acres, gradually increasing the acreage over the next few years.[88]

By deferring and graduating payments at Carlsbad and other projects, Reclamation stretched its budget to the limit. By 1910 Reclamation had assumed control of twenty-four projects, none of which had been completed. And the Reclamation Fund lacked the monies to complete them. Congress loaned the fund $20 million, to be repaid out of the fund income over twenty years beginning in 1920. Congress also stipulated that all new projects had to be approved by the president before work began. By 1924 Reclamation had twenty-seven projects completed or under construction, twenty-one of which began in the early years of the Service,[89] but Reclamation had no more money for Carlsbad.

10

GOD PITY THE WATER USER?

ONE of the major problems at Carlsbad and projects across the West was repaying Reclamation for work. While many people considered reclamation projects as engineering triumphs, many projects were social and economic nightmares. Reclamation failed to anticipate rising costs, the need for irrigators to pay more than original estimates, or the inability of irrigators to pay for projects in ten years. The Reclamation Act required settlers to pay project monies back into the Reclamation Fund, without interest, in ten annual installments. As the time for initial payments got closer, settlers on the majority of projects asked for more flexibility in paying back the government. Disagreement over repayment centered on the discrepancy between Reclamation's estimated project costs and actual construction costs, which were higher and used to calculate repayment.[1]

From the beginning, farmers under the Carlsbad Project had trouble making annual payments. Carlsbad irrigators convinced Reclamation to delay payment from 1908 to 1909, then to 1910, and so on until L. E. Foster recommended postponing the first building charge payment until December 1, 1913.[2] While water users complained that project costs had escalated, Reclamation countered that the water users themselves had requested changes in original project plans.[3] In fact, even while Reclamation sought repayment, water users at Carlsbad urged the Service to expand the project. Newell thought it extremely unwise to extend the irrigation system under the project "until many of the present troubles are solved and a larger amount of money already invested has been returned . . . [and] . . . the Reclamation Service will not advocate the extension of this project until present works have been put in better condition, the seepage reduced, more land cultivated, and a higher degree of success reached by the farmers under the present situation."[4]

Certainly irrigators in the Pecos Valley were not alone in facing repayment problems. Across the West local farmers struggled to make their farms yield enough to pay for Reclamation's labor. Western project demands to extend payments led to conflicts with other western areas that wanted their own projects. Newell argued that the original repayment schedule was just and necessary to provide irrigation for new areas. In 1911 the various project water users banded together to form the National Water Users Association to attack Newell's repayment policy.[5] The National Water Users Association quickly brought its case before new interior secretary, Franklin K. Lane.[6] In 1913 Scott Etter and Francis Tracy went to Washington at Lane's invitation. Results of the conference, according to Pecos Valley water users in their annual report, were positive and showed that the Department of the Interior was shifting from an attitude of "dictatorial authority and fault finding to one of sympathetic cooperation."[7] Lane retained Newell nominally as director, but placed operating power in the hands of a commission. Attacking Reclamation for its preoccupation with building dams and reservoirs and neglecting the human element of projects, Lane backed the water users in their call for extended payments.[8]

In Carlsbad L. E. Foster again recommended postponing the construction payment due December 1, 1913. In late September 1914 Congress issued the Reclamation Extension Act, which extended annual payments over twenty years and urged water users across the West to take advantage of the new law. Washington cautioned that unless water users accepted the act before December 1, 1914, payments due under the old requirements had to be paid by that date.[9]

In April 1915, despite the Extension Act, the matter of repayment on the Carlsbad Project came to a head. Reclamation agreed to the Pecos Water Users Association's demands for a review of project cost issues at Carlsbad by assembling a Central Board of Review. Making up part of the most powerful organization in the engineering world, Reclamation's Central Board consisted of renowned irrigationist Elwood Mead as chairman, William L. Marshall,[10] and I. D. Donnal. Mead, who had submitted reports on Carlsbad to the Department of Agriculture in 1899, had firsthand knowledge of conditions at Carlsbad, including the project's high incidence of speculative and absentee landholders. His history as chief of the Office of Irrigation Investigations in the Department of Agriculture indicated a strong dislike for federal projects like Carlsbad.

Before examining specific issues on a project, the Central Board required input at the local level.[11] The Reclamation Service held hearings in Carlsbad entitled "In the Matter of the Re-evaluation of the Carlsbad Project."[12] The proceedings of Reclamation's Central Board of Review took place from April 13 to April 17 in the Reclamation Service building.

Fulton H. Sears of Chicago represented the local water users association as legal counsel, and P. W. Dent of El Paso represented the Reclamation Service.[13]

L. E. Foster, who became project manager after W. M. Reed in 1906, was one of the key witnesses. When Foster arrived on the project, Reclamation had finished the major construction work. Foster had first come to Carlsbad under the direction of Thomas H. Means and worked to classify lands under the project and select those most suited for agriculture. Foster had little to do with construction work except as it related to operation and maintenance of the project until W. M. Reed left the project.[14]

The crux of the hearing was to determine whether water users in the Carlsbad area should have to pay the assessed construction fees and operation and maintenance charges, which had increased by 50 percent since 1907 as Reclamation tried to improve the valley's irrigation system.[15] Sears, attorney for the water users association, wanted to know how much Reclamation had spent and for what purpose. He asked Foster about an experimental farm and tree nursery, fearing that Reclamation had charged those costs to water users.[16] He was also concerned that Reclamation might have charged preliminary surveys to the project, when such surveys were conducted on other projects without charge. Foster was not aware of any such charges. When asked why construction charges on the project increased from thirty-one dollars to forty-five dollars per acre, Foster pointed to new work done on the project since 1904. Although Reclamation had recommended the work, the water users officially requested it. For example, following recurrent floods, rebuilt spillways on the project caused added expenses, but water users could hardly do without them.[17]

Sears also questioned Francis Tracy along a number of lines. Although Tracy trained as a lawyer, he had gleaned appreciable engineering knowledge working in the valley for over twenty years. Tracy concluded that the Service had never finished work committed to originally. He remarked that spillways at Avalon Dam were insufficient to carry floodwaters from the river even after Reclamation had "fixed" them. Tracy considered Reclamation's first headgates at Avalon inferior, a longtime sticking point between Tracy and Reclamation. He regarded them as symbolic of the ineptitude of project engineers. Tracy held the engineers who installed the gates responsible for the district's water shortage in 1905 and 1906. Tracy argued that water users had had to pay not only for the gates themselves but also for losses that occurred during those years.[18]

Tracy described the gates and how they were supposed to work. Reclamation had originally installed double wooden gates, one tied to another so that when water forced the first one open, the second was supposed to open as well. Reclamation found the gates too weak; so it strengthened them. Through experience Reclamation realized that whenever floodwater of any significance shut the gates, Reclamation's repair work was destroyed. Tracy described a flood in July 1911 when the flood gates were shut and he found running water topping the dam. He remembered seeing superintendent L. E. Foster and other workers trying to open the gates, which were all jammed closed, except perhaps one, which they had managed to open. Foster and others used a rod to suspend a worker above the onrushing water. He hammered mightily with a sledge hammer to open the headgates.[19] Reclamation replaced the original gates in 1911, but Tracy wanted to know why water users should have to pay for the originals.

Dent, attorney for Reclamation, countered that when floods destroyed earlier gates under private investment, the company increased their rental charges for water and tried to raise money to replace them. What was the difference, he asked, between the company charging the water customers for replacing old gates and Reclamation charging for replacement of gates under the present project? Dent stated: "The only difference in the charge of the operation under your company and under the government is that your company charged a profit for the delivering of water, whereas under the government and Reclamation law the settler is supposed to be at least the beneficiary of these works without any profit."[20]

Dent then tried to show that leaking canals and dams existed long before Reclamation entered the valley, and Tracy knew it. Dent referred to a paragraph from a 1910 Army Board of Engineers Report of Reclamation of Arid Lands,[21] in which Tracy described excessive leakage in the reservoir and distribution system. Rehashing issues from the Hondo hearing in 1904, Dent pointed out that the system leaked prior to government purchase. Tracy was trying, in Dent's opinion, to milk the government for problems it did not cause.[22] In trying to ascertain how Tracy and the company had made money, and how much water they wasted, Sears asked him what the company had charged for operation and maintenance. Tracy responded that they charged a flat $1.25 per acre rate for furnishing water, though he admitted that occasionally the company and farmers wasted a tremendous amount of water.[23]

In fact, company practice had been to give property owners all the water they wanted. There were some water users who used as much as sixteen feet an acre on their property. Disbelieving, Dent asked Tracy if he meant sixteen acre-feet of water. Tracy responded in the affirmative. Other water users used what the company called "special water," one and a half feet per acre although they did not have any water contract attached to their lands.[24]

Despite his own company's history of questionable engineering and water waste, Tracy continued to hammer on the incompetency of Reclamation engineers on the project. He claimed that Reclamation furnished the project with men who were absolutely untrained and had no previous experience, and that they had made several mistakes, which cost water users not only water but the money used to pay for their mistakes. Perhaps more true than Reclamation was willing to admit, Tracy suspected that there were a number of men throughout the country on Reclamation projects trying to figure out how to do the necessary work. In reality, a combination of factors—the fact that the original dam was privately designed, Reclamation's own problems with the project, both engineering and political, and nature—all contributed to higher costs at Carlsbad than first anticipated.

Tracy insisted that since Reclamation engineers had increased the cost of projects, the government should absorb the costs of their mistakes. Tracy attacked A. P. Davis, who had served as assistant director of Reclamation and had been in government service thirty-one years with no construction experience. Others in Reclamation, including Frederick Newell and many other men who assumed leadership roles at Reclamation, had no construction experience either. Tracy asked Dent, "Isn't it right that the government should stand some of the lessons these men learned at the expense of the settlers if we have to pay at all?"[25]

At the local level, Tracy cited for incompetency E. W. Myers and W. M. Reed, who was a former employee of the Pecos enterprises in the valley and supporter of the Hondo Project in the early 1900s. Dent asked Tracy if he thought that on the Carlsbad Project there had been more engineering mistakes than would have occurred on the works under private management. Tracy hesitated but still insisted that he thought that Reclamation certainly should have learned from past mistakes and used that experience to place the Carlsbad Project in workable order.[26]

To some degree Reclamation's mistakes occurred as a result of mistakes originally made by private companies in the valley—mistakes and defects that were inherent in the old system under companies for which Francis Tracy had worked. Reclamation, in an attempt to patch up an old imperfect system, inherited such problems.[27] And in 1904 Reclamation's hand was forced by the president of the United States. Reclamation engineers had never wanted to deal with the problems at Carlsbad.

Following the Carlsbad hearing on repayment in April and May of 1915, a local board of review consisting of T. U. Taylor, an engineering professor at the University of Texas for twenty-seven years; Scott Etter, secretary of the water users association; and D. W. Murphy, an engineer in charge of drainage for the Reclamation Service, met in Carlsbad to discuss project issues. Taylor and Etter submitted a majority report to which Murphy dissented strongly. The majority report bitterly criticized management, workmanship, and repayment charges on the project. Taylor and Etter called for significant reductions in project charges. On completion of the local board's report, Taylor and Etter released the majority report before the minority report had been completed and before either of the reports had been submitted to the secretary of the interior. Besides being publicized in the press, the report was reprinted as a circular and distributed among various water users associations on government projects across the West.[28]

The actions by Taylor and Etter at Carlsbad led to similar reports throughout the western states. Local water users on a number of projects asked for similar and equal deductions. Many water users believed that the purpose of local boards was to reduce project costs rather than to determine them, which would reduce the Reclamation Fund by hundreds of thousands of dollars. If that was not enough, Taylor, chairman of the Carlsbad review board, published concurrently with the majority report a bitter criticism of the Reclamation Service.

Taylor, as chairman of the local board and author of the majority report, leveled a number of criticisms at management and construction on the project. Among a litany of arguments, he said "the water user had no voice or vote . . . and had neither the right to hire nor fire and yet he was absolutely compelled to agree to another increase in construction cost per acre or let his land go back to the wilderness."[29] Taylor's comments on the project were dramatic. His concluding paragraph ended with, "God pity the water user on the Carlsbad Project." Taylor later withdrew his comments concerning Reclamation's competency and resulting overcharges on the project. An infuriated Tracy demanded the reason. Taylor responded that "no outside man can afford to serve on a [similar] Board . . . unless he makes up his mind to certify that everything that has been done in the past was 'proper,' otherwise he will put in array against him the most powerful organization in the Engineering world."[30]

Following the hearing and the local board report, on June 23, 1915 the water users association filed a petition protesting the annual maintenance charges with Secretary of the Interior Lane.[31] The association complained that, like the construction charges, the maintenance charge on the project was "unjust, inequitable, confiscatory, wrong in principle," and impossible to pay under Reclamation's public notice of March 2, 1915. When it had signed the Extension Act in early 1915, the association had done so with the belief that it would receive relief from the act and did not realize that Reclamation could increase maintenance charges without local consultation. The water users cited a number of unique factors in arguing for not having to pay the charges, including the climatic conditions of the project, the fact that they grew alfalfa, which used a great deal of water, and the types of porous subsoils underlying a large part of the project.[32] In other words, given that the project was ill suited for high-yield crops, Reclamation should back down on maintenance charges.

Small farmers made an important point when they showed that the schedule of maintenance was now so arranged that the speculator simply let his lands lie idle while the rest of the water users contributed to constructing and cleaning ditches that went directly past the speculator's property. The idle landowners simply paid the minimum of seventy-five cents per acre on uncultivated land, while others trying to develop cultivated lands paid more. Actual resident owners producing crops would now annually pay two dollars per acre, or twice the original estimate for maintenance.[33]

Reclamation's Central Board of Review received the Carlsbad majority report with skepticism. Although the Central Board believed that it had to consider the reasons for the local board's actions and try to determine the fairness and accuracy of criticisms made,[34] all three members detected bitterness by water users toward the Reclamation Service. But most disturbing to the Central Board was the premature and unauthorized publication of Carlsbad's majority report and the ensuing complications for Reclamation. Based on charges of lost crop revenues caused by shoddy and incomplete repairs, poor headgates, and incompetency, two of the local board's recommendations distressed the Central Board most. The first suggested that one-fourth of the total monies expended by the government should be eliminated and that water users should not be required to repay it. The second finding of the local board called for the elimination of another large percentage of federal expenditure, relieving water users of more than half the total project cost.[35]

In March 1916 Reclamation's Central Board of Review submitted to Franklin K. Lane a report dealing with some of the problems at Carlsbad. The Central Board concluded that there were a small number of resident water users having a hard time on the project and that they—but not others—certainly deserved the sympathy of Reclamation.[36] The major problems on the Carlsbad Project, according to the Central Board of Review, included inflated land prices, high freight charges, high interest rates, alien landlordism, and evasion of regulations limiting the size of farm units receiving water, which verged on fraud. It appeared that absentee owners, many of whom had early connections to Pecos Irrigation Company, hired tenants with nominal farming skills to occupy their lands.

The tenants practiced poor agricultural methods because they had neither the reason nor the money nor the equipment needed to cultivate the land as required for profitable irrigated farming. By evading the letter of the Reclamation policy on size of watered property and on absentee ownership, Tracy and speculators were sitting on land, hoping to sell it at a good price.[37]

The project's water users were not the victims of government mistakes, according to the Central Board. Rather, the government had taken over a project of doubtful value and spent a large amount of money to rescue property, which had continuously lost money under private ownership. In the Central Board's estimation, the secretary of the interior and Reclamation had shown great patience in dealing with the Carlsbad Project. The board reminded Lane that according to the water users' own descriptions the district was on the verge of ruin before the government came in. The Central Board noted that fifteen to twenty years previously water users had pleaded with Reclamation for help, afraid of losing 15,000 acres of irrigated property.[38]

A population of 3,000 depended on such property valued in 1904 at $2 million. The Central Board believed that if the government had not taken over the project, farms surrounding Carlsbad would have returned to desert conditions, as they had in other areas in the West. Considering the project's location and private ownership, the board concluded, "The conditions under which the government took over this enterprise were more liberal than sound public policy renders desirable."[39]

Citing an Army Board of Engineers report from 1910, Mead and other Central Board members considered Carlsbad to be the victim of speculation. Therein lay many of the problems on the project. In 1910 Army engineers had questioned the feasibility of the federal government taking on projects under which a large proportion of the land was private: "In undertaking projects where the direct federal interests are small it would seem no more than equalative to add an interest charge to the cost of reclaiming private land or to require that a percentage of private lands be deeded free of cost to the United States."[40] The Central Board felt—as had Pecos Irrigation—that Reclamation was the only authority in the early 1900s that could have saved Carlsbad from abandonment and thus rationalized all repayment schedules under the project.

While various improvements on the project practically doubled the original repayment figure, the board argued that water users were repaying only five cents for each federal dollar actually spent on the project. After initial work in 1906, Reclamation spent $470,000 under the contract, specifying fixed costs of thirty-one dollars an acre. Later, water users consented to spend an additional $65,000 for lining canals and to pay a thirty-five cent per acre increase in the operation and maintenance cost for nine years beginning in 1909. But in 1911 flooding damaged the works and stopped irrigation once again, and water users appealed for more federal money. The government quickly allotted $70,000 worth of emergency funds to the project as well as $120,000 for operation and maintenance, and later another $160,000 was appropriated for canal improvements as requested by the water users association. All told, the government spent some $900,000 for construction on the project, each increment of which was requested and approved by the association.[41]

The Central Board pointed to a number of situations and uncertainties peculiar to the Carlsbad Project that the local board of review neglected to mention in its report. The 20,000-acre project completed in 1912 under the $900,000 allotment continued to be improved by cleaning and lining canals and construction of farm turnouts and other small structures of concrete. The government, according to the Central Board, faced the task of providing enough water to irrigate 20,000 acres. But the perennial flow of the Pecos was not enough for the original area unless Reclamation made canals and distribution systems tighter, which it was proceeding to do.[42]

Problems in the actual reservoirs themselves added to problems in the distribution system. This was due to a long history of silt deposition from the Pecos River and a loss of stored water through sinkholes at the bottom of Lake McMillan. According to the Central Board, the government, which realized these peculiarities too late, was now in the same position as the original investors and companies, which one after another had been unable to master the Pecos River.

Reclamation leveled its strongest argument against speculators under the Carlsbad Project and other projects across the West. But Reclamation saw Carlsbad as excessive in this regard. When the government undertook reconstruction, all land was privately owned except for some state land, later unwisely sold by New Mexico to land speculators in large tracts. "This project stands out therefore as one in which desirable social and economic results of the Reclamation Act have been defeated by the speculative and non-resident ownership of land and by the selfish and unwarranted inflation in land prices."[43] The Central Board might well have included the names Etter, Hartshorn, Tracy, and McLenathen. Etter complained about the lack of local control to hire and fire those considered incompetent, and Tracy vilified Reclamation engineers who failed to expand the project. Yet, they, Hartshorn, and others who lamented the lack of democracy on the project were the same speculative interests denying democracy to small landholders like those near Loving without the wealth or power to do anything about it. The water users association had been controlled by Tracy and absentee proxy votes for a decade. Tracy and large landholders who dictated the agenda of the association had paid lower taxes and assessments than others yet now wanted Reclamation to expand the project to make their lands more valuable.

Reclamation had to include private lands within the Carlsbad Project to make it feasible and practical. But the Central Board also believed the primary objective of the Reclamation Act was the development of public lands to establish homes for the landless farmer. The board thought speculators, under private irrigation projects purchased by Reclamation, had gained improvements intended for settlers at government expense. Mead and the board saw such projects, where Reclamation initially paid for land and water rights and added those costs to construction charges, as benefiting a privileged few. The board saw Carlsbad as an example of severe abuse by speculators. On all projects, according to Mead, the government should buy excess lands in private ownership or fix the selling price of such lands before work began.[44] Absentee landholders, then, were nothing more than promoters relying on the needs of settlers and farmers and on unearned increases in land values to make a profit purely through the fact that such lands were irrigable. The Central Board labeled Carlsbad as symbolic of the worst kind of blemishes on the Reclamation Act that occurred in the West. Board members thought situations like those at Carlsbad should be "eliminated, controlled or abated before any successful system of land settlement in the arid states [could] be carried out."[45]

The Central Review Board concluded that the main problem of the water users under the Carlsbad Project was not high project costs as the local board had reported, but inflated land prices and high interest rates on borrowed money and deferred payments. Farmers cultivated 13,000 of the 20,000 acres under the project, and of 521 farms, farmers irrigated only 345. Out of those 345, tenants operated 200; managers or owners operated only 145. Furthermore, many owners evaded the 160-acre landownership limit set by the Reclamation Act. Ownership of huge areas of land under the project appeared on the records of the water users association in the names of clerks, typists, doctors, laborers, and others whose ownership was only nominal. One person, presumably Tracy, controlled 3,000 acres of unoccupied land. The Central Board pointed out that people neither improving nor cultivating but simply acting as dummy owners for large holders of land worked against the intent of the Reclamation Act. All the while, these same absentee landowners, who refused to pay for construction and other charges, asked farmers and others seeking lands in the area to pay purchase prices anywhere from $50 to $150 an acre. Reclamation's Central Review Board did not see repayment costs as excessive, but suggested that prices for excess lands under the Carlsbad Project be reduced.[46]

Reclamation was willing to help those whom it considered actual farmers under the project. Officials proposed a system of rural credits at low interest rates as a safe and practical direction for those just getting started in irrigation farming. The board also recommended prosecution for those swearing falsely under the Reclamation Act.[47]

The clamoring by water users and the local board had some effect on the Central Review Board's report. First, the board recommended that Reclamation's controller's office adjust any costs made against the Carlsbad Project that should have been made against other projects, such as Hondo upstream. Second, members proposed that administrative expenses incurred through Reclamation as part of the Department of Interior not be charged against landowners. Third, they proposed that the original cost of making and installing the double wooden gates in spillway number one at Avalon Dam not be charged against the water users. To ensure that local water users pay what they owed, the board recommended that the project be limited in size to irrigate 20,000 acres, using the perennial flow of the Pecos River. No further expenditures should be made until farmers brought all acreage under irrigation and Reclamation collected all the construction and maintenance charges. As evidence that Reclamation wanted to wash its hands of the entire situation, the board recommended that the entire project be turned over to the water users as soon as possible. The board stood its ground on charges to the project, noting that the total cost as of December 31, 1914, was $933,840.96, some $500,000 more than what the local board had estimated that water users should pay. The proper annual payment, based on Reclamation's formula, was $47.00. The local board estimated cost at $20.71.[48]

Despite questions about payments to Washington and lingering doubts concerning Reclamation's ability to make the Carlsbad Project viable, improvements on the project continued. In March 1915 Supervising Engineer F. W. Hanna agreed that Carlsbad managers could begin lining the Main Canal, initiate proposals for a drag line excavator to be used for drainage work, and conduct a silt survey at Lake McMillan.[49]

Problems between the local review board and Reclamation's Central Review Board did not stop Reclamation in 1916 from asking for bids for the excavation of spillways for McMillan Reservoir.[50] Reclamation negotiated with the Atchison, Topeka and Santa Fe Railway Company to place a spillway close to their rail line at McMillan.[51] Nor did problems stop local

Cement-lined section of the Main Canal, drop, and lateral gate, Carlsbad Project. Although reluctant to take on the project, Reclamation made numerous repairs and upgrades on the project, including lining canals with concrete.

(Record Group 115, Project Histories, Feature Histories and Reports, 1902–1932, Carlsbad, box 79, entry 10. Department of the Interior, U.S. Reclamation Service, Project History, Calendar Year 1916. Carlsbad Project, Carlsbad, New Mexico. National Archives and Records Administration, Rocky Mountain Region, Denver, Colorado.)

water users from advertising Carlsbad as the greatest natural sanatorium in the Southwest.

In the years following the Central Review Board's reports of 1916, local promoters from the Carlsbad Chamber of Commerce touted Carlsbad's spring, which discharged 4,000 gallons of water per minute according to their booster papers, and two large sanatoriums, one just completed at a cost of $100,000. With paved streets, concrete walks, large shade trees, campgrounds for tourists, large- and small-game hunting, swimming and recreation, telephones, water and lights, a sewer system, fire protection, and two national banks, Carlsbad was a model city. In addition, the newly opened Carlsbad Caverns offered a magnificent underground experience. The Chamber of Commerce promoted Carlsbad as the home of the government's first great irrigation project, "the most successful in the United States."[52] Local boosters promoted the area water supply as inexhaustible, with two great reservoirs storing 50,000 acre-feet of water in McMillan and 7,000 acre-feet for diversion in Avalon. Promoters called the area the greatest breeding section in the Southwest for sheep and cattle.[53]

By the end of the decade, according to boosters, the area grew more Durango cotton than anywhere else in the region.[54] From 1907 to 1922 cotton and alfalfa led the way in crop production. The Carlsbad Project averaged an income of $41.17 per acre each year over the sixteen-year period, with alfalfa averaging $33.55 per acre and cotton averaging $67.99 an acre. By 1922, farmers devoted the bulk of their acreage, some 14,000 acres, to cotton. However, boosters did not acknowledge that average farmers found Reclamation's per acre fee difficult to meet, even if it was extended over twenty years. At $2.35 an acre each year for maintenance, the farmer with 160 acres paid $376 each year.[55]

Francis Tracy continued to accuse Reclamation of incompetency, blaming the Service for mistakes that cost the Carlsbad Project water and money. Typically, Tracy protested the government charges and the repayment schedule and launched a public campaign across the West to question such charges. Reclamation countered that the principal problems at Carlsbad were inflated land prices, alien landlordism, dummy landowners, and watered landholdings that exceeded Reclamation's 160-acre limit. Reclamation engineers had only done what water users requested by improving the irrigation system, and improvements cost money. As Tracy continued his attack on Reclamation, at the same time demanding expansion of the project, a more serious problem loomed on the horizon—new competitors looking for water.

11

AN ELUSIVE RESERVOIR

s Reclamation defended project costs across the West against
growing criticism, water users at Carlsbad faced a threat that caused
all parties under the project to unite in opposition to dangers from
both north and south of the project. To the north, farmers drilled wells
near Roswell to utilize underground aquifers; Carlsbad water users
believed the wells diminished the volume in the Pecos River. Farther
upstream, farmers near Fort Sumner hoped to impound Pecos River waters
to irrigate lands in that area. And downstream from Carlsbad, irrigators
wanted their own portion of the river to water dry lands in Texas.

Francis Tracy was ever vigilant to impending problems to the north of
Carlsbad. Between 1910 and June 1914, the state engineer's office approved
6,742 acres of land for irrigation north of Carlsbad, between the towns of
Roswell and Dayton. Within the same area, permits for an additional 1,221
acres were pending in 1914, and State Engineer James A. French planned to
approve them if no protests were filed.[1] Tracy saw irrigation north of
Carlsbad as a threat to the city's water supply and quickly protested. He
maintained that the Carlsbad District put to beneficial use all the ordinary
flow of the Pecos by direct appropriation during the irrigation season, and
stored all of the winter flow, the ordinary flow having proved insufficient
to irrigate the area.[2] French claimed that since the government had made
no specific attempts to file protests against applications north of Carlsbad,
he saw no justice in holding up applications properly filed and showing
good faith and legality.[3] In May Tracy directed his argument directly to
P. W. Dent, district counsel for Reclamation in New Mexico, Texas, and
Oklahoma, restating remarks made to the state engineer. Tracy did not
understand why the federal government, either by persuasion or injunction
through the state courts, could not prevent the state engineer from
granting further permits except for unappropriated floodwaters.[4] Dent's
hands were tied. With statehood in 1912, state law, not federal territorial

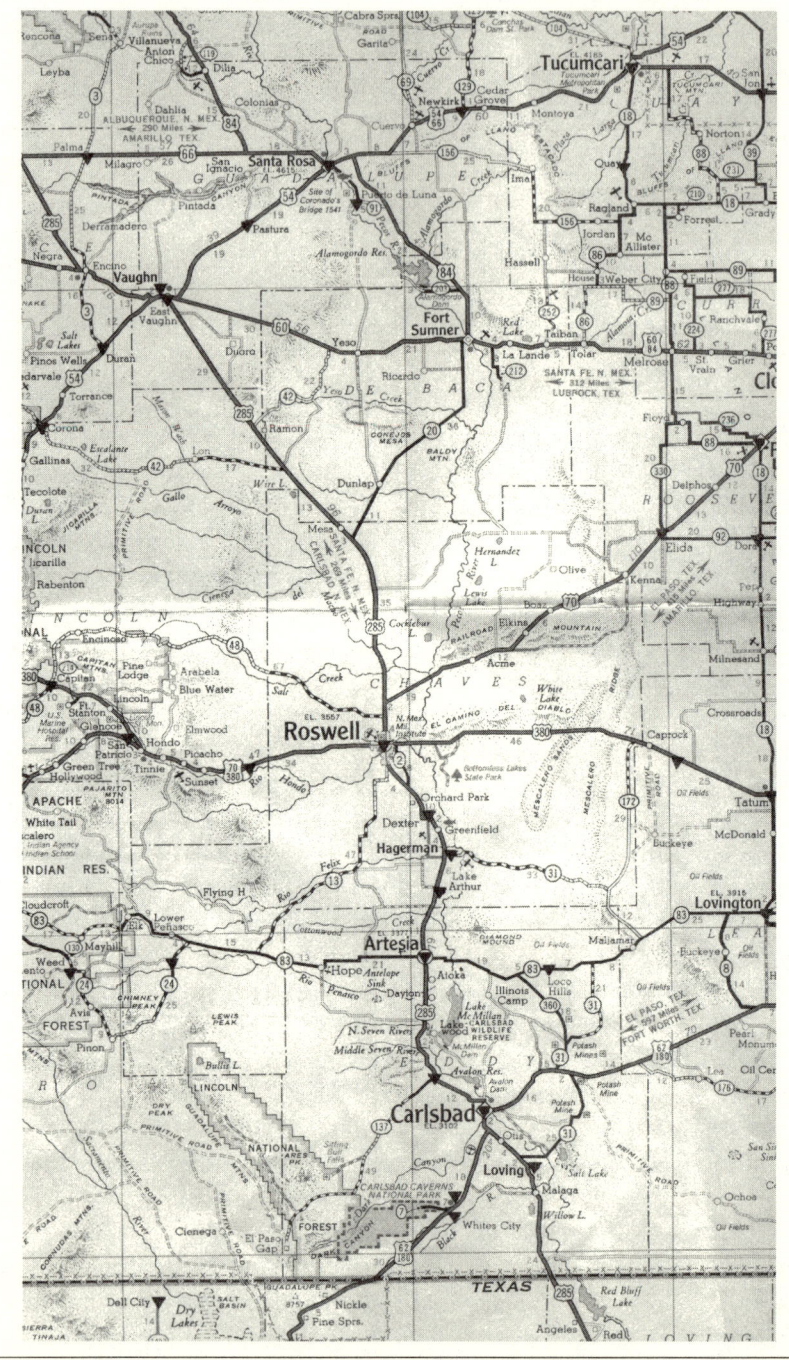

Southeastern New Mexico.
(From Highway Map of Arizona and New Mexico *[Chicago: H. M. Gousha Company, 1950].)*

law, governed New Mexico's water use. The state engineer's office granted permits and offered a thirty-day period for appeal to the state irrigation commission, after which time opponents had no redress against applications except through the courts. Under state law, no lawsuit brought by the water users could be successfully prosecuted unless there was an actual shortage of water causing crop damage.[5]

Dent told Tracy that the Pecos Water Users Association should pass a resolution reciting conditions under the Carlsbad Project. Dent himself was not eager to again take the matter up with the Justice Department, but a resolution from the association would give him a basis for resubmitting the matter at a later time. Dent suggested filing a formal protest against the allowance of any applications north of Carlsbad with New Mexico's state engineer.[6]

Competition for water in the Pecos watershed was not limited to areas north of Carlsbad. At the request of Congressman William Smith of Texas, chairman of the House Committee on Irrigation and Land,[7] Secretary of the Interior Franklin Lane ordered a survey of the entire watershed. Like New Mexicans north of Carlsbad, Texans were insisting on a portion of unappropriated waters from the Pecos River and hoped to get help from Reclamation.[8] This is ironic when one considers that every representative from Texas had voted against the Newlands bill, and that Texas had opposed all Reclamation projects in the West.[9] New Mexicans found it curious that although Texas historically stood against Reclamation in the West, as soon as the federal government started raising money for such projects, Texas promptly made demands for its share of those monies and waters, although it had not one acre of public land.[10] Proceeds from the sale of public lands had funded Reclamation coffers since passage of the Newlands Act in 1902.[11]

Nonetheless, Interior Secretary Franklin Lane assigned P. M. Fogg to investigate and make recommendations to rehabilitate projects along stretches of the Pecos in Texas. After arriving in early March and touring the area, Fogg began the actual survey in April 1914.[12] According to Fogg, in Texas "the doctrine of beneficial use ha[d] hardly made its appearance above the horizon; there is no state engineer and hence . . . the irrigation enterprises have been left to care for themselves."[13] Assuming Texas would develop practicable statutory guarantees of water rights, Fogg recommended construction of a storage dam at Red Bluff, New Mexico, near the Texas line, to serve all ten Pecos irrigation projects south of the line.[14] In 1916 the Texas irrigators organized themselves as the Pecos Valley of Texas Water Users Association, and, on the basis of the Fogg report, they petitioned Interior for aid to build Red Bluff Dam. Until 1906 only states and territories with public federal lands could receive projects. Following a treaty with Mexico over the Rio Grande situation, on May 21, 1906, Congress extended provisions of the Reclamation Act to apply to all of the state of Texas.[15] The association also asked Reclamation to adjudicate intrastate and interstate water rights.

Unfortunately for Texas, 1916 was not the right time to ask the government for money and engineering expertise. By 1916 proceeds from the sale of public lands had dwindled to next to nothing, and Congress only appropriated enough monies from the general fund to rescue Reclamation Service projects already in operation or in the process of construction. Faced with the possibility of involvement in World War I, the federal government cut down its internal improvements and tabled the petition for aid from the Pecos Valley of Texas Water Users Association for ten years. Undeterred by the lack of federal help, the association reorganized into the Red Bluff Water Improvement District. Launching its own studies on the feasibility of projects that it proposed, the group persuaded various state and federal agencies to undertake investigations on their behalf.[16]

While the Texans waited for federal funds, New Mexican irrigators near Carlsbad had their own problems. Leakage and siltation had increasingly reduced the effectiveness of McMillan Reservoir. Experiencing water shortages, the Pecos Water Users Association again requested and received a contract for an investigation and sought federal assistance for construction of a new storage reservoir.[17] Francis Tracy and others had tried to get Reservoir Number Three built since before the turn of the century. Reclamation and other agencies had prepared numerous studies and reports over the years,[18] and while some studies were more or less favorable for building the reservoir, Reclamation had always decided against it.

By the 1920s, Reclamation had slowly made improvements to increase the acreage at Carlsbad to 25,000 acres. The Pecos Water Users Association anxiously wanted to secure some sort of storage reservoir near Carlsbad or several miles away. Consequently on May 2, 1921, Reclamation amended the investigations contract to provide for the investigation of irrigable lands near Fort Sumner and to divide costs between the Pecos Water Users Association and the U.S. government. Irrigators near Fort Sumner wanted a government-built dam on the Pecos to irrigate lands near the old Bosque Redondo reservation. The lands included 8,000 to 10,000 acres, and people in Carlsbad, including Francis Tracy, agreed to this alternate reservoir site to supply water to their own project.[19]

Investigations conducted between the completion of the Carlsbad Project in 1909 and the eventual building of Fort Sumner's Alamogordo Dam point to why Reclamation decided not to build another storage facility close to Carlsbad. Leakage in the irrigation and distribution system had been a vexing problem since its inception. Reclamation reduced leakage in 1909 by constructing an embankment to prevent water from reaching the many sinkholes located in McMillan's coarse sides, where siltation and leakage reduced the effective storage capacity of the reservoir. Salt cedars began appearing at McMillan Reservoir in 1914 on the silt flats where the river entered the reservoir. By 1915, the plant growth obstructed floodwaters and accelerated the rate of sedimentation. Within a few years, silt had built up a delta at the head of the lake, causing the submergence of more land over which the salt cedars then spread. Ironically, the growth of the salt cedars provided an effective silt deposit in the delta area, preserving some of the storage capacity of the reservoir.[20]

Reclamation officials knew that Carlsbad needed additional storage, but feasibility was the big question. Over the course of many investigations, Reclamation reports were not entirely unfavorable, but geologists always reached the conclusion that site number three was too risky because of the area's porous bedrock.[21] As an alternative, geologists had recommended either raising Avalon Dam twenty-three feet, or, in later years, building a new dam at Fort Sumner.[22] The main reason given for not expanding Avalon was that the physical conditions on the northeast and southeast sides of the reservoir were unsatisfactory.[23] Although there were differing opinions as to the merits of increasing the size of Avalon, Reclamation decided against it.[24]

During the course of investigations in the valley, Francis Tracy continued to advocate a third reservoir near Carlsbad. Tracy's insistence on the building of Reservoir Number Three was perhaps legendary at this point. He had advocated the building of the reservoir for at least twenty years. In December 1924 R. F. Walter, acting chief of Reclamation, informed Elwood Mead that on December 10 the Carlsbad superintendent had received a confidential letter from Tracy concerning the Alamogordo Project near Fort Sumner. Tracy preferred, as did the state of Texas, Reservoir Number Three near Carlsbad instead of the Alamogordo site upstream.[25]

Walter reminded Mead that Tracy represented huge interests in the Pecos Irrigation Company and controlled considerable land west of the project. Walter realized that such land could be irrigated only by an extension of the Main Canal at Carlsbad or through construction of a high-line canal from proposed Reservoir Number Three. Reclamation had considered a high-line canal in 1916 to take out water from Avalon Reservoir and lift it forty-five feet above the Pecos River. The canal required flumes, siphons, and other structures, and Reclamation decided that expenses outweighed the benefits. According to Walter, Tracy and Pecos Irrigation

Company's property under the Main Canal was gypsum laden and unirrigable. Walter believed Tracy was concerned that additional storage at the Alamogordo site would be used to irrigate gradually increasing acreage at Fort Sumner, leaving Tracy's lands high and dry. Walter saw Tracy's insistence on Reservoir Number Three as guided by purely selfish motives.[26]

New Mexico's senators, Sam Bratten and Holm Bursum, joined the fray in 1924, equally pursuing further storage and funding for Carlsbad. Bratten communicated regularly with Elwood Mead, the commissioner of Reclamation. In December 1924 Mead informed Bursum that appropriations for additional Carlsbad storage would not be forthcoming at least until the government conducted additional studies in the valley.[27] By early 1925 Reclamation already leaned heavily toward the Alamogordo site as the only feasible storage site available.

In the meantime Tracy contacted Thomas E. Campbell, chairman of Reclamation's Board of Survey and Adjustments in Montrose, Colorado. He exhaustively enumerated reasons why Reclamation should provide Reservoir Number Three and listed past injustices committed on the project.[28] Reclamation conducted hearings on and field inspections of the project during March 18–21, 1925. The inspection team included Campbell, F. M. Goodman, other Reclamation representatives, and Dr. H. L. Kent, New Mexico advisor. Citing the two major problems of the project as shortage of water and impermanency of the water supply, Tracy contended that Reclamation should build the reservoir to recover and store losses from Lake McMillan, arguing that in one year the increased crops generated by the reservoir would be enough to build and pay for the facility.[29] The Board of Adjustments agreed with Tracy as to the serious shortage of water on the Carlsbad Project. It also agreed that a crop failure from lack of water would lead to financial disaster for people living on the project. Furthermore, the board stipulated that the shortage came from leakage at Lake McMillan and that dam site number three—as well as the Fort Sumner site—should again be considered by engineers and geologists. The board recommended immediate authorization by Congress for appropriation from the general Reclamation Fund for studies to take place, noting that the project needed a storage reservoir at the earliest possible time.[30]

Early in January 1924 Miles Cannon, field commissioner for Reclamation, submitted "Government Reclamation in New Mexico, Hondo-Carlsbad Project," dealing with various aspects of the Carlsbad Project. Cannon's report emphasized the inadequacies of project storage, especially at McMillan, where he noted that as early as 1916 silting approached 2,400 acre-feet per year. During the years 1912–1915, the reservoir lost close to

140,000 acre-feet through seepage. In the end, Cannon, like geologists before him, had concluded "there was grave danger of [Reservoir Number Three] not being a success owing to gypsum and broken limestone under the right of way,"[31] and recommended building at the Alamogordo site only. Tracy's dream of a third reservoir remained elusive.

By the early 1920s the federal government had spent close to $2 million rehabilitating and improving the Carlsbad Project. The irrigation works, however, were still besieged by problems of siltation and seepage, salt cedar growth, evaporation, absentee ownership, and additional claims to the waters of the Pecos River. In trying to solve the water needs of both New Mexico and Texas, the states established the Pecos River Compact Commission in 1923. The commission faithfully completed its work, providing a framework for equitable distribution of waters to both New Mexico and Texas. Hesitant to sign the agreement for fear of losing the right to precious waters, New Mexico balked, setting the stage for continued competition over water.

CONCLUSION

ENCOURAGED by grandiose pseudoscientific claims about "rain following the plow," or prodding rain clouds into activity with dynamite, cheap lands, and irrigation, many settlers came west. Settlement of the American West, long linked to the availability of water, focused in eastern New Mexico on surface waters of the Pecos River. Extensive private ventures in the area—at the time reportedly the largest in the world—eventually met failure. Over 90 percent of the private irrigation companies were bankrupt or close to it by 1900 because of poor design and construction, short growing seasons, alkali soils, inadequate drainage, and poor assessment of water availability and flooding.

The advocacy of irrigationists like William E. Smythe combined with new federal support to revive dreams first envisioned in the nineteenth century. In her early work *Desert Passages*,[1] Patricia Limerick discusses a succession of travelers, miners, and appreciators of the desert, finally arriving at irrigationists at the turn of the century. Irrigationists, according to Limerick, unlike those before them, began to "celebrate new opportunities for mastery" of western water through large-scale federal manipulation of river and desert. William Smythe epitomized this celebration of opportunities.

Smythe, editor of *Irrigation Age*, identified seventeen states west of the ninety-seventh meridian as a utopia, awaiting the secret elixir, water. Smythe heralded the climate, resources, and fertility of the lands, claiming that when irrigated, the rich soils of the arid West produced four to ten times as much as eastern soils without irrigation. Whereas others had seen the American West as a safety valve for the teeming populace of eastern cities, Smythe promoted irrigation as a great social movement, which would enhance opportunities for the American middle class.

In New Mexico, the Bureau of Reclamation, a newly created engineering bureaucracy, reluctantly took on the Carlsbad Project. In Carlsbad, one of a handful of the earliest projects, Reclamation found itself facing untold obstacles in engineering, financing, politics, and a harsh natural environment surrounding the place.

At its core, Carlsbad is about the scarcity of water—and the individuals, corporations, and political entities who fought among themselves, established allies, and maneuvered to control it. There was little accommodation for physical, environmental, or economic realities. From its earliest days, the vagaries of the Pecos River often dictated those realities.

Although nineteenth- and twentieth-century histories of the Pecos Valley appear vastly different from each other, the struggle for water is central to both periods. Carlsbad in the nineteenth century is about the heady go-for-broke times in the valley when visionaries and capitalists combined forces to transform the desert. Charles Eddy's prowess as visionary promoter attracted foreign and eastern capital to the desert. Chicagoans, New Yorkers, the English, the Swiss, and displaced easterners in Colorado Springs anted up thousands, then millions, of dollars to develop lands largely acquired through manipulation of the country's land laws. Eastern capital was able to wrest political and economic power from opportunistic Texas cattle rustlers and saloon operators. Easterners who maintained ties with former business associates built an investment structure that largely negated the notion of frontier democracy in the Pecos Valley. Using cheap Mexican labor to build dams and railroads, the various irrigation and railroad enterprises came under the ultimate control of "managers" far removed from James Hagerman's watchful eye. When Charles Eddy departed, Hagerman tried to firmly control his struggling enterprises by appointing new managers, many of whom were unfamiliar with the valley. In the meantime, as costs mounted, nature compounded the problems of salty water, leaky canals and reservoirs, questionable agricultural methods, a needed railroad extension, and less than spectacular settlement. James Hagerman's and, more generally, America's notion of private "can do" initiative washed away with the floods in 1893 and 1904.

In the twentieth century, by the time of the second flood, management had passed to an eastern diehard named Francis Tracy. Obsessed with the notion of expanding the Carlsbad Project, he faced the realities of giving up control to a corps of engineering bureaucrats, cutting their teeth as they went, trying to establish a niche for themselves in the West. Even as Reclamation established itself in the West, Carlsbad represented the conflict between fading private agendas, those of Washington progressives, and those of the local water users, whose agendas were often dictated by corporate and speculative interests, or Washington efficiency experts.

Reclamationists, notably A. P. (Arthur Powell) Davis, the nephew of

John Wesley Powell, Frederick Newell, Elwood Mead, and earlier ideological promoters of the West, including Henry George, Horace Greeley, Nathan Meeker, and Edwin Nettleton, all tried to perpetuate the Jeffersonian myth of small agrarian democracy in the Pecos Valley and across the West. Faced with the reality of corporate speculation in the valley, Reclamation found itself distastefully accommodating local and corporate demands. By a combination of savvy and sheer determination, Francis Tracy, local water users, and one of Roosevelt's Rough Riders caused Reclamation to take on a questionable project. They convinced Reclamation to skirt the 160-acre and residency provisions laid out in the Newlands Act, and they set a precedent for reducing project repayment costs across the West. Although it would be 1987 before the federal government completed a third reservoir, and then only on the basis of multiple use, Brantley Dam and Reservoir culminated a century's worth of badgering and subsequent investigations in the valley.

The learn-as-you-go mentality of the Reclamation Service reflected a wary, unsure measure of its abilities and place in the early days. When the Service began, most irrigation engineers were converted railroad and mining engineers. As the Service practiced on the early projects like Carlsbad, skill and confidence rose to a level enabling engineers to create engineering masterpieces like Hoover Dam. Taking on additional projects, Reclamation responded to local concerns as engineering expertise, appropriations, and political expediency dictated. At Carlsbad, dissatisfied speculators, who controlled the Pecos Water Users Association and other local bodies, demanded improvements and expansion of local projects in order to profit from irrigated land sales. The degree to which Reclamation responded depended on its own set of priorities, the exigencies of each project, and how much political pressure could be brought to bear on the Service by local constituencies.

The result at Carlsbad was a series of crusades for water rights determination and countless surveys and investigations, tempered by hearings and inevitable trips to Washington, D.C. Unlike the "go-go" days of private irrigation, progressive planning took time and prioritization. The slow, churning process of decision making that went against his interests enraged the likes of Frances Tracy, who badgered Reclamation for years. Involved in so many projects needing attention across the West, Reclamation slowly addressed the needs of each project, establishing itself for the long term, setting up local offices and administrations, and trying to resist attempts by corporate and speculative landholders to enrich themselves, sometimes at the expense of small farmers. Contrary to Donald Worster's assertion that the federal government worked in concert with powerful economic elites, the Carlsbad Project nonetheless demonstrates a manipulation of federal law for the benefit of nineteenth-century speculators. Carlsbad reflects the struggle between Reclamation's attempt to stay true to policies adopted under the Newlands Act, given the special nature of the Carlsbad Project, and the political pressure brought to bear by local and national lobby efforts aimed at skirting the law's intent.

Workers finished Brantley Dam in 1987, finally providing for Tracy's dream of a third reservoir. Completion of Brantley marked a milestone in the evolution of the Carlsbad Project. Brantley's concrete center and earthen wings added to the eclectic mixture of features on the project: nineteenth-century canals lined with twentieth-century concrete; Avalon Dam, with its myriad improvements; McMillan, the reservoir built on a sieve; the 1937 Alamogordo Dam; Hondo, a glaring engineering mistake that never held water; and Brantley Dam reflected over one hundred years of irrigation in the Pecos Valley.

EPILOGUE

EARLY in 1923 New Mexico and Texas jointly created the Pecos River Compact Commission following legislation in each state to seek a means for distributing waters equitably among the citizens of the two states. Meeting in El Paso on December 16, 1924, the group agreed on a draft of the Pecos River Compact, and at a later meeting in Santa Fe on February 4, 1925, arbitrators agreed on certain proposed amendments. The commissioners signed the agreement in El Paso on February 10, 1925, and submitted it to the state legislatures for ratification. Major articles of the agreement provided that the compact should not impair existing rights to use the waters of the river and its tributaries. The instrument's basic purpose was to provide for the equitable division of unappropriated flood-waters of the river. The commission disallowed permits for new construction or enlargement of storage facilities within the river's upper basin, the area north of Fort Sumner, in excess of 10,000 acre-feet. Third, and most important to the Carlsbad Project, the middle basin of New Mexico would retain rights in perpetuity to divert and use sufficient water for the irrigation of 76,000 acres of land.[1]

As far as Texas was concerned, the compact's major provision was one that gave Texas the right to divert the natural flow of the river after upstream diversions for agriculture, and to build a reservoir at Red Bluff to serve 40,000 acres in five West Texas counties. The agreement specifically stipulated that nothing in the compact could affect the right of the federal government in the waters of the Pecos River or in the government's Carlsbad Project.[2]

The Pecos Water Users Association and other groups feared the compact might interfere with the right of New Mexico to irrigate. To urge New Mexico to take part in the compact, Texas pushed Congress in 1929 to enact a law (45 Stat. 1502) that emphasized that the consent of Congress was given to the states of New Mexico, Oklahoma, and Texas to negotiate such compacts.

Still New Mexico failed to reconsider the 1925 compact or negotiate a new one. On June 30, 1932, voters of the Pecos Water Users Association voted unanimously for the creation of the Carlsbad Irrigation District. The newly created district had the authority to operate the government's physical plant, issue bonds for improvements to the system, and collect fees from landowners for use of the system.[3]

The Texans next focused their pleas at the national level. Congress provided funds to proceed with Red Bluff's construction, but New Mexicans stipulated in 44 Statute 753 that money be appropriated from the Reclamation Fund, which was depleted, thus blocking the dam again.[4]

By 1933, big-scale government relief spending was well under way for internal improvements. Federal agencies looked on Red Bluff with favor. The Public Works Administration began construction of the Red Bluff Dam and Reservoir. Subsequently, the district bonds were purchased by the Reconstruction Finance Corporation, and over New Mexico's protests construction began in 1934.[5]

All during the controversy of building or not building Red Bluff, the Bureau of Reclamation continued investigating the possibilities for storage of water on the Carlsbad Project. By 1935 definite plans had been prepared for the construction of Alamogordo Dam near Fort Sumner. However, Texas got its chance to block building of the dam, claiming it would adversely effect storage at Red Bluff. New Mexico countered with new efforts to block completion of Red Bluff. In July 1935 Secretary of the Interior Harold Ickes called for a hearing on the matter in Washington, D.C. The secretary appealed to both states to work out their differences, or both New Mexico and Texas would lose the privilege of monies for building any water storage structure.[6]

The result was the Alamogordo Agreement of 1935. Texas withdrew its objection to the Fort Sumner project and New Mexico withdrew its objection to the Red Bluff project. Again, New Mexico failed to make the agreement an official compact. Texas repealed the agreement in 1941 in order to bring suit against New Mexico. Nonetheless, through litigation and compromise the Pecos River Compact was finally signed in 1948 and provided for 76,000 acre-feet of water for the middle Pecos, from Fort Sumner to the Texas line, and 40,000 acre-feet for West Texas.[7]

Immediately following ratification of the second Pecos River Compact, troubles began that continued unabated into the twenty-first century. Unusually dry weather plagued New Mexico in the 1950s and 1960s. In

1988, the U.S. Supreme Court declared that New Mexico had regularly shorted Texas valuable water between 1950 and 1986. The Court decreed that Texas deserved $14 million in cash to make up for the shortfall, and 10,000 acre-feet more water per year than originally stipulated in the compact. Especially vexing was the Court's Special Master's demand that New Mexico repay future underdelivery within six months of the shortfall.[8]

Such demands caused New Mexico to buy available water rights up and down the river. Complicating matters, paying back shortfalls tends to shut down thousands of valuable agricultural acres. And, because of the order in which waters were appropriated over the years, many municipalities stand to lose their water supplies. By retiring enough water rights to supply the increased demand, New Mexico will be forced to reenter the legal arena to make the rules for water sharing more equitable.[9] Although there is much more to say about events and conflict subsequent to 1925 and final adoption of the Pecos River Compact in 1948, I leave that portion of history for a later time.

NOTES

INTRODUCTION

1. Robert M. Utley, *High Noon in Lincoln: Violence on the Western Frontier* (Albuquerque: University of New Mexico Press), 1987.

2. John Wesley Powell, *Report on the Lands of the Arid Region of the United States with a more detailed account of the Lands of Utah* (Boston: Harvard Common Press, 1983), 43–45.

3. Walter Prescott Webb, *The Great Plains* (Boston: Ginn, 1931); Frederick Jackson Turner, "The Significance of the Frontier in American History," in *The Turner Thesis, Concerning the Role of the Frontier in American History*, 3d ed., ed. George Rogers Taylor (Lexington, Mass.: D.C. Heath, 1972); Wallace Stegner, *Beyond the Hundredth Meridian: John Wesley Powell and the Second Opening of the West* (Lincoln: University of Nebraska Press, 1983); Powell, *Report on the Lands;* John Wesley Powell, *The Exploration of the Colorado River and Its Canyons* (New York: Penguin Books, 1987).

4. Webb, *The Great Plains.*

5. Norris Hundley Jr., *The Great Thirst: Californians and Water, 1770s–1990s* (Berkeley and Los Angeles: University of California Press, 1992).

6. William E. Smythe, *The Conquest of Arid America* (New York: Macmillan, 1905).

7. Karen L. Smith, *The Magnificent Experiment: Building the Salt River Reclamation Project, 1890–1917* (Tucson: University of Arizona Press, 1986).

8. Daniel Tyler, *The Last Water Hole in the West: The Colorado–Big Thompson Project and the Northern Colorado Water Conservancy District* (Niwot, Colo.: University Press of Colorado, 1992).

9. Donald Worster, *Rivers of Empire: Water, Aridity, and the Growth of the American West* (New York: Pantheon, 1986).

10. Donald J. Pisani, *To Reclaim a Divided West: Water, Law, and Public Policy, 1848–1902* (Albuquerque: University of New Mexico Press, 1992).

11. William deBuys, *Salt Dreams: Land and Water in Low-Down California* (Albuquerque: University of New Mexico Press, 1999).

12. Dan Flores, *Horizontal Yellow: Nature and History in the Near Southwest* (Albuquerque: University of New Mexico Press, 1999).

13. John Miller Morris, *El Llano Estacado: Exploration and Imagination on the High Plains of Texas and New Mexico, 1536–1860* (Austin: Texas State Historical Association, 1997).

14. James E. Sherow, *Watering the Valley: Development along the High Plains Arkansas River, 1870–1950* (Lawrence: University Press of Kansas, 1990).

15. Donald C. Jackson, *Building the Ultimate Dam: John S. Eastwood and the Control of Water in the West* (Lawrence: University Press of Kansas, 1995).

16. Donald J. Pisani, *Water, Land, and Law in the West: The Limits of Public Policy, 1850–1920* (Lawrence: University Press of Kansas, 1996).

1. VISIONS OF AN IRRIGATION EMPIRE

1. William A. Keleher, *The Fabulous Frontier: Twelve New Mexico Items* (Santa Fe, N. Mex.: Rydal Press, 1945), 180–184; U.S. Department of the Interior, U.S. National Resources Planning Board, *Pecos River Joint Investigation Reports of Participating Agencies* (Washington, D.C.: Government Printing Office, June 1942).

2. U.S. Department of the Interior, Geological Survey, "Hydrographic Survey 243, Report Upon a Reconnaissance of the Pecos Valley (11 March 1889)," by Ralph S. Tarr, U.S. Geological Survey, 1, 2, in U.S. Department of the Interior, Bureau of Reclamation, Record Group (RG) 115, box 443, entry 3, file 651, General Administration and Projects Records, National Archives, Rocky Mountain Region, Denver, Colorado, hereafter cited as Tarr Report. Tarr gives an accurate description of the rivers, which fed the Pecos Valley in the later nineteenth century. Due to climatic changes, oil well drilling, and artesian pumping, these descriptions are important to understand the water supply along the Pecos during the heyday of private irrigation efforts.

3. *Hondo* is a Spanish word meaning "deep." In Spanish the *h* is silent. Long ago the word was spelled *Jondo* and the *j* pronounced as an *h*. T. M. Pearce, *New Mexico Place Names, A Geographical Dictionary* (Albuquerque: University of New Mexico Press, 1968), 71; W. M. Reed, "Reports of Special Agents and Observers, New Mexico Irrigation along Pecos River and Its Tributaries," 1898, 5, Morgan Nelson Papers, Southwest Collection (SWC), Texas Tech University, Lubbock, Texas.

4. Tarr Report, 15–17.

5. The Delaware rises in Texas and joins the Pecos on New Mexican soil. The water is "brackish" as it flows through gypsum. The Peñasco rises in the western part of Eddy County and flows east. Tarr Report, 15–17; Reed, "Reports of Special Agents and Observers," 1, 2, Nelson Papers.

6. W. W. Newcombe, *The Indians of Texas: From Prehistoric to Modern Times*

(Austin: University of Texas Press, 1961), 108, 109, and map 161.

7. Bolton saw the Pecos as the dividing line between the Apache to the west and Comanche to the east, who until the 1870s descended into the Mexican states of Chihuahua, Coahuila, and Durango on raids to gather captives. Herbert Eugene Bolton, *Spanish Exploration in the Southwest* (New York: C. Scribner's Sons, 1916), 320–360.

8. Ernest Wallace and E. Adamson Hoebel, *The Comanches: Lords of the South Plains* (Norman: University of Oklahoma Press, 1951). For more on the Comanche and Apache familiarity with bison hunting lifestyle and how it affected their culture, see Dan Flores, "Bison Ecology and Bison Diplomacy: The Southern Plains from 1800 to 1850," *Journal of American History* 78 (September 1991): 465–485.

9. Randolph Barnes Marcy was born in 1812 in Greenwich, Massachusetts. He graduated from the U.S. Military Academy at West Point in 1832 and spent several years on the Michigan and Wisconsin frontiers. From 1847 to 1859 he was in the West, in 1849 opening a new trail for emigrants from Fort Smith to Santa Fe, and later in Utah fighting against the Mormons. He fought against the Seminoles in Florida and scouted sites for western forts. In 1861 Marcy served under his son-in-law, George B. McClellan, during the Peninsular and Antietam campaigns in the Civil War. Marcy died in 1887 in West Orange, New Jersey. *Dictionary of American Biography,* s.v. "Marcy, Randolph Barnes"; U.S. Congress, House, *Report of Exploration and Survey of Route from Fort Smith, Arkansas, to Santa Fe, New Mexico, Made in 1849, by First Lieutenant James H. Simpson, Corps of Topographical Engineers,* 31st Cong., 1st sess., 1850, Ex. Doc. 45, 1–61; Captain R. B. Marcy, "Journal," in Grant Forman, *Marcy and the Goldseekers* (Norman: University of Oklahoma Press, 1935).

10. U.S. Congress, House, *Report of Exploration,* 1–61; Marcy, "Journal."

11. U.S. Congress, House, *Report of Exploration,* 59–61.

12. Ibid., 59; Marcy, "Journal," quotation on 353.

13. *Report on Exploration,* 60–61; Marcy, "Journal," 355–358.

14. There were no Spanish land grants in the lower part of New Mexico's Pecos Valley as there were in northern New Mexico; thus settlers came to the valley of their own accord. Historians and anthropologists speculated that the Mexicans came from the Manzano Mountains near Albuquerque. This idea is based on oral histories and the fact that some of the earliest settlers claimed to be Penitentes, members of a secret and stringent brotherhood of Catholic men, a group that was very strong in north-central New Mexico. Other Mexican American settlements formed in the valley, including El Berrendo and Rio Hondo. Rio Hondo became a barrio in Roswell known as Chihuahita. Many of these emigrants raised crops to sell at Fort Stanton (forty-five miles west) and at Fort Sumner (fifty-five miles northeast). Many of the Mexicans practiced irrigation through ditches. For more on Missouri Plaza, see James D. Shinkle, *"Missouri Plaza," First Settled Community in Chaves County* (Roswell, N. Mex.: Hall-Poorbaugh Press, 1972), 11–15; Tom Sheridan, *The Bitter River: A Brief Historical Survey of the Middle Pecos River Basin* (Boulder, Colo.: Western Interstate Commission for Higher Education, 1975),

34–36; Malcolm Ebright, "New Mexico Land Grants: The Legal Background," in Briggs and Van Ness, eds., *Land, Water, and Culture*, 281; Ira Clark, *Water in New Mexico: A History of Its Management and Use* (Albuquerque: University of New Mexico Press, 1987), 25; Mario T. Garcia, *Desert Immigrants: The New Mexicans of El Paso, 1880–1920* (New Haven, Conn.: Yale University Press, 1981), 93.

15. Shinkle, *"Missouri Plaza,"* 11–15; Carole Larson, *Forgotten Frontier: The Story of Southeastern New Mexico* (Albuquerque: University of New Mexico Press, 1993), 24–26, 33–34, 59–60, 68–69; Robert J. Rosebaum and Robert J. Larson, "Mexicano Resistance to the Expropriation of Grant Lands in New Mexico," in Briggs and Van Ness, eds., *Land, Water, and Culture*, 281; New Mexico (Territory), Bureau of Immigration, *New Mexico: Its Resources, Climate, Geography, Geology, History, Statistics, Present Condition, and Future Prospects* (Santa Fe, N. Mex.: New Mexican Printing Co., 1894), 245; Keleher, *Fabulous Frontier*, 33; Michael C. Meyers, *Water in the Hispanic Southwest: A Social and Legal History, 1550–1850* (Tucson: University of Arizona Press, 1984), 72; Tarr Report, 26; Smythe, *Conquest of Arid America*, 239.

16. The New Mexico Territorial Legislature adopted numerous memorials that were sent to the U.S. Congress in the 1850s and 1860s detailing the territory's need for transportation, general law enforcement, and military protection against Indians and outlaws, many of the latter coming from Texas. Most of the immigrants into the area were Anglos. Their numbers grew until they composed about one-third of the population. This caused a culture clash. Many Anglos considered the Mexican Americans inferior, lazy, and odoriferous. They described Mexican towns as "dilapidated old adobe buildings, chili, and the smell of garlic . . . everywhere." Whites saw themselves, on the other hand, as progressive, energetic, and superior and were determined to end the Mexican American presence. This conflict led to shootings, assassinations, and other racial violence, including the "Harrell War." The Harrells came to the valley from Lampasas, Texas, after a bloody feud with another family. During a drinking spree that lasted several weeks, family members gunned down several Mexicans near Missouri Plaza for no apparent reason other than their race. Porter A. Stratton, *Territorial Press of New Mexico, 1834–1912* (Albuquerque: University of New Mexico Press, 1969), 24, 132; Sheridan, *Bitter River*, 41.

17. Keleher, *Fabulous Frontier*, 26–27.

18. The first settlers to the area drove off the Indians and called the place "Dogtown" because it was overrun by prairie dogs. Lincoln County was divided in 1889 to form Chaves, Eddy, and Lincoln Counties. Seven Rivers vied with Eddy for the county seat. The Seven Rivers post office closed in 1895, signaling the end of the town. Keleher, *Fabulous Frontier*, 7–23; *Carlsbad Current-Argus*, 20 March 1988; Larson, *Forgotten Frontier*, 27; *History of Eddy County, New Mexico to 1981* (Southeastern New Mexico Historical Society, Carlsbad, N. Mex., Lubbock, Tex.: Craftsman Printers, 1982), 14, hereafter cited as *History of Eddy County*.

19. Apaches, especially the Mescaleros, continued raiding from the Davis and Guadalupe Mountains. By 1855 the United States began a campaign to stop the

raids. For a brief summary of the military operation, see Sheridan, *Bitter River*, 28–32.

20. Charles Goodnight was born 5 March 1836, in Macoupin County, Illinois, moving to Milam County, Texas, in 1846. He entered the cattle business in 1856 and moved to Palo Pinto County in 1857. He served as a scout and guide with the Texas Rangers against the Indians. He and his partner established the Goodnight-Loving cattle trail in 1866 and established a cattle ranch on the Apishapa River in Colorado in 1868. In 1870 Goodnight started a ranch on the Arkansas River four miles above Pueblo. He began a partnership with John Adair of Ireland in 1877 to develop the JA Ranch, which encompassed almost one million acres and fed 100,000 cattle. Goodnight introduced Hereford and Shorthorn cattle into his long-horn herd, and preserved remnants of the Southern buffalo herd by raising calves. *Dictionary of American Biography,* s.v. "Goodnight, Charles"; C. L. Sonnichsen, *The Mescalero Apaches,* 2d ed. (Norman: University of Oklahoma Press, 1973), 113–117; Keleher, *Fabulous Frontier,* 38–39; Paige W. Christiansen, "The Quest for Water in New Mexico," Technical Completion Report Project no. A-026-NMex, New Mexico Water Resources Research Institution and Department of Humanities (Socorro, N. Mex.: New Mexico Institute of Mining and Technology, August 1973), 21; Garcia, *Desert Immigrants,* 112.

21. Sonnichsen, *Mescalero Apache,* 113–117; Christiansen, "The Quest for Water," 21.

22. Christiansen, "The Quest for Water," 21; Sonnichsen, *Mescalero Apache,* 113–117; Keleher, *Fabulous Frontier,* 34; Tarr Report, 22, 24.

23. The Homestead Movement took farmers to places in Texas where only cattlemen had been. Cattlemen began looking west for rangeland. Larson, *Forgotten Frontier,* 27.

24. Ora Brooks Peake, *The Colorado Range Cattle Industry* (Glendale, Calif.: Arthur H. Clark, 1937); For more on the Oliver Loving incident, see J. Evetts Haley, *Charles Goodnight, Cowboy and Plainsman* (New York: Houghton-Mifflin, 1936) and Charles Goodnight, "Starting the Longhorn Westward," *Fort Worth Star Telegram and Sunday Record,* 15 September 1929. For a discussion of Horsehead Crossing, Castle Gap, and other features of the Trans-Pecos in Texas, see Patrick Dearen, *Castle Gap and the Pecos Frontier* (Fort Worth: Texas Christian University Press, 1988).

25. The *Las Vegas Gazette* on 25 November 1875, described Chisum's ranch as stretching one hundred miles, from near Fort Sumner to Seven Rivers. The ranch reached "as far as a man can travel, on a good horse, during a summer."

26. John Simpson Chisum was born 15 August 1824, in Hardeman County, Tennessee. The family moved to Texas in 1837. Chisum became a contractor and built the first courthouse in Paris, Texas. He served as county clerk of Lamar County and in 1854 started in the cattle business, later herding cattle to Concho County. He was one of the first cattlemen to herd cattle into New Mexico, selling to government contractors at Bosque Grande, thirty miles north of Roswell, to feed Indians. Chisum trailed herds along the Pecos from Texas, while partner Charles Good-

night arranged for their sale in Colorado and Wyoming. Chisum established a ranch at South Spring near Roswell in the 1870s and ran as many as 100,000 cattle, although Robert W. Gilbert was issued the first recorded Homestead grant in the area on 2 January 1885. Chisum was one of the first in the area to plant crops. He raised alfalfa and fruit trees. In 1873, following the violence of the Lincoln County War, Chisum was instrumental in helping elect Pat Garrett sheriff. In 1875 Chisum sold out to Hunter and Evans, a St. Louis firm, because he became disgusted with Lincoln County's growing number of cattle rustlers and outlaws, especially from Seven Rivers. Chisum died of cancer in Eureka Springs, Arkansas, on 24 December 1884. He never married and died intestate. His brother, James of Roswell, became the administrator of the estate, which he estimated at $10,000. When he died Chisum owned property and had claims against the U.S. government for Indian depredations. Chisum was survived by brothers James and Pitzer of Paris, Texas, two nephews, Walter and William, and one niece, Sallie Robert. *Dictionary of American Biography*, s.v. "Chisum, John Simpson"; Sheridan, *Bitter River*, 41, 46; David Dary, *Cowboy Culture: A Saga of Five Centuries* (New York: Knopf, 1981), 140; Keleher, *Fabulous Frontier*, 58–60; Tarr Report, 24; Chaves County, Records of Wills, "Petition for Letters of Administration in the Matter of the Estate of John S. Chisum," Roswell, New Mexico.

27. In the Roswell area, three large farms made up the bulk of cultivated acreage. These included the Chisum Ranch, the 1,800-acre Pat Garrett Ranch with 200 to 300 acres in alfalfa and fruit trees, and a third ranch, the Stone Ranch, with 75 acres in alfalfa and 2–3 acres of fruit and vines. Each of the property owners planted 10 acres of timber entirely in cottonwood trees. Sheep rancher Judge Edmund Stone owned the Stone Ranch. Tarr Report, 24; Larson, *Forgotten Frontier*, 78.

28. The Lincoln County War was caused by a combination of Texas outlaws in the Seven Rivers area, cattle rustling, racism, violence, and the transactions of Murphy-Dolan and Company, who held beef contracts at Fort Stanton and the Mescalero Indian Reservation. Murphy-Dolan supplied Lincoln County; they tended to sell their merchandise at inflated prices. The company was reinforced by its association with the Santa Fe Ring, which dominated territorial politics. (The Santa Fe Ring included lawyers, politicians, businessmen, merchants, lawyers, newspapermen, and government officials. The Ring worked together for mutual gain, taking Mexican land grants throughout New Mexico and establishing monopolistic businesses. The leaders of the Ring were Thomas Benton Catron and Stephen Benton Elkins.) In 1877 Dolan-Murphy established a cow camp in Seven Rivers, where they "recruited" beef rustlers. They especially targeted John Chisum's cattle. Chisum himself was also a target of the Santa Fe Ring. John Tunstall, an Englishman who moved to Lincoln County in 1876, threatened the Murphy-Dolan faction by establishing a rival mercantile. The rising competition for cattle, political control, and necessities escalated into violence between the competing groups. Sheridan, *Bitter River*, 40–45; T. Dudley Cramer, *The Pecos Ranchers in the Lincoln County War* (Oakland, Calif.: Branding Iron Press, 1996);

William A. Keleher, *Violence in Lincoln County, 1869–1881* (Albuquerque: University of New Mexico Press, 1957); Maurice Garland Fulton, *History of the Lincoln County War*, ed. Robert N. Mullin (Tucson: University of Arizona Press, 1968); Rosebaum and Larson, "Mexicano Resistance," in Briggs and Van Ness, eds., *Land, Water, and Culture*, 277–280; Malcolm Ebright, *Land Grants and Lawsuits in Northern New Mexico* (Albuquerque: University of New Mexico Press, 1994), 43; Clark, *Water in New Mexico*, 35.

29. Larson, *Forgotten Frontier*, 77.

30. Stretching north to south behind the front ranges of Colorado lie four broad parks encircled by towering peaks. South Park, on the eastern side of the Great Divide, was the former haven of the mountain men and Ute Indians who hunted vast herds of buffalo there. The source of the South Platte River lies here, as did warm mineral salts and springs giving rise to the area's other name, Bayou Salado. At 10,000 feet in elevation, South Park is a near-level plateau and covers 3,000 square miles. One of the first white men to enter the area, James Purcell, from Kentucky, told Lieutenant Zebulon Pike that he had seen gold there in 1803. Pike entered the park in 1806, and in 1844 Lieutenant John C. Fremont crossed the park on his way to California. In 1859 prospectors, moving southwest from Gilpin County, discovered gold in the region and established a handful of mining camps that played out within a few years. The early prospectors were replaced by cattlemen in the 1870s. South Park lies some seventy-five miles west and north of Cañon City and sixty-five miles west of Colorado Springs. Larson, *Forgotten Frontier*, 77; *The Rocky Mountain Directory and Colorado Gazetteer for 1871* (Denver: S. S. Wallihan, 1871), 48–49; Harry Hansen, ed., *Colorado: A Guide to the Highest State*, originally compiled by the Federal Writers' Program of the Works Projects Administration of the State of Colorado (New York: Hastings House, 1970), 6, 383. In the 1860s, during Colorado's gold boom, early prospectors devised a crude wagon road through Ute Pass just west of Colorado Springs to connect the front-range communities with the mining settlements in South Park and California Gulch. Miners increased the use of this road after 1877 when Leadville's silver strikes inaugurated a second boom. Morris Cafky, *Colorado Midland* (Denver: World Press, 1965), 6.

31. Some confusion exists about Amos Bissell. Several of the sources mentioned below say that Amos Bissell was president of Chemical Bank of New York, but the Chemical Bank's official history does not list Bissell as a former president or officer. The president of the Eddy and Bissell Live Stock Company, Amos Bissell, signed the company's Amended Articles of Incorporation in Otsego County, New York. He was listed as a farmer, real estate agent, and produce dealer in "Township Sections of Mini-Biographies," from "History of Otsego County, New York," available at http://www.rootsweb.com/~nyotsego/bios/minimilford.htm. A letter from H. F. Christian (undated letter in H. F. Christian Papers, SWC) says that the Eddy Brothers came from New York with a letter of introduction from George N. Bissell, Amos Bissell's son. Quotation from Eddy Brothers Journal, 1879–1880, Western History and Genealogy Collection, Denver Public Library, Denver

Colorado; Francis G. Tracy Sr., "Pecos Valley Pioneers," *New Mexico Historical Review* 33 (July 1958): 187–204; Keleher, *Fabulous Frontier*, 278; *History of Eddy County*, 26; Lee Myers, *The Pearl of the Pecos: The Story of the Establishment of Eddy, New Mexico and Irrigation on the Lower Pecos River of New Mexico*, compiled from Eddy newspapers between 12 October 1889 and 23 October 1897 (published by the author, c. 1974), v.

32. Harry A. Epperson, *Colorado As I Saw It* (1944?), 48, SWC.

33. Eddy Brothers Journal, 1 January 1880; William McPhee, "The Eddy Brothers," typescript in History File (County) #14, New Mexico State Record Center and Archives, Santa Fe, New Mexico. McPhee gives his own interpretation of journal entries left by John and Charles Eddy during 1879 and 1880. John Eddy was McPhee's grandfather.

34. McPhee, History File #14. See the Eddy Brothers Journal entries for 1 June 1879 to 22 November 1880 in the Denver Public Library.

35. Charles Canterbury, *History of the Fremont County Cattlemen's Association*, 1989, pamphlet located in the Local History Collection, Cañon City Public Library, Cañon City, Colorado; George Everett, *Cattle Cavalcade in Central Colorado* (Denver: Golden Bell Press, 1966), 200.

36. McPhee, History File #14.

37. Paul L. Huntley, *Black Mountain Cowboys and Other Stories* (Cañon City, Colo.: Master Printers, 1976), 167.

38. Everett, *Cattle Cavalcade*, 175.

39. McPhee, History File #14.

40. Huntley, *Black Mountain Cowboys*, 2–8; For more on the mining camps specifically located in South Park, see Norma L. Flynn, *Early Mining Camps of South Park* (published by the author, 1952).

41. According to William Keleher, the brothers had a second Colorado ranch called "El Dorado." Keleher, *Fabulous Frontier*, 278.

42. Eddy Brothers Journal, quoted in McPhee, History File #14.

43. The IM Ranch was located at back of Black Mountain on Badger Creek at the Fremont and Park County line, about eight miles west of the VVN summer headquarters. Ira Mulock and his three sons established the ranch and by 1883 ran 8,000 cattle. The ranch stretched from Cañon City to South Park, then west to Buena Vista, then toward the Arkansas River. The IM was probably the first ranch in the area, while the Eddys' was second. One neighbor was Silas D. Pollock, who ran cattle to north. Pollock was later named postmaster at Alamogordo, New Mexico. M. R. Crylie, "The IM Ranch, 1890–1899, in Everett, *Cattle Cavalcade*, 324–325; Huntley, *Black Mountain Cowboys*, 1–2, 8, 68; Keleher, *Fabulous Frontier*, 286.

44. McPhee, History File #14.

45. Huntley, *Black Mountain Cowboys*, 168; Epperson, *Colorado As I Saw It*, 49.

46. U.S. Congress, House, *Report of the House Committee on the Unlawful Occupancy of the Public Lands*, 48th Cong., 1st sess., 1884, H.R. 1325, *U.S. House Reports*, vol. v., 2, 4, 8; U.S. Congress, Senate, *A communication of the Secretary of*

the Interior to Congress with accompanying papers setting forth the urgent necessity of stringent measures for the repression of the rapidly increasing evasion and violation of the land laws relating to the public land, etc., 47th Cong., 2d sess., Ex. Doc. 61, vol. iii, 29–32; Peake, *The Colorado Range Cattle Industry,* 68–78

47. Huntley, *Black Mountain Cowboys,* 168.

48. Epperson, *Colorado As I Saw It,* 48.

49. Huntley, *Black Mountain Cowboys,* 167.

50. Ibid., 168; Epperson, *Colorado As I Saw It,* 49.

51. Huntley, *Black Mountain Cowboys,* 167.

52. Ibid., 176.

53. Cafky, *Colorado Midland,* 6.

54. Huntley, *Black Mountain Cowboys,* 2.

55. Cafky, *Colorado Midland,* 53.

56. The Eddys involved themselves directly in the Colorado mining industry, although this aspect of their careers occurred after they entered the cattle business. The Eddys owned the Last Chance Mine in the Whitehorn mining district near Salida, which was reached by the Colorado Midland Railway in 1887. In 1893 a 4,000-pound piece of ore taken from the mine was displayed at the World's Fair in Chicago. The Eddys had trouble with partners, and as a result the mine closed for several years until 1902. Huntley, *Black Mountain Cowboys,* 2–8.

57. Both John and Charles Eddy were involved in designing, founding, and building the community of Alamogordo, New Mexico, as a supply center for gold mining activity at White Oaks. Charles Eddy successfully built a rail line from El Paso to Carrizozo following his involvement with James Hagerman in Pecos Valley irrigation. John Arthur Eddy married, had two daughters, and divorced. He lived for many years in Salida, and then ten years in Denver. He died at the age of seventy-six in 1931. *Denver Times,* 11 July 1902; McPhee, History File #14; *Denver Post,* 9 November 1931. For more information on Alamogordo, see Beth Gilbert, *Alamogordo: The Territorial Years, 1898–1912* (Albuquerque, N. Mex.: Starline Printing, 1988).

58. Huntley, *Black Mountain Cowboys,* 176.

59. Huntley, *Black Mountain Cowboys,* 176; Epperson, *Colorado As I Saw It,* 23.

60. Indicative of the Mermods' ineptitude was the fact that they hired far more cowboys than they needed. Many of them apparently hid inside the barn and pretended to work. Still another example is illustrative. The cattle on the Stirrup Ranch were supposed to be counted. The ranch foreman had his men round up about one hundred cattle (out of one thousand) and parade them repeatedly past the brothers' cabin where the brothers sat on the porch, counting the same animals over and over again. Huntley, *Black Mountain Cowboys,* 176; Epperson, *Colorado As I Saw It,* 23.

61. Josephine Hendley Papers, 1862–1977. Microfilm in the SWC.

62. F. Stanley, *The Seven Rivers, New Mexico Story* (Pep, Texas, 1963), 11–12.

63. Ibid.

64. The first Eddy-Bissell headquarters was opposite Seven Rivers on the east

bank of the Pecos. Myers, *Pearl of the Pecos*, v; *Carlsbad Current-Argus*, 23 May 1947; S. I. Roberts, "Fifty Eventful Years," 21 March 1936, 3. Roberts's 3,200-word essay was written as part of the Works Progress Administration's attempt to capture local history during the Great Depression. The essay is located in WPA Files #199, State Records Center and Archives, Santa Fe, New Mexico.

65. Huntley, *Black Mountain Cowboys*, 168.

66. The settlement of John Chisum's estate does not indicate a relationship with Jeff Chisum. For a time the Eddys employed Martin Morose as their trail boss. Morose, and by extension the Eddys, gained a reputation for collecting other people's livestock during the trail drives north. After a time, the Eddys fired him. Morose later organized a successful rustling gang in New Mexico. Lawmen arrested Morose in April 1894 for rustling in Eddy and Lincoln Counties. He was jailed in Juarez, where he filed naturalization papers. Morose's attorney, Thomas Fennessey, had a conflict with "former" gunslinger turned lawyer John Wesley Hardin (who was jailed in Texas for murder in 1878), who filed papers to extradite Morose. Mexican authorities eventually released Morose for a lack of evidence. He was killed by New Mexican lawmen as he crossed the bridge back into the United States in July 1895. Thomas Fennessey left the Pecos Valley after this incident. Hardin became a frequent visitor. He often gambled in nearby Phenix until he accused a fellow player of cheating, grabbed the money, and started to walk off. Another player, Lon Bass, with the help of a six-shooter, persuaded Hardin to put down the money and leave. Epperson, *Colorado As I Saw It*, 48, 50; *Eddy Current,*, 27 April 1895, 13 June 1895, 5 July 1895, 11 July 1895, 15 August 1895; *List of Fugitives from Justice* (Austin: Adjutant General's Office, State of Texas, 1878), in James B. Gillett, *Fugitives from Justice: The Notebook of Texas Ranger Sergeant James B. Gillett* (Austin: State House Press, 1977), 27; Chaves County, Records of Wills, "Petition for Letters of Administration in the Matter of the Estate of John S. Chisum."

67. In the 1880s the cattle industry experienced price declines caused by overproduction. The price per head in 1886 was twelve dollars. By 1889 it was only seven or eight dollars. Coupled with drought and cold winters, the drop in prices forced many small cattlemen out of business, and larger companies restructured. According to Elwood Mead, across the nation many people quit using irrigation as a pretext for land grabbing, and it became real and honest—irrigators used the water to grow feed for cattle. Water, other than from the Pecos, was not readily available. Annual rainfall in the Carlsbad area averages about twelve inches per year, depending on the source. The Bureau of Immigration estimated that the soil needed at least 23.32 inches of rain to make agriculture possible. The Pecos Valley needed close to fifteen inches more. New Mexico (Territory), Bureau of Immigration, *New Mexico: Its Resources,* 10; Elwood Mead, *Irrigation Institutions: A Discussion of the Economic and Legal Questions Created by the Growth of Irrigated Agriculture in the West* (New York: MacMillan Press, 1903), 35–36; Steve Bogener, "Carlsbad Project," in *Project 2000*, forthcoming by the Bureau of Reclamation; Robert W. Larson, *New Mexico Populism: A Study of Radical Protest in a Western*

Territory (Boulder: Colorado Associated University Press, 1974), 52.

68. *Halagüeño* is a Spanish word meaning attractive, charming, bright, rosy, flattering, promising, or encouraging. The ditch at Halagüeño became known as the Southern, or Main, Ditch. Originally, it was to be twenty-five feet wide, four feet deep, and twenty-five miles long. Eddy reduced its size to fourteen feet wide and used it to irrigate his farm. Eddy's ditch later supplied the town of Eddy and some homesteaders. Edwin B. Williams, *Holt Spanish and English Dictionary* (New York: Henry Holt, 1955), 307; "halagüeño," *Larousse Concise Spanish-English Dictionary* (London: Larousse, 1993); Myers, *Pearl of the Pecos*, v, vi; Tarr Report, 30.

69. *Carlsbad Current-Argus*, 23 May 1947, 20 March 1988.

70. Myers, *Pearl of the Pecos*, v; Charles D. Coan, *A History of New Mexico*, vol.1 (Chicago: American Historical Society, 1925), 463; *Carlsbad Current-Argus*, 23 May 1947; F. Stanley, *Seven Rivers*, 9.

71. Joe Nash ran cattle in Pierce Canyon (east of Malaga) in the 1870s. He built a small, temporary irrigation system to take water from the Pecos at Double Crossing (where state highway 128 crosses the river), near what became Harroun Farm. Charles Eddy later acquired the ditch and the land below it and began improvements. It became known as the Hagerman Canal in 1889. Myers, *Pearl of the Pecos*, v, vi; Tarr Report, 30.

72. *History of Eddy County*, 26, 65.

73. The book gave names, the county where the crime was committed, type of crime, date of indictment, and a description of the criminal, if it was available. *List of Fugitives from Justice*, in Gillett, *Fugitives from Justice*; *Carlsbad Current-Argus*, 20 March 1988; Keleher, *Fabulous Frontier*, 26–27.

74. Texas Fever was a disease carried by ticks. It did not affect the longhorns but could kill other cattle.

75. In 1891 Garrett moved to Uvalde, Texas, where he raised racehorses and cultivated a friendship with future vice president John Nance Garner. Garrett moved back to New Mexico at the request of Las Cruces businessmen, and was elected sheriff of Dona Ana County in 1898. He later served as U.S. Collector of Customs at El Paso, using the money earned to return later to raising racehorses and playing poker in and around Las Cruces. On 20 February 1908, Garrett, at the age of fifty-eight, was shot and killed on the road to Las Cruces from his ranch in the Organ Mountains. Keleher, *Fabulous Frontier*, 67–91; Leon C. Metz, *Pat Garrett: The Story of a Western Lawman* (Norman: University of Oklahoma Press, 1974).

76. Maurice Garland Fulton, "Roswell in its Early Years," *Roswell Daily Record*, undated, in Metz, *Pat Garrett*, 150; Keleher, *Fabulous Frontier*, 84.

77. Shinkle, *"Missouri Plaza,"* 11–15; Keleher, *Fabulous Frontier*, 185.

78. A number of irrigation ditches existed in the Roswell area, but relatively few existed south of Roswell toward the state line. These small ditches, including the Pioneer Ditch, with which Pat Garret was associated, irrigated no more than 10,000 acres combined. Irrigators laid out fourteen ditches in the Roswell area, some drawing water from the Berrendo, some from North Spring River, one from

the Hondo, and six from the South Spring River. Lincoln County, Deeds of Record, book G, Carrizozo, New Mexico; Chaves County, Deeds of Record, book 2; *Eddy Argus,* 29 September 1893; Sheridan, *Bitter River,* 55; Tarr Report, 25.

79. Metz, *Pat Garrett,* 152; Clark, *Water in New Mexico,* 88. A flume, or terre plein, is a structure resembling a bridge that conveys diverted river water from one side across to the other.

80. Metz, *Pat Garrett,* 152; New Mexico (Territory), Bureau of Immigration, *New Mexico: Its Resources,* 245.

81. Greene was not popular in circles where he was known. R. A. Kistler of the *Las Vegas Optic* (15 July 1881) called Greene "a stinker," who often borrowed money to "pay his hotel and whiskey bills." Kistler described Greene as "a grinning dwarfish delver in deviltry." *Las Vegas Optic,* 16 July 1881.

82. *History of Eddy County,* 26; *Eddy County News,* 6 June 1947.

83. Myers, *Pearl of the Pecos,* v, vi; Tarr Report, 30.

84. The company incorporated on 31 October 1887. *History of Eddy County,* 26.

85. The new company incorporated under New Mexico Territory's Act of 1887, which allowed companies to use eminent domain to acquire rights-of-way for land or pipelines by condemnation. In *History of Eddy County,* Francis Tracy claimed that the directors were the Eddy brothers, Joseph Stevens, Arthur Mermod, and E. E. Williams. Pisani, *To Reclaim a Divided West,* 105–106; Roberts, "Fifty Eventful Years," 3–5; *Eddy County News,* 6 June 1947; Keleher, *Fabulous Frontier,* 190; *History of Eddy County,* 26, 66; John O. Baxter, *Dividing New Mexico's Water, 1700–1972* (Albuquerque: University of New Mexico Press, 1997), 93.

86. The Pecos Irrigation and Improvement Company derived from the Land and Ditch Company in 1891. *Eddy County News,* 6 June 1947; Sheridan, *Bitter River,* 55; *Lincoln County Leader,* 8 December 1888; New Mexico (Territory), Bureau of Immigration, *New Mexico: Its Resources,* 245; Clark, *Water in New Mexico,* 68, 88.

87. *Eddy Argus,* 25 October 1890.

88. *History of Eddy County;* Keleher, *Fabulous Frontier,* 189–190; *Eddy Argus,* 1 May 1890, 25 October 1890.

89. Keleher, *Fabulous Frontier,* 190.

90. In 1860 only 28,000 acres in New Mexico were irrigated. The number of acres rose to 90,000 by 1879. By 1911, following the Reclamation Act, 450,000 were irrigated, a 400 percent increase from 1879. Christiansen, "The Quest for Water," 27–28; Coan, *A History of New Mexico,* 462.

91. Colorado Springs lay some sixty-five miles from the Eddy's VVN Ranch at Black Mountain, certainly a distance the Eddy brothers traversed easily—especially after the building of the Colorado Midland Railroad through the region in 1887.

92. "History of Otsego County, New York, 1840–1878," 193–196, provided by the Milford Free Library, Milford, New York.

93. *New York Times,* 8 December 1937.

94. Ibid.

95. McPhee, History File #14. McPhee posits that the connection to old-line money in New York and the East Coast allowed Eddy to do what Pat Garrett and other "local" entrepreneurs could not—attract much needed capital to the territory. By extension, Eddy's contacts in the East allowed him to attract capital in Colorado Springs.

96. Tuberculosis is an acute or chronic infection usually caused by inhaling airborne particles expelled from an infected person through coughing. The particles can remain infectious and suspended for a long time. The disease affects more men than women. Neither gender shows symptoms of the disease for up to two years. An undiagnosed or untreated patient may remain in relatively good health for prolonged periods, but is highly infectious. Eventually, symptoms include fever, fatigue, and weight loss, with a morning cough that progressively worsens. The sputum becomes green and purulent, then yellow and mucoid, then blood appears. The disease primarily affects the lungs, but can also involve the gastrointestinal tract, the central nervous system, the genitourinary tract, bones, joints, the mouth, ear, larynx, and bronchi. Medical knowledge and vaccines have virtually eliminated tuberculosis in the United States, but when cases appear the patient is isolated for ten to fourteen days after antibiotics are started and treatment continues for one to two years. *The Merck Manual of Diagnosis and Therapy,* 13th ed. (Rahway, N.J.: Merck, Sharpe, and Dohme Research Laboratories, 1977), 112–124; *The Fortunes of a Decade: A Graphic Recital of the Struggles of the Early Days of Cripple Creek, The Greatest Gold Camp on Earth With Stories of Its Mines, Biographies of the Men Who Made Them, With Many New and Hitherto Unpublished Anecdotes and Incidents of Their Lives.* Written, compiled and published under the Direction of Sargent and Rohrabacher for the *Evening Telegraph,* Colorado Springs, Colorado, October 1900, 82, Western History and Genealogy Collection, Denver Public Library, Denver, Colorado.

97. Marshall Sprague, *Newport in the Rockies: The Life and Good Times of Colorado Springs.* 1961 (Chicago: Swallow Press, 1980), 71–72.

98. William Jackson Palmer was born in 1836 in Kent County, Delaware, later moved to Philadelphia, and worked for coal and railroad companies. From 1858 through 1861, he was private secretary to J. Edgar Thompson, president of the Pennsylvania Railroad. Palmer served in the Civil War although he was a Quaker. After the war, he served as treasurer of the Union Pacific Railroad. Palmer supervised surveys west of the Rio Grande and construction between Sheridan, Wyoming, and Denver. In 1870, he joined the Denver and Rio Grande Railway to work on western and southern connections to Denver. He pushed construction work during the depression of 1870s, and the railroad reached Salt Lake City in 1883. Palmer's railroad connected to the Mexican railroads in the 1880s, and he served as president of Mexican National Railway from 1881 until 1888. Palmer retired from all business interests in 1901. He died near Colorado Springs in 1909.

Palmer's water system continued to provide water for irrigation in Colorado Springs until the mid-1950s. *Dictionary of American Biography,* s.v. "Palmer, William Jackson"; *Colorado Springs Gazette Telegraph,* 10 July 1994.

99. *Colorado Springs Daily Gazette,* 22 January 1885; *Fortunes of a Decade,* 84.

100. Mauly Dayton Ormes and Eleanor R. Ormes, *The Book of Colorado Springs* (Colorado Springs: Dentan Printing, 1933), 268.

101. William A. Otis was the father of Charles A. Otis of Cincinnati, steel magnate and mayor of that city. The Otises were intimately connected to the Pecos Valley irrigation projects through their good friend James J. Hagerman, who supplied iron ore to the Otises from Michigan's Northern Peninsula. Percy Hagerman, *Cheyenne Mountain Country Club: The First Twenty-five Years* (unpublished manuscript, c. 1947), 5–9, Penrose Public Library, Colorado Springs, Colorado; "A History of Cleveland, Ohio," in *Biographical Illustrated,* vol. 2 (Chicago-Cleveland: S. J. Clarke Publishing, 1910), 1026–1027.

102. The first board of directors included B. F. D. Adams, Godfrey Kissel, Arthur Baker, Thomas H. Edsall, Richard H. Hutton, Thomas C. Parrish, Count James Pourtales, E. C. G. Robinson, W. H. Sanford, William R. Yarker, Charles H. White, William J. Wilcox, and Henry L. B. Wills. Cheyenne Mountain Country Club, "Certificate of Incorporation," (Colorado Springs: Out West Printing, 1901), 39–40, Penrose Public Library, Colorado Springs, Colorado; Ormes and Ormes, *The Book of Colorado Springs,* 268; P. Hagerman, *Cheyenne Mountain Country Club,* 8–9, 29.

103. P. Hagerman, *Cheyenne Mountain Country Club,* 22–24.

104. Cheyenne Mountain Country Club, "Certificate of Incorporation," 39–40.

105. Virgil Carrington Jones, *Roosevelt's Rough Riders* (New York: Doubleday, 1971), 35; Edmund Morris, *The Rise of Theodore Roosevelt* (Toronto: Longman Canada, 1979), 620; P. Hagerman, *Cheyenne Mountain Country Club,* 29, 32–33, 36.

106. The story of Stevens's family wealth is fascinating. Stevens family members were among the earliest American colonists, settling in the Boston area about 1670 and participating in the "Boston Tea Party," the American Revolution, and the War of 1812. The elder Stevens attended Columbia Law School in 1861 and 1862 before serving at Harper's Ferry during the Civil War. His mother, Adele Livingston Sampson, born in 1841, met young attorney Frederic Stevens when she was eighteen years old, and married him soon thereafter. The newlyweds moved into a $2 million mansion at Fifty-seventh Street and Fifth Avenue in New York City, where they lived until moving to Newport, Rhode Island, where Mrs. Stevens played a major role in society circles. About 1875 the Marquis de Talleyrand-Perigord, descendant of the prime minister who served under Napoleon, visited New York City. Talleyrand-Perigord had previously married one Bessie Curtis, member of an old Boston family. On meeting the marquis, Mrs. Stevens apparently became infatuated, returning to Paris with him and two of her daughters. In 1882 Mrs. Stevens returned to America seeking a divorce from Frederic, which she finally obtained on the grounds of nonsupport and desertion. In the meantime the

marquis obtained a divorce from his wife, joining the former Mrs. Stevens in Paris where they were married. The marquis's father, apparently pleased with the marriage, ceded to his son the title of the Duc de Dino. The former Mrs. Stevens had secured her vast fortune in such a way that her four children, including Joseph, were legally protected from whatever designs the duke may have had for their fortune. By 1903 the charm of the marriage had worn off, and on 2 April the First Chamber of the Paris Civil Courts pronounced divorce in favor of the duchess, the duke having no representation in court. While married to Stevens, the duchess had four children, three of them girls. Upon the duchess's death of cerebral hemorrhage on July 19, 1912, Joseph Stevens and her son-in-law, Frederick H. Allan, served as executors of her will. *New York Times,* 20 July 1912; *National Cyclopedia of American Biography,* s.v. "Stevens, Frederic William"; *History of the Chemical Bank, 1823–1913* (privately published, 1913), 106–109. For more on Joseph Sampson, see *History of the Chemical Bank.*

107. The company that became Chemical Bank was established in 1823 by six merchants. Originally called the New York Chemical Manufacturing Company, in 1824 the company had its original charter amended to add banking operations to its manufacturing business. The roots of the company go back even further to 1812, when a previous company that would evolve into Manufacturers Hanover—the New York Manufacturing Company—was established in what became New York's financial district. "Chemical Banking Corporation is the leading bank to small business and mid-size companies in New York, New Jersey, and Connecticut . . . and is a world leader in loan syndication and foreign exchange, interest rate and currency swaps and corporate finance service." In January 1996, Chemical gained permission from regulators to merge with Chase Manhattan Bank, making the corporation the second largest banking institution in the country. "History of Chemical Banking Corporation," fact sheet issued by Chemical Banking Corporation, New York, November 1994; *History of Chemical Bank,* 117–162; letter in Christian Papers, SWC.

108. Tracy, "Pecos Valley Pioneers," 187–188.

109. Ibid.

110. Ibid., 188–190.

111. In 1877 the federal government established a plan to grant lands in larger quantities. Through the Desert Land Act, settlers could get 640 acres for $1.25 per acre, but to gain title to the land, the settler had to improve the property through irrigation. For more, see chapter 3.

112. Tracy, "Pecos Valley Pioneers," 187.

113. Dan Flores, *Caprock Canyonlands: Journeys into the Heart of the Southern Plains* (Austin: University of Texas Press, 1990), 55–56.

114. Myers, *Pearl of the Pecos,* 15; Tracy, "Pecos Valley Pioneers," 187–196.

115. John Ganoe, "The Beginnings of Irrigation in the United States," *Mississippi Valley Historical Review* 25 (June 1938): 59–78.

116. The Homestead Law allowed settlers to claim 160 acres of land free, except for filing fees. Settlers had to live on the land and improve it to get clear title.

117. Myers, *Pearl of the Pecos*, 15; *Bolles v. Pecos Irrigation Co.*, 23 N.M. 32 (Supreme Court of New Mexico, 1917); Tracy, "Pecos Valley Pioneers," 196–197; *History of Eddy County*, 27; *Eddy Argus*, 1 February 1890; Eddy County, Records of Patents, book A, 210, Carlsbad, New Mexico.

118. A. T. Andreas, *History of Chicago from the Earliest Period to the Present Time in Three Volumes*, vol. 3, *1871–1885* (Chicago: A. T. Andrews, 1886), 581–582; Keleher, *Fabulous Frontier*, 190.

119. Robert Weems Tansill had two sons, Robert Weems Jr. and Henry Motter Tansill. Robert Jr. died at the age of twenty-six. Sixteen-year-old Henry changed his name to Robert II. The first Robert Jr.'s son was Robert Weems Tansill III. Robert Weems Tansill II, "An Account of the Tansill Family in Eddy County," dictated to J. Evetts Haley, Canyon, Texas, 1965, and later edited by Robert Weems Tansill III, of Ojai, California, for printing in *History of Eddy County*, 172; *Carlsbad Current-Argus*, March 1988; Eddy County, Files and Records of Wills, book 1, 61–65, Carlsbad, New Mexico.

120. Some contend that Greene played almost as important a role in attracting capital to the Pecos Valley irrigation projects as Charles Eddy. S. I. Roberts gives Greene credit for recruiting James Hagerman, R. W. Tansill, Joseph Stevens, William Dominice, Charles Otis, and others. Roberts, "Fifty Eventful Years," 5.

121. *Eddy County News*, 6 June 1947.

122. The *Phoenix Daily Herald*, 6 June 1891, as quoted in *Carlsbad Current-Argus*, March 20, 1988, spoke even more rapturously of Tansill's trip than Tansill did: "As a tenderfoot he [Tansill] plunged into what was three years ago the wild, almost uninhabitable waste of Pecos Valley, New Mexico. When his caravan of eighteen prairie schooners camped at . . . Eddy, Mr. Tansill recognized intuitively that there mingled the land, the waters, and the sunshine which only needed management to create a populous, prosperous country." *Carlsbad Current-Argus*, 20 March 1988.

123. Eddy's sister was married to Edward Campbell Fox, a member of the New York Stock Exchange. Mary Fox, who along with her friend Emma Mermod owned property in the Pecos Valley, served during World War I as president of the Yale Club Auxiliary of the American Red Cross and received a medal and testimonial from President Woodrow Wilson. She died in 1937 in New York at the age of eighty-four. *New York Times*, 8 December 1937; *Eddy County News*, 6 June 1947.

124. *Dictionary of American Biography*, s.v. "Tansill, Robert Weems"; Tansill, "An Account of the Tansill Family," in *History of Eddy County*, 172.

125. Tansill's son contends that his father recruited Joseph Stevens. This has to be a mistake, because Stevens was on board in 1887 and Tansill did not arrive until one year later. It is also questionable whether Tansill had anything to do with recruiting Arthur Mermod, son of the prominent St. Louis jeweler of the same name, given the proximity of the Mermod ranch property to the Eddy cattle operation in Colorado. It is possible that both parties encouraged Mermod to invest his money in the valley. Tansill, "An Account of the Tansill Family," in *History of Eddy County*, 172.

126. Tansill, "An Account of the Tansill Family," in *History of Eddy County,* 172.

127. Ibid. Tansill claims that Stevens invested "$100,000 or so," making him a more important investor than James Hagerman initially. Hagerman soon surpassed everyone connected with irrigation in the valley in terms of capital invested.

128. Tracy, "Pecos Valley Pioneers," 188; Myers, *Pearl of the Pecos,* vi.

129. Ward was an old friend of the family. He owned a large fleet of freight and passenger boats on the Great Lakes, shipping all sorts of material from Buffalo west to Detroit, Milwaukee, Chicago, and other ports prior to rail service to those cities. Hagerman served as a clerk on several of Ward's largest ships, soliciting and arranging to ship passengers and freight from the Vermont Central and Grand Trunk Railroads' railheads to points west on the Great Lakes. Lowry Hagerman, Santa Fe, New Mexico to Mitchell Wilder, Amon Carter Museum, Fort Worth, Texas, 1969, in "Hagerman Family," Vertical File, Colorado Springs Pioneer Museum, Colorado Springs, Colorado; *Denver Republican,* 15 September 1909; Sprague, *Newport in the Rockies,* 103.

130. Hagerman's connection to the Otis family in Cleveland was crucial to supplying such rails. Charles Otis, following a trip to Germany where he learned about the Bessemer-Kelly steel process, opened the largest open-hearth factory in the country in the 1880s. Otis incorporated the American Steel Company and served as president until 1899. Otis and Hagerman became close friends, and Otis was Hagerman's staunchest financial supporter in the gloomy years of the Pecos Valley after 1893. "A History of Cleveland, Ohio," 1027; Alfred D. Chandler Jr., *The Visible Hand: The Managerial Revolution in American Business* (Cambridge: Harvard University Press, Belknap Press, 1977), 259.

131. Hagerman included among his friends and acquaintances Martin Luther Sykes, Albert Keep, Marvin Hughitt, and H. H. Porter of the Chicago and Northwestern Railroad Company, and B. F. Jones, D. J. Morrell, and Tom Scott.

Marcus Alonzo Hanna was born 24 September 1837, at New Lisbon, Ohio. His family moved to Cleveland in 1852. He started out in business working for his father in a grocery and commission firm. In 1867 Hanna entered the Cleveland coal and iron trade. Rapid expansion of his father-in-law's company, Rhodes and Company, led to its reorganization in 1885 under the name M. A. Hanna and Company. In the 1880s he became interested and involved in Ohio Republican politics, and by 1890 Hanna had firmly sponsored William McKinley for Ohio governor. During 1894 and 1895 Hanna devoted his time to electing McKinley president of the United States. As a state and national organizer, Hanna had few equals. In the late 1890s he was an outspoken advocate for railroad and business combinations, which he saw as a natural consequence of business. He also believed strongly in the right of labor to organize and saw negotiating with single representatives advantageous in avoiding work shutdowns. Hanna was touted as a possible Republican candidate for president in 1904. He died in February 1904. *Dictionary of American Biography,* s.v. "Hanna, Marcus Alonzo"; James Hagerman, "Memoirs of His Life,

Written by himself at Roswell, New Mexico," 1908, Hagerman Family Collection, Rio Grande Historical Collections/Hobson-Hutsinger University Archives, New Mexico State University Library, Las Cruces, New Mexico; "Hagerman Family."

132. Hagerman was extremely ill during 1881 and 1882. Commenting on his illness, Hagerman remembered being sick all the time: "I never saw a well day from start to finish. In December, 1881 . . . I went to New York and Philadelphia, [and] I caught a terrible cold. We rushed around trying to finish up our business . . . when I woke one morning I coughed a little and up came a mouthful of blood . . . the next morning in the sleeper I had a dreadful hemorrhage, several of them." Hagerman proceeded to Cleveland where he was cared for by Mrs. Hagerman and Marcus Hanna's wife and a doctor. Hagerman stayed in a hotel for a month before moving into the Hanna home for two more months. To convalesce, Hagerman moved to a small hotel near Kittrel, North Carolina, in the midst of a pine forest. He improved, and in May 1882 traveled to New York where, under advice of "the best doctor [he] could find," he sailed for Europe in July. In London, a Dr. Andrew Clark advised Hagerman and his family to go to Davos, Switzerland, where they remained until September when they went to northern Italy and later to Nice. They returned to the United States in 1884. Hagerman, "Memoirs of His Life," Hagerman Family Collection; "Hagerman Family"; John J. Lipsey, *The Lives of James John Hagerman, Builder of the Colorado Midland Railway* (Denver: Golden Bell Press, 1968), foreword to 27.

133. Larson, *Forgotten Frontier*, 254; Lipsey, *Lives of James John Hagerman*, 27–28; *Denver Republican*, 15 September 1909.

134. Hagerman was a "small pepperish person." He weighed about 120 pounds and loved an argument. In fact, his friends claimed that he felt better when he was angry about something. While Hagerman associated with business titans, he was not known for his tact. He was especially fond of inviting people to "migrate to a climate too hot for comfort." Sprague, *Newport in the Rockies*, 103–107, quotations on pages 104 and 106.

135. F. W. Cragin, *Early Far West Notebook*, vol. 14, 126–128, F. W. Cragin Papers, Colorado Springs Pioneers Museum, Colorado Springs, Colorado; Ormes and Ormes, *The Book of Colorado Springs*, 84, 125, 199, 200.

136. Jay Gould was born on 27 May 1836, in Roxburg, New York. Gould started his business career as a surveyor in several New York, Michigan, and Ohio counties. In 1857 Gould began a leather tanning business in northern Pennsylvania. Around 1860 he began speculating in several small railroads. With $25,000,000 made largely through stock watering of the Erie and New York Central Railroad, Gould turned west. He became director of the Union Pacific in 1874 and eventually forced a merger with the Kansas Pacific at par. He sold his Kansas Pacific stock and made several million dollars. By 1890 he owned the Missouri Pacific system (5,300 miles), the Texas Pacific (1,499), St. Louis-Swerton (1,222), and the International and Great Northern (825), one-half the mileage in the southwest. Gould died of tuberculosis in December 1892, at age fifty-seven. Sprague, *Newport in the Rockies*, 104–107; *Dictionary of American Biography*, s.v. "Gould, Jay."

137. Because of the dangerous conditions, several men died during the two years it took to complete the railway. Sprague, *Newport in the Rockies*, 104–107; Hagerman, "Memoirs of His Life."

138. Hagerman, "Memoirs of His Life"; Cafky, *Colorado Midland*, 8–9.

139. John J. Lipsey, *Colorado's Almost Forgotten Tycoon*, Vertical File, Colorado Springs Pioneer Museum.

140. Miners located the Mollie Gibson in 1880 at the base of Smuggler Mountain, making it one of the first mining locations in Aspen. Ore less than thirty feet from the surface yielded appreciable amounts of silver, and interests in a neighboring mine, the Lone Pine, launched a dispute that lasted two years in the courts. The two claims consolidated into one property, the Mollie Gibson. *Frank Leslie's Illustrated Newspaper*, no. 1797, vol. 70, 22 February 1890, Denver Public Library.

141. Lowe's family had connections to the famous Cripple Creek mining district near Colorado Springs. In April of 1901, several years after the Mollie Gibson controversy, someone in Colorado Springs warned Theodore Lowe, Henry's father, to stay away from the city unless he wanted to be killed—poisoned to be exact. The letter(s) were signed "An Enemy." Theodore Lowe died one year later in Nevada, where he was a civil and mining engineer and a deputy mineral surveyor. He found pay mineral at Cripple Creek in 1881. *Denver Times*, 11 September 1901, 20 February 1902; *History of the Chemical Bank*, 93, 163; *Rocky Mountain News*, 20 October 1880; Ormes and Ormes, *The Book of Colorado Springs*, 116. Hagerman, writing to his sons, tells how Lowe "performed the duties of this trust and how he repaid the help I gave him and his family. . . . [T]he history of this infamy I leave with this paper as it may sometime be necessary for you to refer to it." The note appended by Hagerman's son, Percy, to the bottom of these thoughts on "Lowe's infamy," says "(Note. This paper is also missing. It is better that the whole episode should be forgotten. P.H.)." This is unfortunate, for the full story would be enlightening. Hagerman, "Memoirs of His Life."

142. Hagerman mentions no deception on his part: "When we returned from our trip . . . in 1891, I not only had the Lowe episode to contend with, but also the Wheeler, Gillespie and Shear suits to fight. The wonderful success of the Mollie Gibson Mine had made this trio of conspirators anxious to get back the stock which they had sold me when they thought it would never be valuable." Lipsey, *Lives of James John Hagerman*, 117–152; Hagerman, "Memoirs of His Life."

143. Wheeler intercepted a number of Western Union telegraph messages encoded and sent between the other major investors in the Mollie Gibson, especially between Hagerman and co-equal investor Richard J. Bolles. For more on Hagerman's colorful life as well as a complicated but engrossing study of the Mollie Gibson investment controversy, see Lipsey, *Lives of James John Hagerman*, especially pages 117–152.

144. The Hagerman family retained control of the mine throughout the 1890s. The company's directors for 1898 included Hagerman, his sons Percy and Herbert, William O'Brien, H. A. Burlien, and W. F. Greenwood. In the early 1890s, the Mollie Gibson was the most valuable silver mining property in Colorado. *Facts of*

Colorado Springs, vol. 3, 19 February 1898, 17, Penrose Public Library, Colorado Springs, Colorado; *Colorado Mines* (Denver: Carson, Hurst, and Harper Art Printers, undated), Western History and Genealogy Collection, Denver Public Library, Denver, Colorado.

145. Hagerman believed pro-business government policies benefited society and accepted the class/economic society. He was idealistic, and at times altruistic. He had a distinct personal style and was absorbed by financial details. He was characterized as a man who succeeded at what he set out to do no matter how great the effort or risk involved. Hagerman, "Memoirs of His Life"; Lipsey, *Lives of James John Hagerman,* 16; Larson, *Forgotten Frontier,* 253–254.

146. New Mexico (Territory), Bureau of Immigration, *New Mexico: Its Resources,* 246; Percy Hagerman, *James John Hagerman: A Sketch of His Life,* 1932, in Lipsey, *Lives of James John Hagerman,* 221, and in Hagerman Family Collection.

147. Hagerman, "Memoirs of His Life."

148. Williams worked for Chemical Bank from 1842 until 1903. He served as president from 1878 until 1903. Quotation from James Hagerman, "Statement," c. 1900, Hagerman Family Collection; *History of the Chemical Bank,* 93, 163; *Rocky Mountain News,* 20 October 1880; Ormes and Ormes, *The Book of Colorado Springs,* 116.

149. Hagerman, "Memoirs of His Life"; Lipsey, *Lives of James John Hagerman,* 16.

150. *New York Times,* 27 March 1917.

151. *Fortunes of a Decade,* 114; Ormes and Ormes, *The Book of Colorado Springs,* 91–92; *Eddy Argus,* 7 June 1890; quotation from *Colorado Springs Gazette,* 26 March 1917.

152. P. Hagerman, *Cheyenne Mountain Country Club,* 39–40.

153. Bonbright was part owner of the largest dry-goods wholesaler in Philadelphia, Hood, Bonbright, and Company. The company grew quickly because of its owners' railroad connections and traveling salesmen. John Wanamaker bought the enterprise in the 1870s and began the first department store. *Fortunes of a Decade,* 114; Ormes and Ormes, *The Book of Colorado Springs,* 91–92; *Cripple Creek Weekly Journal,* 8 April 1894; *Eddy Argus,* 7 June 1890. For more on Bonbright, see Norman Scott Brien Gras and Henrietta M. Larson, *Casebook in American Business History* (New York: F. S. Crofts, 1939), 495–496.

2. CREOSOTE AND SAGEBRUSH

1. *History of Eddy County,* 65; Tracy, "Pecos Valley Pioneers," 190.

2. The term *speculator* is usually pejorative, but no precise definition exists. It generally means someone who bought large amounts of land and sold it quickly— for profit. They acted as brokers, buying large tracts and selling smaller ones to individual settlers. Speculation often meant poor use of land and put heavy tax burdens on farmers because many land companies were slow to pay taxes. Some

resisted paying at all and used the courts to stop expenses for building roads or other construction. The companies did not care about tax penalties or tax titles, because they could later settle with poor county boards or have tax titles set aside by courts. This left farmers to build schools, roads, and local railroads without the help of big landowning companies. Taxes never went down, and they drained farmers during economic downturns and forced farming practices that drove prices down and caused soil depletion, erosion, and decreased land values. Most Americans opposed any laws that would restrict their right to accumulate property; so speculation continued because it was legal. Paul W. Gates, *The Jeffersonian Dream: Studies in the History of American Land Policy and Development*, ed. Allan G. and Margaret Beattie Bogue (Albuquerque: University of New Mexico Press, 1996), 17, 108; Peter Temin, *Causal Factors in American Economic Growth in the Nineteenth Century* (London: MacMillan Press, 1975), 22–23.

3. The system did not work well. By 1880 the Preemption Act was a means for speculators to get land—they paid people to swear falsely and obtain land. W. A. J. Sparks established in 1885 that 75–90 percent of all preemption claims were fraudulent. Thomas LeDuc, introduction to *The Public Lands: Studies in the History of the Public Domain*, ed. Vernon Carstensen (Madison: University of Wisconsin Press, 1963), 46; Fremont P. Wirth, "The Operation of the Land Laws in the Minnesota Iron District," in Carstensen, ed., *The Public Lands*, 104, 107; Roy M. Robbins, *Our Landed Heritage: The Public Domain, 1776–1970*, 2d ed. (Lincoln: University of Nebraska Press, 1942), 89–91; Mead, *Irrigation Institutions*, 16; Clark, *Water in New Mexico*, 48; Tracy, "Pecos Valley Pioneers," 189.

4. According to the Northwest Ordinance of 1785, western land was to be divided in a cadastral survey—into ranges, townships, and sections. A township is six square miles, divided into thirty-six sections, each a mile square; the sections are numbered from one to thirty-six beginning in the northeast corner, moving west to section six, then south to section seven, east to twelve, and so forth. Section thirty-six is in the southeast corner. Ranges are numbered east and west from the principal meridian, and townships are numbered north and south. Earl G. Harrington, "Cadastral Surveys for the Public Land of the United States," in Carstensen, ed., *The Public Lands*, 36; Paul Wallace Gates, "The Homestead Law in an Incongruous Land System," in Carstensen, ed., *The Public Lands,* 315; Mark Reisner, *Cadillac Desert: The American West and Its Disappearing Water* (New York: Viking Press, 1986), quotation on 45–46; Fred A. Shannon, "The Homestead Act and the Labor Surplus," in Carstensen, ed., *The Public Lands*, 298; Clark, *Water in New Mexico*, 48; quotation from Tracy, "Pecos Valley Pioneers," 189.

5. Victor Westphall, "The Public Domain in New Mexico, 1854 to 1891," *New Mexico Historical Review* 33 (January 1958): 29–31.

6. Applicant under the Desert Land Act had four ways of watering their land for uses other than grazing purposes. First, they could build a ditch and water their own land. Second, they could join with other owners of land in the area to form a cooperative. Third, they could form a state-sanctioned irrigation district with a

fairly large number of members. Or fourth, they could purchase water from an irrigation company at a specified rate. For applicants who actually planned to settle in New Mexico, it was option four, the purchase of water from an irrigation company, that held sway in the Pecos Valley during the 1890s. Robbins, *Our Landed Heritage,* 219–220; Reisner, *Cadillac Desert,* 43, 44, 46; John Ganoe, "The Desert Land Act in Operation, 1877–1891," *Agricultural History* 11 (1937): 146; Victor Westphall, "The Public Domain in New Mexico, 1854–1891," *New Mexico Historical Review* 33 (April 1958): 132.

7. One of the early Pecos Valley investors later testified that some of the ditches in the area served only "to prove up a lot of the land under the desert entry. . . . [The ditches were used] in the very slightest way." Congress also passed the Timber Culture Act, which required the applicant to plant one quarter (40 acres) of 160 acres (a quarter section) in trees to increase rainfall. The Timber Culture Act was repealed in 1890. Mead, *Irrigation Institutions,* 16, 17; Reisner, *Cadillac Desert,* 44, 45; *Hearing in the Matter of the Protest of the Pecos Irrigation Company to the Honorable Secretary of the Interior against the Construction of the Hondo Reservoir by the Reclamation Service,* Roswell, New Mexico, 6 September 1904, 8, U.S. Department of the Interior, Bureau of Reclamation, RG 115, box 1, entry 4.

8. Paul Gates sees the opportunity for abuse not only in the Desert Land Act, but also in the Timber Culture and Timber and Stone Acts, and encroachment onto Indians lands. Laws allowed dummy entrymen to enter thousands of acres and later assign the lands to corporate interest and speculators across the West. See Gates, *History of Public Land.*

9. In 1877 alone, the General Land Office disposed of 4,850,000 acres. In 1884 it disposed of 27,530,000 acres. The average between 1877 and 1884 was 2,000,000 per year. Harold H. Dunham, "Some Crucial Years of the General Land Office, 1875–1890," in Carstensen, ed., *The Public Lands;* Gates, "Homestead Law," in Carstensen, *The Public Lands,* 318; Mead, *Irrigation Institutions,* 16; Clark, *Water in New Mexico,* 49–51.

10. *Eddy County News,* 6 June 1947; Westphall, "The Public Domain" (April 1958), 130–131.

11. Westphall, "The Public Domain" (April 1958), 130.

12. Reisner, *Cadillac Desert,* 46; Mead, *Irrigation Institutions,* 36–37.

13. Westphall, "The Public Domain" (January 1958), 30–31; Michael C. Robinson, *Water for the West: The Bureau of Reclamation, 1902–1977* (Chicago: Public Works Historical Society, 1979), 8; Larson, *Forgotten Frontier,* 76–79.

14. Westphall, "The Public Domain" (January 1958), 30–31.

15. Clark, *Water in New Mexico,* 48–49.

16. Westphall, "The Public Domain" (January 1958), 33.

17. Ibid., 39–40.

18. Ibid., 42–45.

19. Tarr Report, 8–9.

20. U.S. Congress, House, *Aliens Owning Lands in the United States,* 49th Cong., 1st sess., 1885–1886, *U.S. House Reports,* no. 1951.

21. Gates, "The Homestead Law." See also Allan G. Bogue and Margaret Beattie Bogue, "'Profits' and the Frontier Land Speculator," in Carstensen, ed., *The Public Lands,* 373–374. The Bogues tend to reject the concept of speculator as evil bogeyman bent on making huge monetary profits and question the impreciseness of the words "profits" and "speculator." Speculation in the West usually was a losing business. Ventures were a gamble involving much risk. Figures for speculation profits are sketchy at best. See also U.S. General Land Office, *Annual Report, 1883* (Washington, D.C.: Government Printing Office, 1883), 6–8.

22. Robbins, *Our Landed Heritage,* 271–272.

23. Frost owned 1,840 acres in the Pecos Valley, which he bought in 1885. Frost was later forced to resign his position. U.S. Congress, Senate, *Fraudulent Acquisition of Titles to Lands in New Mexico,* 48th Cong., 2d sess., 1884–1885, Ex. Doc. 106, vol. 2, serial 2263; Westphall, "The Public Domain" (April 1958), 138–139; Ebright, *Land Grants and Lawsuits,* 214; Eddy County, Deeds of Record, book B, 299.

24. Clark, *Water in New Mexico,* 50–53.

25. Westphall, "The Public Domain" (January 1958), 45–56.

26. Westphall, "The Public Domain" (April 1958), 136–137; Clark, *Water in New Mexico,* 50.

27. U.S. Congress, House, *Adjustment of Private Land Claims in New Mexico,* 49th Cong., 1st sess., 1885–1886, Ex. Doc. 209, vol. 33, serial 2401.

28. *White Oaks Golden Era,* 25 December 1884.

29. Prosecutors tried 637 cases of fraud between 1883 and 1889 in New Mexico. The number should have been greater, but the territory could not locate the defendants. *White Oaks Golden Era,* 10 January 1884; Clark, *Water in New Mexico,* 54.

30. Paul W. Gates, *History of Public Land Law Development* (Washington, D.C.: Government Printing Office, 1968), 477; Robbins, *Our Landed Heritage,* 291–295.

31. George Washington Julian was born in 1817, near Centerville, Indiana. An attorney, in 1845 he was elected to the Indiana state legislature. Julian was an abolitionist and was elected to Congress in 1848 as a Free Soiler. Elected as a Republican to Congress in 1860, he was reelected four times. Julian served as chairman of the Committee on Public Lands and played a major role in passage of the Homestead Act in 1862. Julian joined other radical Republicans after the Civil War and supported the right of women to vote. After 1870 he devoted his time to various reform measures, including land reform in New Mexico under President Grover Cleveland. He died in 1899. *Dictionary of American Biography,* s.v. "Julian, George Washington."

32. George W. Julian, "Land Stealing in New Mexico," *North American Review* 145 (1887): 2, 28–29.

33. Julian, "Land Stealing," 2, 28–29; Clark, *Water in New Mexico,* 49.

34. Julian, "Land Stealing," 29.

35. *White Oaks Golden Era,* 15 October 1885.

36. Ibid.

37. Ibid., 5 November 1885.

38. U.S. Congress, House, *Annual Report of Governor of New Mexico, 1885,* 49th Cong., 1st sess., 1885–1886, Ex. Doc. 1, vol. 12, serial 2379, 1005.

39. *Annual Report of Governor of New Mexico, 1885,* quotation on 1008.

40. Hundreds of ditch companies formed with eastern capital to reclaim the land. Few survived more than ten years. Robert Tudor Hill, *The Public Domain and Democracy: A Study of Social, Economic, and Political Problems in the United States in Relation to Western Development* (New York: Columbia University Press, 1910), 185–186; Clark, *Water in New Mexico,* 54; Reisner, *Cadillac Desert,* 113; Robbins, *Our Landed Heritage,* 326–327.

41. In 1884 the Eddy-Bissell Cattle Company was one of the going concerns along the middle Pecos River valley of New Mexico. Owning a ranch of 20,000 acres, in the spring of 1884 the company branded 9,000 head of cattle near Seven Rivers. These cattle Charles Eddy turned loose on the east side of the Pecos, while rounding up some 2,500 head and sending them north to his Colorado ranch in April. *White Oaks Golden Era,* 15 May 1884; Larson, *Forgotten Frontier,* 77.

42. Myers, *Pearl of the Pecos,* 141.

43. Deposition of Marshal A. Upson, in *White Oaks Golden Era,* 17 December 1885; "Roosevelt Hearings," in McPhee, History File #14.

44. *White Oaks Golden Era,* 17 December 1885; Keleher, *Fabulous Frontier,* 90; "Roosevelt Hearings," in McPhee, History File #14.

45. *White Oaks Golden Era,* 17 December 1885.

46. Ibid.

47. Ibid.

48. Ibid.

49. Ibid.

50. William Slaughter, Charles H. Slaughter, and John B. Slaughter were all active in Eddy County land trades. See Eddy County, Records of Patents, book A, Deeds of Record, books B, 1, 7, Warranty Deeds, books A and B; *White Oaks Golden Era,* 17 December 1885.

51. *White Oaks Golden Era,* 17 December 1885.

52. Eddy County, Deeds of Record, book A, 2–10, 78–86, 110.

53. Westphall, "The Public Domain" (April 1958), 129–130.

54. U.S. General Land Office Records, *Reports of Registrar Edmond G. Shields and receiver James Brown, Las Cruces Land Office,* 50th Cong., 2d sess., 4 October 1888, H. Ex. Doc. 1, serial 2636, 76.; quotation from Westphall, "The Public Domain" (April 1958), 130.

55. Ganoe, "The Desert Land Act in Operation,"146; Westphall, "The Public Domain" (April 1958), 132. According to some sources, the Pecos Valley immediately felt the impact of the "Big Die." See New Mexico (Territory), Bureau of Immigration, *New Mexico: Its Resources,* 10; Mead, *Irrigation Institutions;* Larson,

New Mexico Populism, 52. Other sources indicate that the immediate economic impact of the "Big Die" was not as great in New Mexico and losses of cattle not quite so heavy. A cattle depression hit New Mexico in the 1890s, causing Desert Land entries to drop. Westphall, "The Public Domain" (April 1958), 132; Ganoe, "The Desert Land Act since 1891," *Agricultural History* 11 (1937): 267.

56. Westphall, "The Public Domain" (April 1958), 132.

57. Ibid.; Ganoe, "Desert Land Act in Operation," 146–147. See Eddy County, Records of Patents and Deeds of Record.

58. Ganoe, "Desert Land Act in Operation," 149–150; Westphall, "The Public Domain" (April 1958), 135. See Eddy County, Deeds of Record, book 1, for entries from 1888 through the 1890s.

59. Tarr included the history and general description of the valley, who owned the land, and abuses of federal lands laws. He detailed the water supply of the Pecos and its tributaries. He described proposed water usage by the Pecos Irrigation and Improvement Company and other irrigation companies along the Pecos and Hondo Rivers, and discussed the difficulty of controlling the Pecos and possible reservoir sites in the valley. "Synopsis of Tarr Report," I-IV, in Tarr Report.

60. See Eddy County, Records of Patents, books A and B, Deeds of Record, books 1–4; Tarr Report, 9.

61. Witnesses who verified the claim were B. A. Nymeyer, G. W. Witt, and F. W. Stevens of New York. *Eddy Argus,* 19 October 1889, 2 November 1889, 28 November 1889. See Eddy County, Records of Patents, books A and B, and Deeds of Record, books 1–4.

62. Fennessey later served as Eddy County clerk, a convenient position given his connections to Charles Eddy's land company.

63. See Eddy County, Records of Patents, books A and B, Deeds of Record, books 1–4; Tarr Report, 9.

64. Tarr Report, 9–10.

65. Tarr Report, 10.

66. Hagerman, "Statement," c. 1900, Hagerman Family Collection. See Eddy County, Records of Patents and Deeds of Record.

67. *Eddy Argus,* 13 December 1890.

68. *Eddy Argus,* 13 December 1890.

69. *Eddy Argus,* 13 December 1890.

70. *Eddy Argus,* October 1889.

71. In 1887 the New Mexico Territory allowed irrigation companies the right to use ditch rights-of-way for their operations. *Eddy Argus,* 28 February 1891.

72. See Eddy County, Deeds of Record, books 1 and 2.

73. Eddy County, Records of Patents, book A, 74, 358, 406, 544; Deeds of Record, book 1, 14–50; book 3, 476; book 5, 273, 280, 332; book 6, 122; book 7, 443, 550, 554, 573, 597, 599, 601; book 8, 190.

74. Ibid.

75. Eddy County, Deeds of Record, book A, 2–10, 78–86, 110.

76. Eddy County, Deeds of Record, book A.

77. Eddy County, Mortgage Records, book 1, Water Rights, 1891 and 1892.

78. Powell, *Report on the Lands;* Gates, "Homestead Law," in Carstensen, ed., *The Public Lands,* 318; Mead, *Irrigation Institutions,* 16; Clark, *Water in New Mexico,* 49–51.

79. Powell, *Report on the Lands;* Reisner, *Cadillac Desert,* 47.

80. Everett W. Sterling, "The Powell Irrigation Survey, 1888–1893," *Mississippi Valley Historical Review* 27 (December 1940): 421–434.

81. Surveys increased the selling price of land tenfold. Mead, *Irrigation Institutions,* 20.

82. Sterling, "The Powell Irrigation Survey," 422, 423; Robbins, *Our Landed Heritage,* 327; *U.S. Statutes at Large,* Act of 2 October 1888, vol. 25, sec. 1, 526.

83. Sterling, "The Powell Irrigation Survey," 423.

84. Ralph Tarr to Captain C. E. Dutton, U.S. Geological Survey in Embudo, New Mexico, 11 March 1889, in Tarr Report.

85. U.S. General Land Office, *Annual Report, 1890* (Washington, D.C.: Government Printing Office, 1890), 71–74.

86. Sterling, "The Powell Irrigation Survey," 425; U.S. Congress, Senate, 51st Cong., 1st sess., 2 July 1890, S.R. 1466, 33, shows a large increase in land patents issued in 1889 and 1890.

87. See U.S. Department of the Interior, U.S. Geological Survey, *Eleventh Annual Report, 1890,* pt. 2, 111–200; *Twelfth Annual Report, 1891,* pt. 2, 1–212; *Thirteenth Annual Report, 1892,* pt. 3, 451–478, for complete descriptions of approved sites.

88. The U.S. Geological Survey isolated 147 possible sites in the West, which would water 1.8 million acres of land. U.S. General Land Office, *Annual Report, 1891* (Washington, D.C.: Government Printing Office, 1891), 51.

89. Clark, *Water in New Mexico,* 61.

90. Charles Eddy to Governor L. Bradford Prince, 7 July 1890, in Governor L. Bradford Prince, Papers, Record Group: Letters Received, New Mexico State Records Center and Archives, Santa Fe.

91. Hagerman to Florence(?), 21 September 1890, Hagerman Family Collection.

92. Senator Stewart, as head of a select committee on irrigation, accused Powell of misappropriating funds and sought to turn other western senators against him. Leading the fight to protect Powell and the survey was John H. Reagan of Texas, minority leader on the committee, who saw the incorporation of irrigation companies in New Mexico as a portent of landlordism and a threat to the homesteading concept on the public lands. Others, however, saw no danger in corporate reclamation. Pisani, *To Reclaim a Divided West,* 163; Sterling, "The Powell Irrigation Survey," 432; *U.S. Statutes at Large,* XXVI, 391; Gates, *History of Public Land Law Development,* 484; Gates, *The Jeffersonian Dream,* 108.

93. Powell, *Report on the Lands,* iii–xxiii.

94. The law to strengthen the Desert Land Act was 26th U.S. Statute, 1095–1103. Harry N. Scheiber, *U.S. Economic History, Selected Readings* (New York: Knopf, 1964), 262–264; Robinson, *Water for the West,* 8.

95. New Mexico (Territory), Bureau of Immigration, *Eddy County, New Mexico, the Most Southeastern County in the Territory* (Santa Fe, N. Mex.: J. S. Duncan, 1903), 41.

96. Ganoe, "The Beginnings of Irrigation," 75–77.

97. Tracy, "Pecos Valley Pioneers," 189–190.

98. Ganoe, "The Desert Land Act in Operation," 146–149.

99. Ganoe, "The Desert Land Act since 1891," 266.

3. JACKRABBITS, CANALS, AND COTTONWOODS

1. Nymeyer immigrated to the United States from Holland. He came to New Mexico in 1887 as a partner to J. O. Cameron, a lawyer from Texas. Nymeyer served as county surveyor for thirty years, and he, along with Cameron, operated law, land, insurance, and collection businesses in Eddy. Nymeyer died in 1925. Cameron served as a U.S. commissioner. He left Carlsbad in 1912. *Eddy Argus*, 12 October 1889; 4 June 1890; Charles William Lewis Jr., "Early Settlers in Carlsbad, New Mexico and Vicinity," compiled for Southwestern New Mexico Historical Society, November 1976.

2. Pisani, *To Reclaim a Divided West*, 58–159; *United States of America v. Hope Community Ditch, et al.*, District Court, District of New Mexico, #712, Equity, vol. 1 and Hearings at Carlsbad, New Mexico, 8–10 June 1931, 3257.

3. *History of Eddy County*, 26; *Carlsbad Current-Argus*, 20 March 1988.

4. Charles Eddy, Joseph C. Lea, Arthur A. Mermod, Pat Garrett, and Edgar B. Bronson incorporated the Pecos Valley Town Company. *Meeting the Train: Hagerman, New Mexico, and Its Pioneers* (Hagerman, N. Mex.: The Society, c. 1975), 16.

5. *Eddy Argus*, 28 May 1891.

6. Ibid.

7. *Eddy Argus*, 14 February 1891, 22 March 1890, 6 April 1894.

8. *Carlsbad Current-Argus*, 20 March 1888.

9. *Eddy Argus*, 23 November 1889.

10. The foundation for the new hotel was laid in January 1890. *Eddy Argus*, 12 October 1889, 25 January 1890.

11. Finlay came to the valley as the confidential clerk for the construction company of Ward and Courtney. *Eddy Argus*, 15 March 1890.

12. The first paper was the company-owned *Argus*, established to promote investment and settlement in the valley. Between 1879 and 1900, 283 new papers were added to the 11 existing in New Mexico Territory. Almost every town had a newspaper, and politicians usually controlled them. The first paper in the lower Pecos Valley was the *Roswell Valley Register*, established as a Republican paper in 1888. It became the *Roswell Register* in 1890. The *Democratic Roswell Record* was formed in 1891 and lasted until 1899. The Pecos Irrigation and Investment Company, owned largely by Republican Charles Eddy, established the *Argus* in Eddy in 1889. Five other papers also operated in Eddy during the 1890s. They included the

Democratic Current (1892) and the *Daily Eddy Current* (1894). Stratton, *Territorial Press,* 24, 32–33; *Eddy Argus,* 12 October 1889.

13. Fall worked on a number of cases before moving to Las Cruces. He later became secretary of the interior under Warren G. Harding, was implicated in the Teapot Dome Scandal, and served one year in the federal penitentiary. *Eddy Argus,* 25 June 1892; Myers, *Pearl of the Pecos,* 147.

14. *Eddy Argus,* 13 December 1890.

15. *Eddy Argus,* 28 July 1891.

16. *Rocky Mountain News,* 5 June 1892.

17. La Huerta was a private suburb to the north of Eddy. Deed records indicate that John Arthur Eddy and Charles Eddy sold land in La Huerta to A. A. Freeman, Charles McLenathen, Ed Motter, and others. Residents of La Huerta spent over $250,000, used over 8 million feet of lumber in house building, and had over $200,000 worth of improved livestock. *Eddy Argus,* 13 January 1893, 17 February 1893; New Mexico (Territory), Bureau of Immigration, *New Mexico: Its Resources,* 247.

18. Even after the company established the park, citizens continued to cross the footbridge and walk in the area that became known as Hagerman Heights. *Eddy Argus,* 22 March 1892.

19. The Hotel Hagerman was sold to the Schlitz Brewery in 1908, then to the Bates Brothers in 1912. It burned in 1916. *History of Eddy County,* 1, 28; *Eddy Weekly Current,* 21 April 1893.

20. At least until January 1891 Pat Garrett played a role in the Pecos Valley enterprises. Although Garrett seemed to have been more closely associated with Charles Eddy before the arrival of James Hagerman, he still maintained an active business interest in the valley. Garrett, although in a reduced capacity, still played a role in bringing investors to the valley—literally. Garrett and James Brent opened a livery stable in Eddy in December 1889 when the partners moved their stock into a brick stable, forty by eighty feet long, furnished with "modern appliances" for the care of horses and carriages. Garrett provided transportation from Pecos, Texas, to Eddy, Seven Rivers, and Roswell, New Mexico, from January 1890 until January 1891. Garrett and Brent, as well as independent travelers, frequented a way-station at the Delaware River. There the proprietress had two tents and an empty wagon bed for refuge. Each tent was equipped with two wide beds. Travelers could rent space in the beds. Eddy was also a stopover, with main destinations of Seven Rivers and Roswell. Garrett and partner James Brent provided hacks (coaches or carriages for hire) that ran daily, except Sundays, until 13 January 1891, when the railroad reached Black River some seventeen miles south of Eddy. With the coming of the railroad, the partners sold their lines in Eddy to the Pecos Valley Town Company, owned by Charles Eddy. After selling the livery, Garrett and James Brent left the valley in April 1891 for Uvalde. *Roswell Register,* 18 April 1891; *Eddy Argus,* 14 December 1889, 23 November 1889, 15 February 1890; James M. Day, "Buggy Days in the Permian Basin," *Permian Historical Annual* 17 (December 1977): 3–20.

21. *Eddy Argus,* 29 March 1890, 23 April 1892, 6 February 1892, 2 December 1892.

22. *Eddy Argus,* 25 January 1890, 2 March 1890, 8 March 1890, 20 August 1892.

23. The First National Bank of Eddy listed as officers Charles Eddy, president, and William Bonbright, Hagerman's old mining partner, as vice president. Eddy's brother John Arthur, Bonbright, and Charles H. McLenathen, one of the first major real estate dealers in the valley, and Eddy's attorney, W. A. Hawkins, served as directors. *Fortunes of a Decade,* 114; *Eddy Argus,* 3 October 1891; Lewis, "Early Settlers."

24. *Eddy Argus,* 5 July 1890, 12 July 1890, 18 July 1891.

25. *Eddy Argus,* 18 July 1891.

26. *Eddy Weekly Current,* 3 March 1893.

27. Charles Eddy to Governor L. Bradford Prince, 6 July 1890, Prince Papers.

28. Ibid.

29. *Carlsbad Argus,* 16 May 1947.

30. *Eddy Argus,* 18 January 1890.

31. *Eddy Argus,* 3 May 1890.

32. *Eddy Argus,* 1 March 1890.

33. Sheridan, *Bitter River,* 50–51, 60; Metz, *Pat Garrett,* 155; Coan, *History of New Mexico,* 566; *Pecos Valley Register,* 18 April 1889.

34. Sheridan, *Bitter River,* 60; Larson, *Forgotten Frontier,* 28, 29, 70–71, 82; Elvis E. Fleming and Minor S. Huffman, eds., *Roundup on the Pecos* (Roswell, N. Mex.: Chaves County Historical Society, 1978), 16–18, 20.

35. Ruth Kessler Rice, *Letters from New Mexico, 1899–1904,* ed. Margaret W. Reid (Albuquerque, N. Mex.: Adobe Press, 1981), 5, 8, 9.

36. Westphall, "The Public Domain" (January 1958): 28–29.

37. Ibid.

38. *Eddy Current,* 2 February 1895.

39. *Eddy Argus,* 30 January 1892.

40. Hagerman seriously underestimated the impact of artesian drilling in the valley. *Eddy Argus,* 31 January 1891; *Roswell Daily Record,* 7 October 1937; Robert Lingle and Dee Linford, *The Pecos River Commission of New Mexico and Texas* (Santa Fe, N. Mex.: Rydal Press, 1961), 98–99; Larson, *Forgotten Frontier,* 29.

41. *Eddy Weekly Current,* 23 December 1892; *Carlsbad Current-Argus,* 20 March 1988; *History of Eddy County,* 27.

42. *Eddy Argus,* 31 January 1891.

43. *Eddy Argus,* 12 October 1889.

44. *Eddy Argus,* 18 January 1890.

45. *History of Eddy County,* 68.

46. *Eddy Argus,* 28 February 1891, 8 March 1890.

47. *Eddy Argus,* 23 August 1890, 30 August 1889.

48. Mermod never permanently lived in New Mexico. He and his wife, Emma, maintained a residence in St. Louis. Mermod died in 1908, an apparent suicide. A

chambermaid found his body in a St. Louis hotel, a bottle containing carbolic acid at his bedside. His mouth and right hand were burned. At the time of his death, the Mermods were separated. Emma Mermod died in 1916. *Eddy Argus,* 21 December 1889; undated, uncredited newspaper clipping, Missouri Historical Society, Necrology Scrapbook IIC, St. Louis Public Library, St. Louis, Missouri; *St. Louis Republic,* 30 May 1916, Missouri Historical Society, Necrology Scrapbook IX.

49. *Eddy Argus,* 6 June 1891.

50. He eventually purchased Hagerman's residential property, the "Heights," and the power dam that supplied water and electricity to the Heights. *Eddy County News,* 25 July 1947; *Carlsbad Current-Argus,* various 1892.

51. The mile-long racetrack was reportedly one of the finest and fastest in the Territory. The first races were held in October 1890. Horse racing was not the only sport locals enjoyed. The town boasted a baseball team, which played teams from Roswell, White Oaks, Georgetown, Deming, Silver City, Piños Altos, and Silver City. *Eddy Argus,* 3 May 1890, 18 October 1890, 9 May 1991; *Eddy Weekly Current,* 14 April 1893.

52. *Eddy Weekly Current,* 14 April 1893.

53. *Eddy Weekly Current,* 14 April 1893, 21 April 1893.

54. *Eddy Argus,* 31 January 1891.

55. Charles Eddy convinced Hagerman to build a residence on the east side of the river. Charles McLenathen contracted to build the house. When completed in 1892, the new house contained twelve rooms, including library and billiard and smoking rooms. Including a power dam, pump, and other developments, Hagerman eventually spent $225,000 on the place. *Eddy Argus,* 7 May 1892; Percy Hagerman, *The Great Illusion: Notes on the Hagerman Experience in the Pecos Valley,* unpublished manuscript, May 1934, Hagerman Family Collection. In 1934, Percy Hagerman looked through his father's vast accumulation of office files, selected and retained a few printed circulars, letters, and other papers from which to compose an account of what he called "The Pecos Valley Adventure." Subsequent to recording what Hagerman called "The Great Illusion," nearly all of the documents, correspondence, and other files stored at either Colorado Springs or Roswell were burned, including the account books of numerous defunct companies in the Pecos Valley. Hagerman was exasperated by his father's blind faith in turning the desert of the Pecos Valley into an agricultural mecca. Percy Hagerman said, "No get rich scheme or hair-brain mining promotion was ever founded on a more unsound basis than the Pecos Irrigation and Improvement Company, which was the original illusion from which everything else naturally evolved."

56. Roberts, "Fifty Eventful Years," 7–9; *Eddy Current Weekly,* 16 December 1892, 23 December 1892.

57. The nine-foot-high dam later became known as Tansill Dam. The 1894 Eddy Light and Ice Company became the Carlsbad Public Utilities Company. The Roswell Electric Light Company was organized in 1904 and was the predecessor of Southwestern Public Service Company, now Xcel Energy. In 1910 the Otis Company of Cleveland, Ohio, bought controlling interest in the Roswell company. Otis

consolidated the light company with the community's gas utility and called the new organization the Roswell Gas and Electric Company, which was managed by D. W. Lowe. The Otises, staunch supporters of Hagerman's irrigation scheme in the valley, apparently launched their own company. In 1912 the Otis group purchased the Milan Bush Ranch north of Roswell and divided it into small farms, digging forty wells to irrigate them. Investors ran trains from the East to Roswell, hoping to attract settlers with free irrigation water and to increase the electricity usage. The land promotion "led to a great waste of water which flooded not only the growing areas, but the surrounding countryside." Roswell Gas and Electric was strapped financially in 1921 and was sold to Roswell Public Service Company. The Paul C. Dodge Company of Chicago purchased this company in 1925 and changed the name to Southwestern Public Service Company (SPS). The Dodge group bought two competing companies in Carlsbad, one the Public Utilities Company, as well as an electric plant and distribution system in Artesia. Today, Xcel Energy (the former SPS) serves a 52,000-square-mile area, ninety-six communities, and provides wholesale power to four municipal systems and several rural electric cooperatives. The company provides services and/or operates plants in seventeen states, including the entire Southwest. "History of the Southwestern Public Service Company," Public Information Offices, Southwestern Public Service Company, Amarillo, Texas, and *The Southwesterner*, 90th anniv. ed., November/December 1994, both provided by Bill Crenshaw, Amarillo office, SPS; *Eddy Argus*, 30 September 1892; Myers, *Pearl of the Pecos*, 147.

58. The flume was five feet by six feet. *Eddy Argus*, 30 September 1892; *Eddy Weekly Current*, 23 December 1892.

59. *Eddy Weekly Current*, 23 December 1892.

60. *Eddy Argus*, 4 May 1894.

61. Hagerman, "Statement of Assets and Liabilities, Year ending December 31, 1893," Hagerman Family Collection.

62. *Eddy Argus*, 24 August 1894, 14 October 1892; New Mexico (Territory), Bureau of Immigration, *New Mexico: Its Resources*, 247, 267–268.

63. Little is known of Frost's origins. He reportedly came from Vienna, Austria, New York, or New Orleans. Herbert H. Lang, "New Mexico Bureau of Immigration, 1880–1912," *New Mexico Historical Review* 51 (July 1976): 195–208.

64. Lang, "New Mexico Bureau of Immigration," 195.

65. Otis was a farming community located six miles south of Eddy. It was named for T. E. Otis, director of the Atchison, Topeka and Santa Fe Railroad. *Eddy Weekly Current*, 21 April 1893; Pearce, *New Mexico Place Names*, 115.

66. *Eddy Argus*, 18 June 1892; *Carlsbad Current-Argus*, 20 March 1988.

67. *Eddy Weekly Current*, 23 December 1892.

68. *Eddy Argus*, 28 February 1891.

69. *Eddy Argus*, 7 December 1889.

70. *Eddy Argus*, 23 November 1889.

71. Lewis, "Early Settlers."

72. Ibid.

73. During the latter half of the nineteenth century, New Mexico gained the reputation as a natural sanatorium based on its climate. Soldiers and government workers in the territory and travelers all extolled the virtues of the arid territory. By the end of the century, the New Mexico Territory Bureau of Immigration enthusiastically advertised not only agriculture and irrigation in the region, but also the healthful aspects of the territory. For example, Jay Gould spent his final years operating in the Southwest, traveling in a personal coach with his family, largely because of his tuberculosis. Jake W. Spidle Jr., "Coughing, Spitting, and New Mexican History," in *Essays in 20th Century New Mexico History,* ed. Judith Boyce DeMark (Albuquerque: University of New Mexico Press, 1994), 169–171.

74. William Thorton Parker, M.D., Munich, "Shall We Have a National Sanitarium for Consumptives?" Billy M. Jones, Papers, Collection of Misc. Articles, 1838–1906, SWC, microfilm; Spidle, "Coughing, Spitting," 169–171.

75. Spidle, "Coughing, Spitting," 172.

76. *Eddy Weekly Current,* 27 January 1893.

77. In Lincoln County, Fort Stanton was abandoned as a military post in 1898, and in 1899 the government designated the fort a tuberculosis hospital for merchant marines. "Germans Sit Out War in Desert Drydock," *New Mexico Magazine,* November 1992, 91; Spidle, "Coughing, Spitting," 172–173.

78. Spidle, "Coughing, Spitting," 173.

79. Sharlot M. Hall, "The Burden of the Southwest," *Out West Magazine* 28 (January 1908): 3–4.

80. *Eddy Weekly Current,* 9 December 1892.

81. Charles Eddy to Governor L. Bradford Prince, 3 September 1890, Prince Papers.

82. *Eddy Weekly Current,* 23 December 1892.

83. Ibid.

84. *Eddy Weekly Current,* 30 December 1892.

85. *Eddy Weekly Current,* 13 January 1893.

86. *Eddy Weekly Current,* 13 January 1893, 20 January 1893.

87. *Eddy Weekly Current,* 27 January 1893.

88. *Eddy Weekly Current,* 9 December 1892.

89. Ibid.

90. *Eddy Weekly Current,* 3 March 1893.

91. *Eddy Weekly Current,* 30 June 1893.

92. *Eddy Argus,* 28 February 1891.

93. Ibid.

94. *Eddy Weekly Current,* 21 April 1893.

95. Frost served the Bureau of Immigration for twelve years—until President Theodore Roosevelt appointed Herbert J. Hagerman (son of J. J. Hagerman) the territorial governor. Hagerman did not like Frost because he exaggerated too much and produced only glowing reports of the economy. *Eddy Weekly Current,* 21 April 1893; Lang, "New Mexico Bureau of Immigration," 203–204, 208.

96. King was born January 6, 1842, in Newport, Rhode Island, and in 1859

entered the Yale Scientific School, graduating in 1862. In 1863 King and a geologist took a horseback trip across the United States. After working and studying Nevada's Comstock Lode, they crossed the Sierra Nevada by foot. During 1865–1866, King worked under General McDowell in surveying the desert of Southern California. In the fall of 1866, King convinced Congress to appropriate funds to survey the region from eastern Colorado to the California boundary, and to place him in charge. King worked on this survey until 1877. His results were published in seven volumes as *Report of the Geological Exploration of the Fortieth Parallel 1870–1900*. In 1878 Congress made King the head of the newly established U.S. Geological Survey, where he remained until going into the private sector in 1881. King lost heavily in the depression following 1892, "had an attack of nervous prostration," and was confined to Bloomingdale Asylum during 1893 and 1894. In 1901 King contracted pneumonia, followed by tuberculosis. (Like Hagerman, Tansill, and others, King headed west for his health, dying in obscurity in Arizona.) *Eddy Argus*, 14 November 1892; P. Hagerman, *The Great Illusion*, 8, Hagerman Family Collection; *Dictionary of American Biography*, s.v. "King, Clarence"; Keleher, *Fabulous Frontier*, 192–194; Larson, *Forgotten Frontier*, 222; Thurman Wilkins, *Clarence King: A Biography* (Albuquerque: University of New Mexico Press, 1988).

97. Circular, Pecos Irrigation and Improvement Company, 15 December 1891, in P. Hagerman, *The Great Illusion*, 9, Hagerman Family Collection.

98. P. Hagerman, *The Great Illusion*, 9, Hagerman Family Collection.

99. Ibid.; Keleher, *Fabulous Frontier*, 192–194.

100. P. Hagerman, *The Great Illusion*, 9, Hagerman Family Collection; Keleher, *Fabulous Frontier*, 192–194; Lipsey, *The Lives of James J. Hagerman*, 222.

101. *Eddy Argus*, 28 February 1891.

102. *Eddy Argus*, 28 February 1891, 30 May 1891.

103. *Eddy Argus*, 30 May 1891.

104. *Eddy Argus*, 30 May 1891, 19 December 1891.

105. *Eddy Argus*, 28 February 1891, 30 May 1891.

106. Following up on Gaullieur's visit, in June 1891 William A. Hawkins, Charles Eddy's personal attorney, wrote a letter to the governor of New Mexico Territory. Hawkins told Governor L. Bradford Prince about the inspection visit by Gaullieur, but instead of mentioning Switzerland as Gaullieur's homeland, he said that Gaullieur represented Norway and Sweden on behalf of a philanthropic association composed largely of leading officials from those countries, including the minister of the interior, two senators, and others. Hawkins noted that officials had organized the association, with quasi-official recognition from the governments of Norway and Sweden, to protect emigrants. Hawkins concluded that since Gaullieur gave it his unqualified endorsement, he was highly pleased with southeastern New Mexico. Before he left in the spring of 1891, Gaullieur asked Eddy if the governor of the Territory might provide official endorsement of activities in the lower Pecos Valley, especially those of Pecos Irrigation and Improvement Company and the standing of the men who headed the company. Hawkins and Eddy

recognized that such an endorsement from the governor's office gave Gaullieur authority to bring a large group of settlers to the valley. By way of informing the governor, Hawkins detailed the activities of three large enterprises underway in the valley: the irrigation system, the railroad, and town developments. According to Hawkins, canals would irrigate 360,000 acres of land. Hawkins requested nothing less from Governor Prince than a general letter endorsing all merits of southeastern New Mexico—its climate, soil, water, and the standing of men "at the helm" of the various companies. Prince complied with Hawkins's request. William A. Hawkins to Governor L. Bradford Prince, Santa Fe, 18 June 1891, Prince Papers; *Eddy Argus,* 30 May 1891.

107. *El Paso Herald,* 19 December 1891.

108. Ibid.

109. Eddy County, Deeds of Record, books 4, 8, 14; *El Paso Herald,* 19 December 1891; *Eddy Argus,* 14 November 1891.

110. The New Mexico town was established in 1891, renamed Florence in 1894, and again renamed Loving in 1908, after Oliver Loving. Myers, *Pearl of the Pecos,* 146; *Webster's New Geographical Dictionary,* s.v. "Vaud"; Angelo Heilprin and Louis Heilprin, eds., *A Complete Pronouncing Gazetteer or Geographical Dictionary of the World,* s.v. "Vaud"; *Eddy Argus,* 14 November 1891; Pearce, *New Mexico Place Names,* 93.

111. Malaga was originally known as Lookout, then Kirkwell. The town became known as Malaga in 1891, named for a variety of grape grown by the Swiss. Malaga is also the name of a province on the Mediterranean in Spain, once held by the Romans. It had a large trade in wines, especially muscatels like Dulce and Lagrima. Eddy County, Deeds of Record, book 4, 19, 27; *Eddy Weekly Current,* 13 January 1893; Heilprin and Heilprin, *A Complete Pronouncing Gazetteer,* s.v. "Malaga"; *Webster's New Geographical Dictionary,* s.v. "Malaga"; *Eddy Argus,* 6 August 1892; *History of Eddy County,* 14; Pearce, *New Mexico Place Names,* 90.

4. DAMS AND VIOLENCE IN THE PECOS VALLEY

1. *U.S. v. Hope,* vol. 1, 3237.

2. Ibid., 3258.

3. Ibid.

4. Joe Nash started the Harroun Ditch in the 1870s. Dan Harroun later improved it.

5. *U.S. v. Hope,* vol. 1, 3259–3260.

6. Roberts, "Fifty Eventful Years," 4; *Carlsbad Current-Argus,* 16 May 1947.

7. Engineering students were gaining some practical experience in water engineering by 1888. At Powell's insistence, students had to use the Rio Grande near Santa Fe to learn how to gauge stream flow from a makeshift raft. Baxter, *Dividing New Mexico's Water,* 79.

8. H. H. Cloud, Biographical File, Colorado Historical Society, Denver, Colorado; L. D. Blauvelt, Biographical File, Colorado Historical Society.

9. Nettleton, who had studied the irrigation systems of the Nile, the Euphrates, and in India, China, Italy, and Spain, was, by all accounts, a "thoroughly trained water expert." He estimated the thirty-six-mile-long, thirty-two-foot-wide Greeley, Colorado, ditch would cost $20,000. By 1878, the colony had spent $87,000. Edwin S. Nettleton, Biographical File, Colorado Historical Society, Denver, Colorado; Robert Dunbar, *Forging New Rights in Western Waters* (Lincoln: University of Nebraska Press, 1983), 21–23; Marshall Sprague, *Colorado: A Bicentennial History* (New York: Norton, 1976), 61–63; *Carlsbad Current-Argus*, 20 June 1947; Betty Marie Daniels and Virginia McConnell, *The Springs of Manitou* (Denver: Sage Books, 1969), 15; Sprague, *Newport in the Rockies*, 30, 31, 35.

10. Hagerman, looking back over the decade, remembered, "[t]he irrigation scheme had been examined and favorably reported on by Engineer Nettleton, at that time having a high reputation. I had him and another engineer make a second examination and their report was extremely favorable. After I went into the company, a number of others followed me, a large sum of money was raised and work went on rapidly." Nettleton was the state engineer of Colorado and served as Hagerman's chief advisor and consulting engineer. He measured the river and estimated its possibilities, assuring Hagerman of an adequate water supply. Tracy, "Pecos Valley Pioneers," 203; P. Hagerman, *The Great Illusion*, 3, Hagerman Family Collection; Hagerman, "Statement," c. 1900, Hagerman Family Collection.

11. Nathan Jaffa and W. S. Prager of Roswell helped finance the business. Long and First New Mexico Reservoir officials planned a primary distribution reservoir half a mile long and 40 feet deep on the Hondo. A second reservoir would cover ten square miles, with a dam 60 feet deep, 800 feet wide on the bottom, and three-eighths of a mile wide on top. First New Mexico officials estimated that together the reservoirs would store enough water to reclaim 100,0000 acres of land. The Pecos Irrigation and Improvement Company spent $30,000 preparing the Hondo site before it abandoned the project, partly because of a scarcity of money and partly because builders feared that the reservoir would not hold water. Percy Hagerman was certain that Charles Eddy profited personally from this deal, whereby Pecos Irrigation and Improvement had to purchase the water rights from First New Mexico Reservoir. Percy Hagerman further asserted that Eddy was involved in holding up plans to build the reservoir near the Seven Rivers area until such time as the company acquired Eddy's title to certain lands that would be submerged by the Seven Rivers Reservoir (this reservoir later became known as McMillan Reservoir). Hagerman told Reed that Charles Eddy bought the land first, then sold it to Pecos Irrigation and Investment Company. Annual Stockholder's Report, in P. Hagerman, *The Great Illusion*, 12, 13, Hagerman Family Collection; Sheridan, *Bitter River*, 70; Coan, *History of New Mexico*, 465; C. A. Hundertmark, "Reclamation in Chaves and Eddy Counties, 1887–1912," *New Mexico Historical Review* 47 (1972): 306; *Eddy Argus*, 15 March 1890; Tarr Report, 28, 31–32; Hagerman to Reed, 1 July 1904, U.S. Department of the Interior, Bureau of Reclamation, RG 115, box 1, entry 4, file 25; St. Louis City Directories for 1880, 1885–1886, 1895–1896, 1889–1890, St. Louis Public Library; New Mexico (Terri-

tory), Bureau of Immigration, *Compilation of Facts Concerning the Pecos Valley: Resume of the Improvement Being Made in Southeastern New Mexico, Embracing the Magnificent Pecos Valley.* 1891 (Santa Fe, N. Mex.: New Mexican Printing, 1978), 8; Herbert M. Wilson, "Pecos Valley Canals," *Engineering News* 26 (17 October 1891): 350–351; *Eddy Argus,* 12 October 1889.

12. *Eddy Argus,* 12 October 1889, 13 December 1890; New Mexico (Territory), Bureau of Immigration, *New Mexico: Its Resources,* 249; Tarr Report, 30; Eddy to Prince, 3 September 1890, Prince Papers.

13. The number of acres the company planned to irrigate changed from year to year. Eddy and Hagerman thought big and then bigger. As reality settled in, the acreage decreased.

14. William C. Bradbury, born in Massachusetts in 1849, moved to Colorado in 1871. He married Hattie Howe, daughter of B. H. and Elizabeth Howe, and then worked as a "cow puncher." Because his health was failing (possibly from tuberculosis), in 1880 Bradbury moved to Tres Piedra, New Mexico. His company built over 700 miles of railroad track, including the Colorado Midland. The company also constructed numerous irrigation systems in Colorado, New Mexico, Idaho, and Wyoming. The contractor, builder, and rancher died on October 3, 1925, at the age of seventy-six. *Ballenger and Richards 25th Annual Denver City Directory,* 1897. The directory lists the company as "Bradbury W C and Company (W C Bradbury), railroad and canal contractors, 55 Railroad Building, 1515 Larimer"; *The Trail,* vol. 18, no. 22 (25 November 1935), Denver Public Library; *Eddy Argus,* 19 October 1889; "Biographical data to accompany portrait of Mr. William C. Bradbury" and "Brief Incidents of Special Historical Interest" in Clippings File, Denver Public Library, Denver, Colorado.

15. *Eddy Argus,* 12 October 1889, 15 November 1890.

16. *Eddy Argus,* 12 October 1889.

17. *Eddy Argus,* 4 January 1890.

18. The Southern Canal was also known as the Southwestern Canal and the West Side Canal.

19. Wilson, "Pecos Valley Canals," 351; William E. Gibbs, "History, the Farmer, and the Pecos Valley Past," *Southwest Heritage* 11 (fall 1981): 2–7.

20. Irrigation companies often excavated canals and ditches using scrapers pulled by horses. Most of the ditches were wide and shallow, promoting evaporation. Pisani, *To Reclaim a Divided West,* 120–121; Ash Upson, *Lincoln County Leader,* 8 December 1888, from Keleher, *Fabulous Frontier,* 187.

21. Wilson, "Pecos Valley Canals," 351; Gibbs, "History, the Farmer, and the Pecos Valley Past," 2–7.

22. Witt, known to locals as "Boston," came to Eddy in its infancy. He later built the first Eddy County courthouse and purchased lands along the Black River southwest of Eddy, where he planted hundreds of cottonwoods and a fruit orchard watered by a constant water supply called Blue Springs. Witt was active in local affairs, serving for years as a member of the Eddy County Board of Commissioners. U.S. Department of the Interior, Bureau of Reclamation, RG 115, box 443,

entry 7, folder 651; *Eddy Argus,* 19 October 1889, 23 November 1889; *History of Eddy County;* Lewis, "Early Settlers."

23. *Eddy Argus,* 12 October 1889; U.S. Department of Agriculture, Office of Experiment Stations, Bulletin 215, *Irrigation in New Mexico,* by Vernon Sullivan (Washington, D.C.: Government Printing Office, 1909), 20; New Mexico (Territory), Bureau of Immigration, *Compilation of Facts,* 11.

24. *Eddy Weekly Current,* 2 December 1892; Wilson, "Pecos Valley Canals," 350–351.

25. *Eddy Argus,* 5 April 1890.

26. James Dix Schuyler, *Reservoirs for Irrigation, Water Power and Domestic Water Supply. With an Account of Various Types of Dams and the Methods and Plans of Their Construction. Together with a Discussion of the Available Water-Supply for Irrigation in Various Sections of Arid America; the Distribution, Application, and Use of Water; the Rainfall and Run-off, the Evaporation from Reservoirs; the Effect of Silt Upon Reservoirs, etc.* (New York: John Wiley and Sons, 1901), 49; *Eddy Argus,* 29 March 1890, 15 April 1890.

27. Myers, *Pearl of the Pecos,* 24; *History of Eddy County,* 27; *Eddy Argus,* 22 March 1890.

28. During the dam's construction officials and workers alike referred to it as Rock Dam. Later it was referred to as Six-Mile Dam, Reservoir No. 2, the Eddy Dam (in honor of Charles Eddy), and finally Avalon. *Eddy Argus,* 29 March 1890, 5 April 1890.

29. Mark Hufstetler and Lon Johnson, *Watering the Land: The Turbulent History of the Carlsbad Irrigation District* (Denver: National Park Service, Rocky Mountain Region and Bureau of Reclamation, 1993), 28; Wilson, "Pecos Valley Canals," 351–352.

30. Schuyler, *Reservoirs for Irrigation,* 49, 51; New Mexico (Territory), Bureau of Immigration, *Compilation of Facts,* 11; Wilson, "Pecos Valley Canals," 352.

31. *Eddy Argus,* 12 October 1889.

32. *Eddy Argus,* 28 February 1891.

33. *Eddy Argus,* 30 August 1890.

34. *Eddy Argus,* 9 May 1891.

35. L. B. Howell, "Pecos Valley Irrigation System," *Engineering News and American Railway Journal* 36 (17 September 1896): 181–182.

36. Mann did not leave. He quit his job with Bradbury and Company and stayed in Eddy. *Eddy Argus,* 1 March 1890, 14 February 1891, 28 February 1891; Coan, *History of New Mexico,* 465.

37. *Eddy Weekly Current,* 2 December 1892.

38. *Eddy Argus,* 28 February 1891.

39. *Eddy Argus,* 1 February 1890.

40. *Eddy Argus,* 13 December 1890.

41. *Eddy Argus,* 28 February 1891; Tarr Report, 28–30; Eddy to Prince, 3 September 1890, Prince Papers.

42. Tracy, "Pecos Valley Pioneers," 203; Pecos Irrigation and Improvement

Company Annual Report, December 1891 and 1891, circular, as cited in Hagerman, *The Great Illusion*, 5–7, Hagerman Family Collection; Clark, *Water in New Mexico*, 88; Sheridan, *Bitter River*, 58; Howell, "Pecos Valley Irrigation System," 181–82; Larson, *Forgotten Frontier*, 269; Reisner, *Cadillac Desert*, 56; U.S. Department of Agriculture, Office of Experiment Stations, Bulletin 86, *The Use of Water in Irrigation: Report of Investigations Made in 1899*, by Elwood Mead and C. T. Johnson (Washington, D.C.: Government Printing Office, 1900), 88; New Mexico (Territory), Bureau of Immigration, *New Mexico: Its Resources*, 245; Schuyler, *Reservoirs for Irrigation*, 54.

43. *Eddy Argus*, 12 October 1889, 25 October 1890; Coan, *History of New Mexico*, 465.

44. *Eddy Argus*, 28 February 1891.

45. *Eddy Argus*, 28 February 1891; Myers, *Pearl of the Pecos*, vi; New Mexico (Territory), Bureau of Immigration, *New Mexico: Its Resources*, 250; New Mexico (Territory), Bureau of Immigration, *Compilation of Facts*, 12; Schuyler, *Reservoirs for Irrigation*, 56.

46. Tracy, "Pecos Valley Pioneers," 203; Pecos Irrigation and Improvement Company Annual Report, December 1891 and 1891, circular, as cited in P. Hagerman, *The Great Illusion*, 5–7, Hagerman Family Collection; Clark, *Water in New Mexico*, 88; Sheridan, *Bitter River*, 58; Howell, "Pecos Valley Irrigation System," 181–82; Larson, *Forgotten Frontier*, 269; Reisner, *Cadillac Desert*, 56; U.S. Department of Agriculture, Office of Experiment Stations, *The Use of Water*, 88; New Mexico (Territory), Bureau of Immigration, *New Mexico: Its Resources*, 245; Schuyler, *Reservoirs for Irrigation*, 54.

47. *Eddy Argus*, 28 February 1891.

48. *Eddy Argus*, 28 October 1892.

49. Ibid.; Coan, *History of New Mexico*, 465; Sheridan, *Bitter River*, 58; *History of Eddy County*, 14.

50. Schuyler, *Reservoirs for Irrigation*, 55; Howell, "Pecos Valley Irrigation System," 182.

51. *Eddy Argus*, 1 March 1890; Myers, *Pearl of the Pecos*, 4.

52. *Eddy Argus*, 25 October 1890.

53. Wilson, "Pecos Valley Canals," 351; Gibbs, "History, the Farmer, and the Pecos Valley Past," 2–7.

54. *Eddy Argus*, 27 January 1893.

55. *Eddy Argus*, 28 February 1891.

56. *Eddy County News*, 27 June 1947.

57. *Twelfth Census of Population, 1900, New Mexico*, vol. 3, *Eddy County to Grant County to Guadalupe County* (Washington, D.C.: Government Printing Office, 1900); *Eddy Weekly Current*, 23 April 1893.

58. Carolyn Zeleny, *Relations between the Spanish-Americans and Anglo-Americans in New Mexico* (New York: Arno Press, 1974), 157.

59. Christiansen, "The Quest for Water," 5.

60. The Newlands Act provided for federal reclamation in the western states.

Robert N. McClean, *That Mexican! As He Really Is North and South of the Rio Grande* (New York: Fleming H. Revell, 1928), 134; Juan Gómez-Quiñones, *Mexican American Labor, 1790–1990* (Albuquerque: University of New Mexico Press, 1994), 151; Zeleny, *Relations between the Spanish-Americans and Anglo-Americans,* 77–78, 162–163.

61. *Eddy Argus,* 30 August 1890.

62. *History of Eddy County,* 68.

63. Rice, *Letter from New Mexico,* 46.

64. Epperson, *Colorado As I Saw It,* 48.

65. *Eddy Argus,* [?] April 1890; *Eddy Weekly Current,* 23 December 1892.

66. *Eddy Argus,* 19 October 1889.

67. *Eddy Argus,* 23 November 1889, 8 March 1890.

68. Mexicans were often recruited in large groups because they worked, legally or not, cheaply. Alan M. Kraut, *The Huddled Masses: The Immigrant in American Society* (Arlington Heights, Ill.: Harlan Davidson, 1982), 25; Carey McWilliams, *North from Mexico: The Spanish Speaking People of the United States,* updated by Matt S. Meier (New York: Praeger, 1990), 195.

69. *Eddy Argus,* 20 December 1890.

70. *Eddy Argus,* 28 December 1889.

71. Jacales were one-room Mexican huts (usually about twelve feet by twelve feet) made of mesquite or yucca posts stuck into the ground, with mud in between. The roof was thatched with tule or cane. The door, if one existed, was usually a blanket. Many Anglos believed that Mexicans were content living in a jacal with two meals (biscuits made on a tin and baked in ashes) a day. Some supposed that eight or ten Mexicans could live happily with five acres, goats, old horses, two or three cows, and perhaps a donkey. The *Argus* reported that "Chihuahua on the east side of the river is growing a little. Another stately jacal was erected there this week." Interestingly, while the town was segregated, the schools were not. The Eddy school had 181 pupils in 1891, including 20 Mexicans and several blacks. *Eddy Argus,* 30 August 1890, 7 February 1891, 13 June 1891, 5 September 1891; Arnoldo DeLeón, *The Mexican American Image in Nineteenth Century Texas* (Boston: American Press, 1983), 31–32; McWilliams, *North from Mexico,* 196.

72. Gómez-Quiñones, *Mexican American Labor,* 45, quotation on 40.

73. *El Paso Daily Times,* 14 January 1902; *El Paso Herald,* 4 December 1915.

74. Garcia, *Desert Immigrants,* 17, 33–35, 243.

75. Railroad building provided a significant source of employment for Mexican nationals and Mexican Americans. Railroad employment was not stable, since companies hired men to meet specific needs, then let them go. In New Mexico, villagers worked part-time on railroads and farmed part-time. By the 1900s, on some lines Mexicans constituted 70 percent of section crews and 90 percent of extra gangs, making one dollar a day. Many rail companies recruited directly in the interior of Mexico. McClean, *That Mexican!* 134; Gómez-Quiñones, *Mexican American Labor,* 151; Zeleny, *Relations between the Spanish-Americans and Anglo-Americans,* 77–78, 162–163; Garcia, *Desert Immigrants,* 37, 40.

76. Many Mexicans in turn believed that Anglos were arrogant, insolent, heretics, and barbarians. Garcia, *Desert Immigrants*, 106–07.

77. The temperance movement began in the 1830s and blossomed in the 1870s. Many towns had only a few hundred citizens and dozens of saloons. Many people came to see alcohol and saloons as evil and menaces to society. Drinkers spent money needed for their children's education, clothing, and food. Alcohol put men in a world of whores, drug fiends, pimps, thieves, and gamblers. Such company taught men to abuse their wives and molest their children. Temperance, on the other hand, promoted domestic tranquility, sobriety, frugality, and industry. Norman H. Clark, *Deliver Us from Evil: An Interpretation of American Prohibition* (New York: W. W. Norton, 1976), 2–6; *Carlsbad Current-Argus*, 20 March 1988.

78. *Eddy Argus*, 7 December 1889.

79. *Eddy Argus*, 5 April 1890.

80. *Eddy Weekly Current*, 13 January 1893.

81. Most Catholics at the time were Mexican Americans and were considered an unwanted element. The *Argus*'s competitor, the *Weekly Current*, encouraged Catholics to relocate to the upper valley. *Eddy Weekly Current*, 2 December 1892.

82. Before Lone Wolf and Phenix, Seven Rivers, sixteen miles north of Eddy, and Lookout, south of town on the Black River near Malaga, earned reputations as havens for trail-weary cowboys, rustlers and whiskey drinkers. Myers, *Pearl of the Pecos*, iv, 111.

83. The Wolf, for which Lone Wolf was known, was a tent saloon with a stuffed wolf on display. *Eddy Argus*, 8 March 1890, 21 June 1890; Myers, *Pearl of the Pecos*, 111.

84. *Eddy Argus*, 15 February 1890.

85. D. R. Harkey, *Mean As Hell* (Albuquerque: University of New Mexico Press, 1948), 49–50.

86. This implies some hypocrisy. Although the *Eddy Argus*, controlled by Eddy and Hagerman, was purported to be temperate, it carried Lone Wolf advertisements and news from Wolftown. One ad noted that the Lone Wolf had "a stock of pure liquors . . . their 'Belle of Bourbon' is guaranteed to be the real article." *Eddy Argus*, 15 February 1890, 8 March 1890.

87. Harkey, *Mean As Hell*, 49–50; *Eddy Argus*, 17 October 1889, 19 September 1891, 23 April 1892, 30 April 1892, 7 May 1892, 2 July 1892, 7 August 1892, 5 May 1893, 16 February 1894; *Eddy Weekly Current*, 2 December 1892; Lewis, "Early Settlers."

88. The *Argus* suggested that Charles Eddy and the Pecos Valley Town Company tried to root out saloons in the town of Eddy by establishing Phenix. Phenix, a town of mostly wooden shacks, opened for business on April 23, 1892. It featured a saloon, gambling, and women. H. C. Bennett opened the first business. A hack service between Eddy and Phenix provided transportation. The nickname "Jagville" derived from an English word meaning "load," intimating that Phenix had "as much liquor as one can carry." No remnants of the town remain today. It was located near the 1800 block of South Canal Street on the east side of the road.

When the county surveyor laid out the road between Eddy and Malaga, he positioned it to run through Phenix. *Eddy Argus*, 23 April 1892, 7 May 1892, 2 July 1892, 7 August 1892, 5 May 1893, 16 February 1894; Myers, *Pearl of the Pecos*, 146; *History of Eddy County*, 27; Roberts, "Fifty Eventful Years," 13.

89. The few blacks in town "were treated as a joke," according to Sheridan, *Bitter River*, 61. Most lived on the east side of the river and worked in menial jobs such as dishwashers and hotel porters. *Eddy Argus*, 28 April 1893.

90. *Eddy Weekly Current*, 6 January 1893.

91. *Eddy Argus*, 6 January 1893.

92. *Eddy Weekly Current*, 27 January 1893.

93. *Eddy Weekly Current*, 31 March 1893.

94. *Eddy Weekly Current*, 23 December 1892; *Eddy Argus*, 30 December 1892.

95. *Eddy Argus*, 7 April 1893.

96. *Eddy Argus*, 16 August 1890.

97. *Eddy Argus*, 24 November 1893.

98. *Eddy Argus*, 23 March 1894.

99. *Eddy Weekly Current*, 26 May 1893.

100. The town rebuilt. *Eddy Weekly Current*, 2 June 1893; *Carlsbad Current-Argus*, 20 March 1988; *Eddy Argus*, 2 June 1893.

101. *Eddy Argus*, 14 February 1891.

102. While a lot of violence occurred in and near Eddy, only one incident was considered a capital case. In 1894 James Barnett, a killer from Texas wanting to prove his reputation, shot two sleeping workers at Avalon Dam. Judge Freeman sentenced Barnett to hang. *Eddy Argus*, 30 March 1894; *Eddy Argus*, 2 November 1889.

103. *Carlsbad Current-Argus*, 17 July 1966.

104. Ibid.

105. *Eddy Argus*, 26 April 1890.

106. *Eddy Argus*, 12 October 1889; *Eddy Weekly Current*, 2 December 1892.

107. *Eddy Weekly Current*, 14 April 1893.

108. *Eddy Argus*, 17 February 1893.

109. The paper contended that there were some saloons operating in Eddy behind "spurious business fronts." When Mrs. M. J. Borden lectured a group on temperance, several men got up, went outside, and hid their hip flasks under the house—not because they were embarrassed or ashamed, but because the flasks were uncomfortable to sit on. *Eddy Argus*, 7 August 1892, 19 September 1891, 12 October 1889.

110. *Eddy Argus*, 26 September 1891.

111. *Eddy Argus*, 9 December 1892.

112. *Eddy Argus*, 30 April 1892, 17 March 1893.

113. *Eddy Weekly Current*, 2 December 1892.

114. Cameron was a business partner of Nymeyer, who owned the land where Phenix was located and surveyed the town. Freeman was a justice of the New Mexico Supreme Court in 1890 and became the first judge of the Fifth Judicial

District in 1891. Freeman held court in Socorro, but lived in Eddy. He had the reputation as a "fearless judge" and tried many notorious outlaws. *Eddy Weekly Current*, 2 December 1892; Lewis, "Early Settlers"; *Eddy Argus*, 30 April 1892.

115. *Eddy Argus*, 8 September 1893; *Eddy Weekly Current*, 21 April 1893.

116. Roberts, "Fifty Eventful Years," 13.

117. Ibid.

118. Harkey, *Mean As Hell*, 50.

119. The Edmunds Act was known as the federal antipolygamy act and was enacted on 2 March 1882. The *Argus* noted that men usually served jail time for the offense, and women got a nominal penalty. The paper noted that the law "has well nigh put a stop to the loose and immoral custom" of polygamy and fornication. *Eddy Argus*, 16 May 1895; *U.S. Statutes at Large*, ch. 47, sec. 22, 30.

120. *Eddy Argus*, 13 January 1893, 5 May 1893.

121. Harkey, *Mean As Hell*, 56–69.

122. *Eddy Current*, 19 September 1895.

123. *Twelfth U.S. Census of Population, 1900, New Mexico*, vol. 3, *Eddy County to Grant County to Guadalupe County*.

5. COMPANIES, CANALS, AND CAPITAL

1. *U.S. v. Hope*, vol. 1, 3257.

2. Incorporation laws empowered the directors to levy assessments in money or labor for contracts, and to maintain the ditches and employ "ditch riders" to oversee the distribution of water. Capital stock is all of a company's securities or common stock, preferred stock, and bonds. Dunbar, *Forging New Rights*, 29; Roberts, "Fifty Eventful Years," 3, 4; Coan, *History of New Mexico*, 463; *U.S. v. Hope*, vol. 1, 3259.

3. *U.S. v. Hope*, vol. 2, 565.

4. Ibid.

5. Roberts, "Fifty Eventful Years," 4, 5; *Eddy County News*, 6 June 1947; P. Hagerman, *The Great Illusion*, Hagerman Family Collection; Pecos Irrigation and Investment, Articles of Incorporation, 18 July 1888, New Mexico State Records Center and Archives, Santa Fe, New Mexico; Sheridan, *Bitter River*, 55; Coan, *History of New Mexico*, 463, Dunbar, *Forging New Rights*, 146; Eddy to Prince, 3 September 1890, Prince Papers; *U.S. v. Hope*, vol. 1, 3260.

6. *Eddy County News*, 6 June 1947; P. Hagerman, *The Great Illusion*, 2, Hagerman Family Collection.

7. For more on Greene, see chapter 2.

8. Many irrigation companies overestimated the rate of settlement. Greene started his own project to attract eastern money. He bought 640 acres three miles south of Eddy and turned it into a vineyard. He improved another section and parceled it out to small farmers. Greene remained a stockholder in Pecos Irrigation and Improvement until he sold his stock during the Panic of 1893. Dunbar, *Forging New Rights*, 26; Pecos Irrigation and Investment, Articles of Incorpora-

tion; P. Hagerman, *The Great Illusion,* 2, 3, Hagerman Family Collection; Hundertmark, "Reclamation in Chaves and Eddy Counties," 305; Tracy, "Pecos Valley Pioneers," 193; George Anderson, *The History of New Mexico: Its Resources and People* (Los Angeles: State Publishers, 1907), 770.

9. The figures pertaining to Hagerman's initial investment in the Pecos Valley vary between $30,000 and $40,000. Alan Gilpin, *Dictionary of Economic Terms,* 2d ed. (New York: Philosophical Library, 1970), 9; Coan, *History of New Mexico,* 463; P. Hagerman, *The Great Illusion,* 2, Hagerman Family Collection; Percy Hagerman, *James John Hagerman: A Sketch of His Life,* unpublished manuscript, 1932, Hagerman Family Collection.

10. George W. Campbell, the editor of the *Las Vegas Gazette,* and various other editors stressed the need for railroads in southeastern New Mexico, arguing that poor roads and poor mail service hampered growth. The first railroad in New Mexico, built in 1880, ran along the Santa Fe Trail, through Kansas, Colorado, then through Raton, New Mexico, to Las Vegas and into Santa Fe. Southeastern New Mexico was isolated from the rest of the territory. Lincoln County encompassed 27,000 square miles; the trip from Seven Rivers to Lincoln was between 125 and 150 miles. The trip from Lincoln to Santa Fe took another five days. *Las Vegas Gazette,* 4 May 1873; Keleher, *Fabulous Frontier,* 25–26, 33.

11. When Hagerman joined the company, Garrett was forced out, since he had little monetary interest in the company. Sheridan, *Bitter River,* 57; Hagerman, *The Great Illusion,* 2, 3, Hagerman Family Collection.

12. Percy Hagerman described Charles Eddy as "one of the most remarkable men I have ever met, full of enthusiasm, magnetism, and breeziness, a most interesting personality, a plausible and convincing talker, in short, a wonderful promoter." Eddy, according to a note in the Christian Papers (SWC), had jet-black hair with fierce blue eyes and a personality full of energy and magnetic force. Francis Tracy described Eddy as having "piercing black eyes." He noted that Eddy was "nervous, high strung, and impetuous, with a full resonant voice, an impressive manner, [and] a great personal magnetism." Tracy, "Pecos Valley Pioneers," 188. Several sources note that Mrs. Hagerman likened Eddy to a Svengali (a hypnotic character in George du Maurier's novel *Trilby*) because of his magnetic personality. Larson, *Forgotten Frontier,* 253, 254; Sheridan, *Bitter River,* 57; Keleher, *Fabulous Frontier,* 194–195; P. Hagerman, *The Great Illusion,* 12, Hagerman Family Collection.

13. P. Hagerman, *The Great Illusion,* 3, Hagerman Family Collection.

14. Common stock is a portion of a company's capital. Stock represents shares of ownership. Common stockholders share distributive earnings after prior claims of bondholders and owners of preferred stock. Preferred stocks are also shares of ownership, but have prior claim on distributed earnings over common stock. Owners of preferred stock also have prior claim if a company dissolves. The risk and cost is higher for preferred stock, but it usually has higher returns. Bonds are securities issued by government, public bodies, corporations, and companies carrying fixed rates of interest. Terms and repayment schedules are printed on the

note. They may be secured or not, and maturity varies. Bonds are usually offered for sale through an issuing house, which buys the securities and sells them to the public. Gilpin, *Dictionary of Economic Terms*, 22, 42, 189, 205, 229, 231, 242.

15. Hundertmark, "Reclamation in Chaves and Eddy Counties," 304; P. Hagerman, *The Great Illusion*, 3, Hagerman Family Collection.

16. When the Bureau of Reclamation finally came into the Pecos Valley several years later, they spent anywhere from $30 to $150 an acre to improve the region. Wilson, "Pecos Valley Canals," 350; P. Hagerman, *The Great Illusion*, 8, 9, Hagerman Family Collection.

17. P. Hagerman, *The Great Illusion*, 3, Hagerman Family Collection.

18. Reorganization of a company was not unusual in the nineteenth century. Reorganization allowed owners to reform financial and administrative systems and often lowered fixed charges on bonds by converting them into preferred stock. Chandler, *Visible Hand*, 184.

19. Water "rights" gave subscribers a perpetual right to *use* water from the ditches; water-right fees paid for ditch construction. The annual rental fee was a service fee to pay for repairs and operation. In California, some irrigation companies charged a royalty and an annual rental fee—they did not sell any water rights. State of Colorado, Office of Secretary of State, Articles of Incorporation for Pecos Irrigation and Improvement, 19 May 1890, in *Bolles v. Pecos Irrigation Company*, Docket 992, 23 December 1908, District Court of Eddy County, New Mexico; Dunbar, *Forging New Rights*, 24–25, 31; Keleher, *Fabulous Frontier*, 196.

20. P. Hagerman, *The Great Illusion*, 4, Hagerman Family Collection; Bogener, "Carlsbad Project," 9.

21. P. Hagerman, *The Great Illusion*, 4, Hagerman Family Collection; Bogener, "Carlsbad Project," 9; "Carlsbad Project: Annual Project History," through 1913 [1 and 2] and 1910–1912, vol. 1, U.S. Department of the Interior, Bureau of Reclamation, RG 115, box 3, vol. 1, p. 21; Sheridan, *Bitter River*, 58; *History of Eddy County*, 27; *U.S. v. Hope*, vol. 1, 3261.

22. Gardner T. Sanford and Charles S. Kelley, bankers and brokers, teamed up to form the firm in 1884. Kelley visited the Pecos Valley in 1890 and subsequently wrote twenty columns about his trip for the *New Bedford Standard*. *Eddy Argus*, 9 August 1890; Leonard Bolles Ellis, *History of New Bedford and Its Vicinity: 1602–1892* (Syracuse, N.Y.: D. Mason, 1892), 520.

23. Otis was born in Bloomfield, Ohio, on 26 June 1835. In 1853, following his father's lead, Otis organized the firm of Ford and Otis, establishing the first forge in Cleveland and manufacturing axles and bar iron. Following the Civil War, Otis traveled to Germany where he learned how to refine the making of steel, returned to this country and founded the Otis Iron and Steel Company, later, Otis Steel Company, and built the largest open-hearth steel plant in the United States. Otis remained president of the company until 1899, when he sold it to an English syndicate. In 1872, Otis was elected mayor of Cleveland. In 1894 he was named president of the Commercial National Bank, holding office until 1904. Otis and others founded what became the American Steel and Wire Company, and he was associ-

ated with the American Steel Screw Company, the Cleveland Electric Railway Company, and Otis Elevator Company. "A History of Cleveland, Ohio," 1027.

24. Sidney Dillon was born 7 May 1892, in Northampton, New York. He started work at the age of seven as a "waterboy" for the Mohawk and Hudson Railroad. Later, he oversaw construction of several railroad projects in New England and began bidding on such projects. Over the next thirty years, Dillon built thousands of miles of railway across the country, including the Union Pacific Railroad. He served as a director of the road from 1864 to 1892 and president from 1874 until 1884, and again from 1890 to 1892. After 1870 Dillon served largely as a financier, having already made a fortune. Managing railroad securities took the bulk of his time. He associated with Jay Gould, managing many of his properties, and served as a director of the Western Union Telegraph Company and other transportation companies. *Dictionary of American Biography*, s.v. "Dillon, Sydney"; P. Hagerman, *The Great Illusion*, 4, Hagerman Family Collection.

25. *Eddy County News*, 13 June 1947; *Eddy Argus*, 5 July 1890; P. Hagerman, *The Great Illusion*, 3, 4, Hagerman Family Collection; Coan, *History of New Mexico*, 465; "Carlsbad Project: Annual Project History," through 1913 [1 and 2] and 1910–1912, vol. 1, U.S. Department of the Interior, Bureau of Reclamation, RG 115, box 3, vol. 1, p. 20; U.S. Department of the Interior, National Park Service, *Historic American Engineering Record: Carlsbad Irrigation District, HAER No. NM-4*, vol. 2, *Historical and Descriptive Data* (Washington, D.C.: National Park Service, 1991), 30; John Steele Gordon, "The Farthest Fall," *American Heritage*, July–August 1997, 14, 18; Gilpin, *Dictionary of Economic Terms*, 121; John Mack Faragher et al., *Out of Many: A History of the American People*, 2d ed. (Upper Saddle River, N.J.: Prentice Hall, 1997), 589; Richard B. Morris, ed., *Encyclopedia of American History, Updated and Revised* (New York: Harper and Row, 1965), 261; Chandler, *Visible Hand*, 145–48, 176–77, 184, 491, 509; Hagerman to Reed, 1 July 1904, U.S. Department of the Interior, Bureau of Reclamation, RG 115.

26. *Eddy County News*, 13 June 1947; *Eddy Argus*, 5 July 1890; P. Hagerman, *The Great Illusion*, 3, 4, Hagerman Family Collection; Gordon, "The Farthest Fall," 14, 18.

27. Joseph Lea, Charles Eddy, Arthur Mermod, Patrick Garrett, and Edgar Bronson organized the Pecos Valley Town Company on April 8, 1889, for the purpose of establishing the town of Eddy and selling lots in the vicinity. Roberts, "Fifty Eventful Years," 5.

28. P. Hagerman, *The Great Illusion*, 4, Hagerman Family Collection.

29. U.S. Department of Agriculture, Report No. 64, *Field Operations of the Division of Soils, 1899*, by Milton Whitney (Washington, D.C.: Government Printing Office, 1900) 18; Eddy County, Deeds of Record; P. Hagerman, *The Great Illusion*, 5–7, Hagerman Family Collection; Hagerman, "Statement," c. 1900, Hagerman Family Collection; Tracy, "Pecos Valley Pioneers," 203; Clark, *Water in New Mexico*, 68; Tarr Report, 17–30.

30. The cost of land and irrigation waters seemed high to many small farmers. Thus, they used as much water as possible to ensure they got their money's worth.

This led to salinization and overirrigation. That farmers tended to overwater comes as no surprise since the New Mexico Bureau of Immigration advised them to "let the water run a month if need be, rather than turn it off before the ground is thoroughly soaked." The company originally proposed fixing the price of "water rights" for the first 20,000 acres at $10.00 per acres, the next 10,000 acres at $12.50 per acre, and after that the company would sell no "rights" for less than $15.00 an acre. The company wanted one-tenth in cash down and the balance to be paid in ten annual installments at 8 percent interest. In addition, an annual water rent was fixed at $1.25 per acre for cultivated land, and $0.25 an acre for uncultivated land. Pecos Irrigation and Investment required half the amount on June first, the balance due in December of each year. Observers feared that average farmers could not take advantage of the water rates since numerous friends of the company took up the first 20,000 acres and speculators took up the next 10,000. Therefore, average farmers were forced to pay $15.00 an acre for water usage. Without a railroad and markets for their crops, farmers could not pay the fees. Ralph Tarr, in his 1889 evaluation, suggested that much of the land under ditch would eventually become the property of the company through failure to pay taxes. And Tarr concluded, "it is possible that the company looks on the matter in the same light." New Mexico (Territory), Bureau of Immigration, *New Mexico: Its Resources*, 46; U.S. Department of Agriculture, Report No. 64, *Field Operations of the Division of Soils, 1899*, 18; Eddy County, Deeds of Record; P. Hagerman, *The Great Illusion*, 5–7, Hagerman Family Collection; Hagerman, "Statement," c. 1900, Hagerman Family Collection; Tracy, "Pecos Valley Pioneers," 203; Clark, *Water in New Mexico*, 68; Tarr Report, 17–30.

31. Hagerman, "Statement," c. 1900, Hagerman Family Collection.

32. P. Hagerman, *The Great Illusion*, 5, Hagerman Family Collection; Tracy, "Pecos Valley Pioneers," 204.

33. P. Hagerman, *The Great Illusion*, 5, 6, Hagerman Family Collection.

34. Pecos Irrigation and Improvement Company, Annual Report, December 1891, in P. Hagerman, *The Great Illusion*, 6, Hagerman Family Collection; Wilson, "Pecos Valley Canals," 350–51.

35. P. Hagerman, *The Great Illusion*, 5–7, Hagerman Family Collection.

36. P. Hagerman, *The Great Illusion*, 6, Hagerman Family Collection; *Eddy County News*, 18 July 1947.

37. P. Hagerman, *The Great Illusion*, 6, Hagerman Family Collection.

38. Ibid., 7.

39. Wilson, "Pecos Valley Canals," 350; P. Hagerman, *The Great Illusion*, 8, Hagerman Family Collection.

40. Hagerman to John R. Holland, 28 December 1889, in P. Hagerman, *The Great Illusion*, 8, Hagerman Family Collection.

41. P. Hagerman, *The Great Illusion*, 10, Hagerman Family Collection; U.S. Congress, Senate, 48th Cong., 1st sess., 1884, S. Doc.181.

42. Most capital for irrigation came from the East Coast or from British

investors. British money led to the organization in the 1880s of the North Colorado Irrigation Company, which built the eighty-five-mile-long High Line Canal, a diversion of the Platte, for $650,000. The stock Hagerman offered to the British was not entirely his own. It was, according to son Percy, "all of which father [Hagerman] and his immediate friends could furnish." Dunbar, *Forging New Rights*, 24; P. Hagerman, *The Great Illusion*, 10, 11, Hagerman Family Collection; U.S. Department of the Interior, National Park Service, *HAER*, 33–34.

43. P. Hagerman, *The Great Illusion*, 11, Hagerman Family Collection.

44. Wheeler was born 9 October 1842, in Hopkinton, Massachusetts. In July 1862 he was made a cadet at the U.S. Military Academy. In 1866, as a second lieutenant, Wheeler did survey work in California (possibly alongside Clarence King, who was there during that time). In 1871 Wheeler was selected to survey lands west of the one hundredth meridian for the Army Corps of Engineers. Originally topographical in nature, the survey expanded its role to include geology, zoology, and ethnology of the region. From 1871 to 1879, on fourteen trips of between three and eight and a half months each, Wheeler traversed the mountains and inter-montane parts of the country, accumulating information for forty partial volumes entitled *Report upon United States Geographical Surveys West of the One Hundredth Meridian*. Wheeler's work was compiled into eight complete volumes plus atlases between 1875 and 1889. Wheeler, as chief of the Army Corps' survey organization, supervised publication of his reports in 1879 when the U.S. Geological Survey took over all survey tasks in the West. Exposure and fatigue resulting from his explorations in Europe led to Wheeler being declared permanently incapacitated for active service in 1889, although he continued to work for the army as he was able until 1888. He died 3 May 1905, in New York, where he spent the last years of his life. P. Hagerman, *The Great Illusion*, 11, Hagerman Family Collection; *Dictionary of American Biography*, s.v. "Wheeler, George Montague."

45. The Baring Bank failed because of overspeculation in securities prior to the Panic of 1893. The bank financed and speculated in Argentina, then could not sell the securities. It had to sell other stocks, causing rumors and eventually distrust. The bank closed on 15 November 1890. W. Jeff Lauck, *The Causes of the Panic of 1893* (Boston: Houghton, Mifflin, 1907), 63–65; P. Hagerman, *The Great Illusion*, 11, Hagerman Family Collection.

46. P. Hagerman, *The Great Illusion*, 11, Hagerman Family Collection.

47. Ibid., 10.

48. Ibid., 12.

49. Ibid., 12; New Mexico (Territory), Bureau of Immigration, *New Mexico: Its Resources*, 46.

50. New Mexico (Territory), Bureau of Immigration, *New Mexico: Its Resources*, 46; P. Hagerman, *The Great Illusion*, 12, Hagerman Family Collection.

51. P. Hagerman, *The Great Illusion*, 12–13, Hagerman Family Collection.

52. Eddy County, Deeds of Record, book 6, 88–89.

53. *Eddy Argus*, 28 February 1891.

54. The only outlet was through Pecos and the Texas and Pacific Railroad. P. Hagerman, *The Great Illusion*, 12–14, Hagerman Family Collection; Hagerman, "Statement," c. 1900, Hagerman Family Collection.

55. P. Hagerman, *The Great Illusion*, 4, Hagerman Family Collection.

56. The Pecos Company was a holding company—a company that existed to hold securities of other corporations. When corporations' stock was pyramided on top of one another, the holding company allowed someone with little actual investment to control assets. A person holding 51 percent of a holding company would control all companies that the holding company had 51 percent interest in and all the subsidiaries of those second-tier companies. These holding companies often became top heavy, since they depended on income from subsidiaries to pay for investments. The holding companies were usually highly leveraged—with borrowed money—but they decided the policies of their subsidiary companies. The purpose of a holding company was to extend prices and control output to monopolistic conditions to circumvent the Sherman Anti-Trust Act of 1890. The Sherman Act outlawed "every . . . combination . . . in restraint of trade." It was an ineffective attempt to stop large companies from controlling certain segments of markets. Directors of Pecos Irrigation and Improvement included Hagerman, Henry C. Lowe, R. W. Tansill, Lewis H. Jackson, Charles Eddy, and W. A. Hawkins. The company's headquarters were in Colorado Springs, with a branch in Eddy. Charles Eddy served as general manager. *Eddy County News*, 13 June 1947; *Eddy Argus*, 5 July 1890; P. Hagerman, *The Great Illusion*, 3, 4, Hagerman Family Collection; Coan, *History of New Mexico*, 465; "Carlsbad Project: Annual Project History," through 1913 [1 and 2] and 1910–1912, vol. 1, U.S. Department of the Interior, Bureau of Reclamation, RG 115, box 3, vol. 1, p. 20; U.S. Department of the Interior, National Park Service, *HAER*, 30; Pecos Irrigation and Improvement, Articles of Corporation, in *Bolles v. Pecos Irrigation;* Gordon, "The Farthest Fall," 14, 18; Gilpin, *Dictionary of Economic Terms*, 121; Faragher et al., *Out of Many*, 589; R. Morris, ed., *Encyclopedia of American History*, 261; Chandler, *Visible Hand*, 145–148, 176–177, 184, 491, 509; Hagerman to Reed, 1 July 1904, U.S. Department of the Interior, Bureau of Reclamation, RG 115.

57. U.S. Department of the Interior, National Park Service, *HAER*, 33–34; Tracy, "Pecos Valley Pioneers," 199; P. Hagerman, *The Great Illusion*, 13, 14, 17, Hagerman Family Collection; *Eddy Argus*, 12 May 1893.

58. P. Hagerman, *The Great Illusion*, 14, Hagerman Family Collection.

59. *Eddy Argus*, 11 August 1893; *History of Eddy County*, 28.

60. *Eddy Argus*, 11 October 1893; Schuyler, *Reservoirs for Irrigation*, 51.

61. The dam had thirty-one floodgates, each five by seven feet. *Eddy Argus*, 11 August 1893; Schuyler, *Reservoirs for Irrigation*, 51.

62. *Eddy Argus*, 11 August 1893.

63. Ibid.; *History of Eddy County*, 28.

64. Just a year before the flood, the *Eddy Argus* stated that the Avalon Dam "is one of the strongest structures of that kind in the United States." The *Argus* on 1 September 1893, reported the damage as $125,000. In remembering the event in

1900, James Hagerman stated that the loss to the irrigation company and the rail-road amounted to $500,000. Schuyler estimated that the damaged amounted to $776,000, or seven dollars per acre for the 110,000 acres covered. Whatever the cost, the flood wiped out the wooden flume, two bridges, and the dam. U.S. Department of the Interior, Bureau of Reclamation, RG 115, box 3, vol. 21; *Eddy Argus*, 9 July 1892, 11 August 1893, 1 September 1893; P. Hagerman, *The Great Illusion*, 14, 16, Hagerman Family Collection; Hagerman, "Statement," c. 1900, Hagerman Family Collection; Schuyler, *Reservoirs for Irrigation*, 56.

65. *Eddy Argus*, 11 August 1893.

66. Ibid.

67. From 1887 through 1890, high speculation by British investors in North America, Central America, South Africa, and Australia led British trust and invest-ment companies to organize, promote, underwrite, and speculate in securities of other companies. From 1891 through 1893 the financial world turned to conser-vatism and caution and away from the reckless speculation of the three previous years. The change was due to the discovery of rash and corrupt financial practices and liquidation of trust companies, which had absorbed public savings in Britain. Growing apprehension over the currency situation—fear that the U.S. Treasury would not have enough gold to cover all the treasury notes—and other worldwide financial factors caused the absence of speculation. Companies and the stock market collapsed. Lauck, *Causes of the Panic of 1893*, 40–41, 97–113; *Eddy Argus*, 16 February 1894; Hagerman, "Statement," c. 1900, Hagerman Family Collection; P. Hagerman, *The Great Illusion*, 14, 15, Hagerman Family Collection; Larson, *Forgotten Frontier*, 263.

68. Keleher, *Fabulous Frontier*, 198; Faragher et al., *Out of Many*, 632.

69. While the flood caused a financial setback, Hagerman's financial statement for 1893 is still impressive. He had $80,000 available in cash on hand, with accounts receivable of over $500,000, including $32,000 from Pecos Irrigation and Improvement. He owned $530,000 worth of Pecos Valley Irrigation and Improve-ment and Pecos Valley Railway Company bonds. He held $119,000 worth of sub-scriptions in the Pecos Valley, plus other assets worth $3.7 million. His gross income for 1893 was $456,525. Net income amounted to $291,348. Hagerman owned 25,000 acres in Colorado. His property in the lower Pecos Valley totaled 2,872 acres. The statement also listed liabilities of $3.7 million, including a "profit and loss account" of $3.2 million. Bogener, "Carlsbad Project," 10; P. Hagerman, *The Great Illusion*, 14–16, Hagerman Family Collection; *Eddy Argus*, 20 October 1893; Hagerman, "Statement," c. 1900, Hagerman Family Collection.

70. The firm of Ward and Courtney rebuilt the dam. The contract was com-pleted by January 1894. *Eddy Argus*, 20 October 1892, 1 September 1893.

71. Schuyler, *Reservoirs for Irrigation*, 50.

72. *Eddy County News*, 4 July 1947; Clark, *Water in New Mexico*, 88.

73. Ibid.

74. *Carlsbad Current-Argus*, 16 May 1947; Larson, *Forgotten Frontier*, 269.

75. Myers, *Pearl of the Pecos*, vi; *Eddy County News*, 4 July 1947.

76. *Eddy Argus*, 9 November 1894.

77. P. Hagerman, *The Great Illusion*, 19, Hagerman Family Collection.

78. When Eddy left he was the vice president and manager of Pecos Irrigation and Improvement. *Eddy Argus*, 27 April 1894; Bogener, "Carlsbad Project," 11; P. Hagerman, *The Great Illusion*, 19, Hagerman Family Collection; *History of Eddy County*, 28.

79. *Carlsbad Current-Argus*, 16 May 1947.

80. First Annual Report, Pecos Company, 20 March 1894, in P. Hagerman, *The Great Illusion*, 16, 17, Hagerman Family Collection.

81. Bogener, "Carlsbad Project," 3–4; P. Hagerman, *The Great Illusion*, 6–7, Hagerman Family Collection.

82. New Mexico (Territory), Bureau of Immigration, *New Mexico: Its Resources*, 250.

83. P. Hagerman, *The Great Illusion*, 6–7, Hagerman Family Collection.

84. Bogener, "Carlsbad Project," 3–4.

85. P. Hagerman, *The Great Illusion*, 19, Hagerman Family Collection; John Middaugh, *Frontier Newspaper: The El Paso Times* (El Paso: Texas Western Press, 1958), 59–63.

86. W. A. Hawkins was born 6 April 1861, in Tennessee. His uncle, Tennessee governor Alvin Hawkins, raised him. Hawkins attended Vanderbilt University, then went west. He was admitted to the New Mexico bar in 1885 and joined the prestigious firm of Conway and Posey in Silver City. When Charles Eddy moved to New Mexico, he needed a lawyer, as his attorney in Colorado Springs could not leave his practice in the Springs. He recommended Hawkins. Hawkins subsequently moved his practice to Eddy. Blodgett was the former private secretary to Hagerman. He came to the valley to work for Pecos Irrigation and Investment as a stenographer and typist. *History of Eddy County*, 27; *Eddy Argus*, 29 March 1890, 27 April 1894; David Townsend, *You Take the Sundials and Give Me the Sun* (Alamogordo, N. Mex.: Alamogordo Daily News, 1984 [Sun Country Printing], 50.

87. Blodgett, according to Percy Hagerman, was weak and easily led and was finally "ruined by drink." Hawkins rejoined his friend Charles Eddy and promoted the El Paso and Northeastern Railroad. P. Hagerman, *The Great Illusion*, 20, 21, Hagerman Family Collection; Townsend, *You Take the Sundials*, 50.

88. P. Hagerman, *The Great Illusion*, 20, Hagerman Family Collection.

89. Many people successfully planted orchards, including John Chisum; others, including Hagerman, failed. One orchard, established with trees from France and run by Frenchmen Henri Bole, a noted French horticulturist, failed because terrible sandstorms destroyed leaves and small trees. The storms also filled in ditches and laterals. Yet, the Bureau of Immigration claimed that only Santa Fe rivaled Eddy County in fruit production. Eddy County growers raised apples, peaches, pears, plums, prunes, cherries, and grapes, all large in size, with perfect flavor. The bureau also claimed that one acre of land would support 108 peach trees and that local apple orchards produced from 1,000 to 1,500 pounds of fruit each year—with

some of the apples having a circumference of fifteen inches! While the size of the apples is questionable, now it is not unusual to plant seventy-five apple trees per acre. Each tree should produce between 35 and 40 pounds of fruit. Thus an acre of apple trees could produce up to 3,000 pounds of fruit. Sheridan, *Bitter River*, 54; New Mexico (Territory), Bureau of Immigration, *Eddy County, New Mexico, the Most Southeastern County*, 28, 32; Tracy, "Pecos Valley Pioneers," 200; U.S. Department of the Interior, National Park Service, *HAER*, 47; P. Hagerman, *The Great Illusion*, 25, 38, Hagerman Family Collection.

90. Larson, *Forgotten Frontier*, 269; Robert A. Meeker, "Pecos Valley Railway and Proposed Extension," report submitted to Harvey Fisk and Sons, No. 24, Nashua Street, New York, New York, c. 1896, 4, Hagerman Family Collection. Meeker undoubtedly arrived before the crop was fully harvested. He reported that the crop was successful, with the highest percentage of sugar of any beets in the world.

91. Hagerman's beets probably suffered from curly-top disease or root rot. The soil's salt content probably hurt the crop too. The salt content of the Pecos River equaled 240 parts soluble matter per 100,000 parts water, the limit of plant endurance. That Hagerman attempted to raise beets was not so unusual. The Bureau of Immigration promoted the area for sugar beets based on their great size, high sugar content, and purity. U.S. Department of Agriculture, Report No. 64, *Field Operations of the Division of Soils*, 17–19; New Mexico (Territory), Bureau of Immigration, *Eddy County, New Mexico, the Most Southeastern County*, 27; *Eddy Argus*, 16 February 1884; U.S. Department of Agriculture, Office of Experiment Stations, *The Use of Water*, 98; Keleher, *Fabulous Frontier*, 199; P. Hagerman, *The Great Illusion*, 11, 24, Hagerman Family Collection; *Denver Field and Farm*, 21 August 1897.

92. Lipsey, *The Lives of James John Hagerman*, 227; P. Hagerman, *The Great Illusion*, 15, 25, Hagerman Family Collection.

93. *Hearing in the Matter of the Protest . . . against the Construction of the Hondo Reservoir*, 24–25, U.S. Department of the Interior, Bureau of Reclamation, RG 115, box 1, entry 4, loc. 42/4/3:2.

94. As to Hagerman Heights, in 1890 or 1891 Eddy had interested Hagerman in building a residence or plantation on the east side of the river. The house that Hagerman built also included a reservoir, water pump, and waterpower to supply electricity to the house, orchard, and other farming developments. All of this eventually cost Hagerman some $225,000. When Hagerman eventually left the lower valley in 1900, moving to the Roswell area, he sold the entire outfit to Robert Weems Tansill for $10,000, reflecting a $215,000 loss. P. Hagerman, *The Great Illusion*, 15, 16, Hagerman Family Collection.

95. P. Hagerman, *The Great Illusion*, 15, 16, Hagerman Family Collection; *U.S. v. Hope*, vol. 1, 3257.

6. RAILROADS AND RECEIVERSHIP

1. Eddy to Prince, 3 September 1890, Prince Papers.

2. *Eddy Argus*, March 1890 [date unreadable].

3. *Eddy Argus*, April 1890 [date unreadable].

4. *Eddy Argus*, 30 August 1890.

5. *Eddy Argus*, 13 December 1890.

6. *Eddy Argus*, 1 November 1890.

7. *Eddy Argus*, 17 January 1891; *History of Eddy County*, 27.

8. *Eddy Argus*, 17 January 1891.

9. Ibid.

10. Keleher, *Fabulous Frontier*, 200.

11. Elvis E. Fleming, "J. J. Hagerman and the Pecos River Railroad," *Permian Historical Annual* 23 (December 1977): 25.

12. A number of men who had made their names in the railroad business served as directors of the railroad company, including Hagerman, who had built the Colorado Midland; William McMillan, the namesake of the storage reservoir and president of the St. Louis Car and Iron Company; Charles Otis, who had founded Otis Steelworks in Cleveland; Richard Bolles, who had been involved with Hagerman in mining operations and extended his investments later to the Florida Everglades; B. F. Ham of Cranford, New Jersey; and W. A. Otis, W. P. Bonbright, and Thomas Edsall, all of Colorado Springs. Local resident directors included Charles Eddy and Frederick Dominice, who represented the Swiss colonists who had moved to the valley two years before. Arriving for the celebration were company president James Hagerman, vice president Charles Eddy, and directors Charles Otis, Daniel Ells, and Richard Bolles. *Eddy Weekly Current*, 14 April 1893; *Eddy Argus*, 12 October 1894.

13. Richard H. Gravel, "The Pecos Valley Railroad, 1889–1906" (master's thesis, Eastern New Mexico University, Portales, New Mexico, 1924), 148.

14. *Eddy Weekly Current*, 3 February 1893, 10 March 1893.

15. *Roswell Record*, 12 January, 3 February, 6 April 1894.

16. Fleming, "J. J. Hagerman and the Pecos River Railroad," 23.

17. Gravel, "The Pecos Valley Railroad," 31–32; Fleming, "J. J. Hagerman and the Pecos River Railroad," 23.

18. *Roswell Record*, 12 January, 3 February, 6 April 1894.

19. *Roswell Record*, 24 April 1894.

20. Hagerman also chartered a special train from Eddy, adding to the crowd that day of 4,500. Several speeches were made, including one by Governor W. T. Thornton. Hagerman praised the region's elevation and climate, irrigation, cattle, and resources. While dignitaries gave lofty speeches, the crowd consumed 20 head of cattle, 500 watermelons, and other foods prepared for the occasion. Fleming, "J. J. Hagerman and the Pecos River Railroad," 24, 25.

21. *Eddy Argus*, 12 October 1894.

22. *Eddy Argus*, 7 October 1894.

23. *Eddy Argus*, 19 October 1994.

24. *Eddy Argus,* 12 October 1894.

25. *The Pecos Valley, the Fruit Belt of New Mexico* (Eddy, N. Mex.: Pecos Irrigation and Improvement Company, 1891), microfilm, SWC, Western Americana, reel 415, no. 4200; Larson, *Forgotten Frontier,* 269.

26. *Eddy Argus,* 1 November 1890.

27. By 1887 the Santa Fe was the largest rail system in the world—perhaps too big at 8,000 miles. The company had been on the verge of bankruptcy in 1888, but had recovered and by 1893 had 9,232 miles of track and had a capital worth of $647,000,000 because Jay Gould refinanced the Santa Fe. Gould was also involved in the Missouri Pacific, Texas and Pacific, St. Louis Southwestern, and the International and Great Northern. Gould died in December 1892, but the Santa Fe continued to grow. By 1917 it boasted assets of $847,000,000 and 11,291 miles of track. Chandler, *Visible Hand,* 160, 168, 171, 513; Lipsey, *The Lives of James John Hagerman,* 235–36.

28. *Roswell Record,* 20 May 1892.

29. In 1893 Hagerman and Eddy tried unsuccessfully to sell $5 million worth of Pecos Irrigation and Improvement company bonds to extend the road to a point thirty miles south of Las Vegas, where the Atchison, Topeka and Santa Fe crossed the Pecos River. *Roswell Record,* 28 April 1893; *Eddy Argus,* 20 February 1892; Gravel, "The Pecos Valley Railroad," 25, 27, 28.

30. The floating debts (debts that involve short-term borrowing) included the collateral trust notes and loans from banks and individuals. Gilpin, *Dictionary of Economic Terms,* 106.

31. The company's bonded debt included $300,000 on the 8 percent and $500,000 on the 6 percent Pecos Irrigation and Improvement bonds, $2,346,000 at 5 percent for the Pecos Valley Railway Company. The floating debt stood at $687,039.16. The company listed assets other than cash ($72,816) and accounts receivable ($299,591) at $922,553.03. It needed $411,000 to meet maturing interest obligations in 1895. Percy Hagerman claimed the company needed an additional $230,000 to stay afloat; however, based on these figures, the amount was closer to $196,000. P. Hagerman, *The Great Illusion,* 17–18, Hagerman Family Collection.

32. Ibid., 18.

33. A trust is an arrangement where property is transferred to a trusted third party by a grantor. The trustee holds the property for the benefit of another—in this case, stock and bondholders. Ibid., 18–22; Jack Friedman, Jack Harris, and J. Bruce Lindeman, *Dictionary of Real Estate Terms* (Woodbury, N.Y.: Barron's, 1984), 285.

34. P. Hagerman, *The Great Illusion,* 22, Hagerman Family Collection.

35. Ibid., 22, 23.

36. Ibid., 16, 23, 24.

37. Ibid., 16, 23, 24.

38. By the end of 1896, Hagerman owned 72 percent (10,200 shares, which cost him $1,230,000), and Otis owned 28 percent of the company. At first Otis and Hagerman kept putting money into Pecos Irrigation and Improvement because

they thought the success with the extended railroad would rehabilitate the irrigation company. They continued to pour money into the company because they feared that a receivership for the irrigation company would ruin the railway deal. Finally a receivership for Pecos Irrigation and Improvement became imminent. Sheridan, *Bitter River*, 64; P. Hagerman, *The Great Illusion*, 26, Hagerman Family Collection; Larson, *Forgotten Frontier*, 270.

39. Charles Eddy visited White Oaks several times in 1900. He was convinced that coal, gold, and other minerals warranted a rail line into town. By May 1900 Eddy had completed ten miles of track from El Paso. Progress was slow because of opposition from the Southern Pacific and Santa Fe. Eddy continued building, but not through White Oaks. The rail line asked White Oaks for a right-of-way, a flat vacant area near the west edge of town for a depot, shops, and a siding, plus a $50,000 cash bonus. The town balked. The railway dropped the demand for a bonus. The town refused to compromise. It did not want to spend money or make concessions for the railroad. Consequently, the railroad passed White Oaks by, and the town died. F. Stanley, *The White Oaks, New Mexico Story* (Pep, Texas, undated), 21–22; Morris B. Parker, *White Oaks: Life in a New Mexico Gold Camp*, ed. with intro. by C. L. Sonnichsen (Tucson: University of Arizona Press, 1971), 117–21; Townsend, *You Take the Sundials*, 47–67.

40. Eddy and Hawkins campaigned to have the surrounding county named for the governor, Miguel Otero. Eddy and Hawkins were also instrumental in establishing the resort town of Cloudcroft by building a rail line, the "Cloud Climbing Route," to the mountains. Townsend, *You Take the Sundials*, 54–61.

41. Ibid.

42. Dawson was named for John Barkley Dawson, born in Kentucky in 1830. Dawson trailed a herd of cattle from Arkansas to California in the early 1850s, ran herds to Colorado for Charles Goodnight in the late 1860s, and served off and on with the Texas Rangers from 1864 to 1867. Passing through northern New Mexico, Dawson purchased a tract of land from Lucien Maxwell in 1869. Maxwell owned the largest piece of real estate in the United States—2 million acres. Dawson bought 3,700 acres for $1,000. The Maxwell lands had passed to Maxwell through marriage ties dating back to the sixteenth century. After the grant passed to the Maxwell Land Grant Company in 1880, the latest owners started examining the property to see who lived on the grant, for coal deposits, worth $1 billion to railroads in the region, had been discovered. The Maxwell Land Grant Company initiated ejection lawsuits in the 1880s and 1890s to get rid of what officials considered squatters, including Dawson, who hired Las Vegas attorney Andrieius A. Jones. Jones became the first U.S. senator from New Mexico after statehood in 1912. Dawson won in a jury trial in San Miguel County District Court and on appeal in the U.S. Supreme Court. The Court decided Dawson owned closer to 20,000 acres than 3,700. Charles Eddy watched all this legal activity intently, consulting his attorney, W. A. Hawkins, who assured him Dawson would win. Toby Smith, *Coal Town: The Life and Times of Dawson, New Mexico* (Santa Fe, N. Mex.: Ancient City Press, 1993), 2–5.

43. Larson, *Forgotten Frontier*, 268.

44. *Bolles v. Pecos Irrigation Co.*, 23 N.M. 32; P. Hagerman, *The Great Illusion*, 26, Hagerman Family Collection.

45. Hagerman, like Charles Eddy, had talked to the Rock Island Railroad about a connection with the company's Liberal, Kansas, branch and another connecting rail to El Paso. According to his son Percy, Hagerman believed the company had no intention of making an El Paso connection, and Eddy would fail. Discussions with the St. Louis and San Francisco Railroad Company also fell through. P. Hagerman, *The Great Illusion*, 34–35, Hagerman Family Collection.

46. Meeker, "Pecos Valley Railway and Proposed Extension," 1, 3, Hagerman Family Collection.

47. In 1894 the company had 121 miles of main canals, 273 miles of laterals, and 900 miles of farm and sublaterals, for a total of 1,294 miles of ditch. New Mexico (Territory), Bureau of Immigration, *New Mexico: Its Resources,* 250; Meeker, "Pecos Valley Railway and Proposed Extension," 3, Hagerman Family Collection.

48. Meeker, "Pecos Valley Railway and Proposed Extension," 8, Hagerman Family Collection.

49. Ibid., 19.

50. The Pecos Valley and Northeastern Railway Company took control of the bankrupt Pecos Valley Railway on 30 April 1898. Gravel, "The Pecos Valley Railroad," 48–50; Fleming, "J. J. Hagerman and the Pecos River Railroad," 27.

51. Subscribers to the $1,000,000 worth of preferred and common stock included the Pullman Palace Car Company, $200,000; B. P. Cheney of Boston, $168,000; Frederick Ayer of Boston, $50,000; Charles Head and Company of New York, $50,000; Edgerly and Crocker of Boston, $25,000; Clement Parker and Company of Boston, $25,000; C. C. Converse of Boston, $10,000; Lee Higginson and Company, $8,000; Maitland, Copple and Company, New York, $50,000; Charles Otis and other investors in Cleveland, $150,000; James M. Ham of New York, $10,000; Harriman and Company of New York, $200,000; and James Hagerman, $54,000.

George Mortimer Pullman was born in Brocton, New York, in 1831. For a while he was a cabinetmaker, moving to Chicago in 1855. There he contracted to raise street and building levels in the city. In 1858 he contracted with the Chicago and Alton Railroad to remodel two-day cars into sleeping cars. Pullman then went to the Colorado mining fields where he ran a general store from 1859 to 1863. In 1867 Pullman and Ben Field organized the Pullman Palace Car Company and established plants in several cities, including St. Louis and Chicago, to produce their patented sleeping car. Pullman died in 1897, survived by his widow and four children.

Benjamin Pierce Cheney was born 12 August 1815, in Hillborough, New Hampshire. Cheney early on worked in his father's blacksmith shop, a tavern, a store, and later worked with stage lines in New England, establishing business with the American Express Company. Through the express business, Cheney amassed a fortune. He died in 1895. *Dictionary of American Biography*, s.v. "Cheney, Benjamin Pierce."

P. Hagerman, *The Great Illusion*, 27, 28, Hagerman Family Collection; *Dictionary of American Biography*, s.v. "Pullman, George Mortimer"; Meeker, "Pecos Valley Railway and Proposed Extension," 15, Hagerman Family Collection.

52. P. Hagerman, *The Great Illusion*, 27, 28, Hagerman Family Collection.

53. Keith L. Bryant Jr., *History of the Atchison, Topeka, and Santa Fe Railway* (Lincoln: University of Nebraska Press, 1974), 192.

54. Gravel, "The Pecos Valley Railroad," 52–53, 55; *Pecos Valley Argus*, 4 February 1898.

55. P. Hagerman, *The Great Illusion*, 20, 28, 29, Hagerman Family Collection.

56. Perhaps fantasizing, or anticipating the interests of the Santa Fe, in October 1898 Hagerman also announced plans to extend his Pecos Valley Northeastern Railroad south to Boquillas on the Rio Grande in Brewster County, Texas. From that point the line would connect with the Mexican Northern, which was being pushed north to the same point. Hagerman realized that if the Pecos Valley road connected at Boquillas with the Mexican Northern, the result would be a through route for Chicago and Kansas City, all the way to Mexico City. The route would be several hundred miles shorter than any other. *Denver Times*, 12 October 1898.

57. *Meeting the Train*, 24; Fleming, "J. J. Hagerman and the Pecos River Railroad," 28; Larson, *Forgotten Frontier*, 269.

58. P. Hagerman, *The Great Illusion*, 33, Hagerman Family Collection.

59. Eddy organized a company, the Dawson Fuel Company, which for $400,000 purchased the mineral rights and a 23,000-acre town site from Dawson and the Springers, who had obtained a portion of the coal field. Eddy's savvy was uncanny. Having discovered that Dawson's coal was of a superior nature—the best coking coal in the Southwest—and a good enough suppply for one hundred years, he realized the impact on the Phelps Dodge Corporation smelters in southeast Arizona. Phelps Dodge, which contemplated buying Eddy's coal and rail system, built a rail line from Bisbee, Arizona, to El Paso. In 1905, Eddy offered his entire system, including the now bankrupt El Paso and Northeastern line, for sale to Phelps Dodge. The corporation threatened to pull out of negotiations with Eddy in favor of building a line to the San Juan coalfields in the Four Corners area. Eddy, who had tested the San Juan coal, apprised the company that the coal would not coke. In July 1905 the directors of Phelps Dodge agreed to pay Eddy $16 million for his line. Smith, *Coal Town*, 5–7.

60. P. Hagerman, *The Great Illusion*, 35, Hagerman Family Collection.

61. Bryant, *History of the Atchison, Topeka, and Santa Fe*, 192.

62. P. Hagerman, *The Great Illusion*, 35, Hagerman Family Collection.

63. The management of Hagerman's companies stayed under E. O. Faulkner until 1898. D. N. Nichols, the former general superintendent, became general manager in 1899. Hagerman expected the railroad to solve his financial woes. It did not, according to son Percy. The elder Hagerman took a substantial loss relative to his overall investment in the valley. Lipsey, *Lives of James John Hagerman*,

227–28; Fleming, "J. J. Hagerman and the Pecos River Railroads," 29; P. Hagerman, *The Great Illusion*, 20, 28, 29, Hagerman Family Collection.

64. *U.S. v. Hope*, vol. 2, 565.

65. "Plan for the Settlement of the Affairs of the Pecos Irrigation and Improvement Company of New Mexico and for the Reorganization of a New Company by the Bondholders," 25 May 1898, 7, U.S. Department of the Interior, Bureau of Reclamation, RG 115, General Records, Oversize Records, 1902–1910, box 1, entry 4.

66. *Eddy County News*, 18 July 1947; Hagerman to Reed, 1 July 1904, U.S. Department of the Interior, Bureau of Reclamation, RG 115; "Plan for the Settlement," 6, RG 115.

67. Tracy, an early settler of the town of Eddy and cousin of Joseph Stevens, was one of the minor investors in land and irrigation ventures in the valley during its early days.

68. *Eddy County News*, 25 July 1947.

69. *Bolles v. Pecos Irrigation Co.*, 23 N.M. 32; P. Hagerman, *The Great Illusion*, 30, Hagerman Family Collection; *Eddy County News*, 25 July 1947.

70. *Eddy County News*, 25 July 1947; Hagerman to Reed, 1 July 1904, U.S. Department of the Interior, Bureau of Reclamation, RG 115.

71. Hagerman to Reed, 1 July 1904, U.S. Department of the Interior, Bureau of Reclamation, RG 115; "Plan for the Settlement," 5–6, RG 115; *Eddy County News*, 18 July 1947, 25 July 1947.

72. *Hearing in the Matter of the Protest ... against the Construction of the Hondo Reservoir*, 6–7, U.S. Department of the Interior, Bureau of Reclamation, RG 115; *Eddy County News*, 25 July 1947, 1 August 1947; "Plan for the Settlement," 9, 10, RG 115.

73. During the skirmish between 6 percent and 8 percent bondholders, the irrigation company's payrolls fell way behind schedule, and employees were understandably disgruntled. Hagerman knew that certain past balances were due to workers from the Pecos Company reaching back several months. He realized that such expenses had to be paid prior to such a receivership. *Eddy County News*, 25 July 1947, 1 August 1947; "Plan for the Settlement," 9, 10, U.S. Department of the Interior, Bureau of Reclamation, RG 115; *Hearing in the Matter of the Protest ... against the Construction of the Hondo Reservoir*, 6–7, RG 115.

74. "Plan for the Settlement," 11, U.S. Department of the Interior, Bureau of Reclamation, RG 115.

75. In a 1904 letter to W. M. Reed, Hagerman claimed that the reorganization committee and receiver found it impossible to raise the money to reorganize the company, and on 26 May 1899, they offered to sell the Northern Canal to Hagerman for $50,000. Tansill, Tracy, Percy Hagerman, G. B. Shaw of Chicago, and I. W. Rogers made up the committee to carry out the agreement. In any plan of reorganization the 8 percent bondholders would get 60 percent and the 6 percent bondholders were to receive 40 percent on any future dividends returned from the new investment company. At the time of the reorganization, Hagerman, Otis,

Kelley, Tracy, and A. C. Campbell (attorney for Hagerman) controlled 90 percent of the securities in the valley. *Bolles v. Pecos Irrigation*, 23 N.M. 32; *Eddy County News*, 25 July 1947; P. Hagerman, *The Great Illusion*, 30, 31, Hagerman Family Collection; "Plan for the Settlement," 11–12, 14–15, U.S. Department of the Interior, Bureau of Reclamation, RG 115; Hagerman to Reed, 1 July 1904, RG 115.

76. Next to Hagerman, Tansill was the largest individual holder of securities in Pecos Irrigation and Improvement Company. Tansill also bought Hagerman's private residence in Eddy. Hagerman spent more than $225,000 improving the property and building a house. Tansill paid Hagerman $10,000 for the Heights. *Eddy County News*, 25 July 1947, 1 August 1947; "Proposal of J. J. Hagerman to Buy Northern Canal," in Minutes of the Meeting of Stockholders of Pecos Irrigation Company, Carlsbad, New Mexico, 27 December 1900, Tracy Papers; *Roswell Daily Record*, 29 July 1904; *History of Eddy County*, 28; "Plan for the Settlement," 11, U.S. Department of the Interior, Bureau of Reclamation, RG 115; P. Hagerman, *The Great Illusion*, 15, 16, Hagerman Family Collection.

77. Hagerman spent well over $2 million in the valley. When he left he owed $234,336 in notes. After he moved to South Spring Ranch near Roswell, which he purchased in 1898, Hagerman sold all but the 6,000-acre Chisum ranch. He had a railroad siding there and his personal car. After he sold the railroad, Hagerman had $250,000 clear and in cash, plus some land, stocks, and bonds. He built the three-story, twenty-room house in 1902. In Roswell, Hagerman and his wife became friends with lawyer Harold Hurd. Anna Hagerman was godmother to Hurd's son, Peter, the artist. Hagerman donated forty acres to the New Mexico Military Institute and helped pay for Roswell's Carnegie Library. Hagerman, except for a few incidents, was finally able to live his life in peace and pleasure.

Other than irrigation woes, one troubling incident involved Hagerman's son, Herbert. Herbert James Hagerman was born in Milwaukee on 15 December 1871. After attending Cornell University, he practiced law in Colorado Springs. In 1898, under President William McKinley, Hagerman represented the United States in the American Embassy in Russia (1898–1901). From 1901 to 1905, Hagerman joined his father, James, in the Pecos Valley, growing fruit near Roswell. In 1906 Theodore Roosevelt appointed Hagerman territorial governor of New Mexico. Hagerman served until Roosevelt asked him to resign in April 1907 following his alienation of prominent New Mexico power brokers. During his short stint as governor Hagerman tried to root out corruption and gambling and initiated irrigation and tax laws. The incident clouded the Roosevelt/Hagerman friendship.

James Hagerman eventually sold the Northern Canal, which he acquired through purchase, to the federal government. He died in 1909. An inventory of his property, as of 1910, indicated that he had less than $1,500 in the bank. His heirs appraised his estate (not including real estate) at less than $170,000. At the time of his death, Hagerman still owned property in New Mexico at Carlsbad, Hagerman, and in Chaves County. He also had land in various places in Colorado and Michigan.

Sheridan, *Bitter River*, 64; P. Hagerman, *The Great Illusion*, 32, 36, Hagerman Family Collection; Larson, *Forgotten Frontier*, 269–275; Fleming and Huffman, *Roundup on the Pecos*, 19; Robert Larson, *New Mexico's Quest for Statehood, 1846–1912* (Albuquerque: University of New Mexico Press, 1968), 258; Thomas A. McMullin and David Walker, *Biographical Directory of American Territorial Governors* (Westport, Conn.: Meckler, 1934), 257; *National Cyclopedia of American Biography*, s.v. "Hagerman, Herbert James"; Hagerman, "Statement," c. 1900, Hagerman Family Collection; "The South Spring Ranch, Located in the Famous Pecos Valley Four Miles Southeast of Roswell, Chaves County, New Mexico" (Colorado Springs: Out West Printing and Stationary Company, undated), pamphlet in Hagerman Family Collection; Chaves County, Records of Wills, "Inventory Containing the Property of James J. Hagerman, Deceased, Commended this 21st Day of January A.D. 1910," Roswell, New Mexico.

78. *Eddy County News*, 1 August 1947; *History of Eddy County*, 28.

79. Percy Hagerman gives the Amarillo line credit for reviving the region's economy "for several years," allowing Pecos Irrigation Company to survive until 1904. P. Hagerman, *The Great Illusion*, 32, Hagerman Family Collection.

80. *Eddy County News*, 5 September 1947, 26 September 1947; *U.S. v. Hope*, vol. 2, 571–572.

81. Karlsbad, Czechoslovakia, seventy miles west of Prague, is known as a health resort for its sulfur springs. *Eddy County News*, 1 August 1947; *Webster's New Geographic Dictionary*, "Karlsbad."

82. *Eddy County News*, 25 July 1947.

83. Percy Hagerman to Herbert J. Hagerman, 26 February 1899, Hagerman Family Collection.

84. *Eddy County News*, 26 September 1947; *U.S. v. Hope*, vol. 2, 573.

85. *U.S. v. Hope*, vol. 2, 572; *Eddy County News*, 26 September 1947; Pecos Irrigation Company, Certificate of Incorporation, 19 September 1900, State Records Center and Archives, Santa Fe, New Mexico; Coan, *History of New Mexico*, 465; Minutes of the Meeting of Stockholders of Pecos Irrigation Company, 27 December 1900, Tracy Papers.

86. *Eddy County News*, 26 September 1947; Minutes of the Meeting of Stockholders of Pecos Irrigation Company, 27 December 1900, Tracy Papers; Pecos Irrigation Company, Certificate of Incorporation, 19 September 1900.

87. *Eddy County News*, 26 September 1947; Minutes of the Meeting of Stockholders of Pecos Irrigation Company, 27 December 1900, Tracy Papers; Pecos Irrigation Company, Certificate of Incorporation, 19 September 1900; *U.S. v. Hope*, vol. 2, 572.

88. Ibid.

89. *Eddy County News*, 26 September 1947.

7. PRIVATE IRRIGATION DROWNS IN DITCHES
OF ITS OWN MAKING

1. Gates, *History of Public Land,* 651; Ganoe, "Desert Land Act since 1891," 266; Reisner, *Cadillac Desert,* 116; Mead, *Irrigation Institutions,* 344, 346.

2. The Carey Act did not apply to territories, including New Mexico. By 1902 only 11,321 acres had been patented under the Carey Act. Gates, *History of Public Land,* 650; Samuel P. Hays, *Conservation and the Gospel of Efficiency: The Progressive Conservation Movement, 1890–1920* (Cambridge: Harvard University Press, 1959), 9–10; *U.S. v. Hope,* vol. 1, 3264.

3. Hays, *Conservation and the Gospel of Efficiency,* 9–10; Reisner, *Cadillac Desert,* 114–115; Clark, *Water in New Mexico,* 78.

4. By 1903 Roosevelt had appointed a Public Land Commission to report on conditions, operations of land laws, disposal, and settlement of public land. Roosevelt wanted to close the loopholes and get rid of speculators. Hays, *Conservation and the Gospel of Efficiency,* 67–68; Reisner, *Cadillac Desert,* 116–117; Ganoe, "Desert Land Act since 1891," 268; Clark, *Water in New Mexico,* 78; Gates, *History of Public Land,* 652; Pisani, *To Reclaim a Divided West,* 116.

5. U.S. Congress, Senate, *Extracts on Irrigation of Arid Lands,* 57th Cong., 1st sess., 1901–1902, S. Doc. 446, vol. 30, serial 4249, 16–19; Gates, *History of Public Land,* 652–653.

6. Newlands was born in Mississippi in 1848. The Yale and Columbia graduate moved to California to practice law in 1870. Newlands's father-in-law owned interest in several silver mines, and when he died, in 1889, Newlands managed the estate. In doing so, Newlands moved to Nevada where he became involved in politics, economics, and irrigation. He used the Newlands Act as a way to propel himself into the Senate, where he served from 1903 until his death in 1919. Reisner, *Cadillac Desert,* 116; Hays, *Conservation and the Gospel of Efficiency,* 11–12; Robbins, *Our Landed Heritage,* 332; Pisani, *To Reclaim a Divided West,* 116.

7. Gates, *History of Public Land,* 654; Clark, *Water in New Mexico,* 79.

8. The Newlands Act provisions included:

> Section 1—All monies from sale and disposal of public lands in the sixteen
>   western states with semiarid or arid land went into the Reclamation Fund
>   to be used for planning, construction, maintenance of dams, and irrigation
>   works.
> Section 2—The Secretary of the Interior would authorize construction projects as funds permitted.
> Section 3—The Secretary of the Interior must authorize withdrawal from
>   entry lands needed for such projects and lands susceptible to irrigation.
> Section 4—Users of water had to pay for it—sufficiently enough to return
>   the cost of building the works, but with no interest.
> Section 5—The Reclamation Fund would provide for operation and maintenance until local project users paid for the cost of work. The government

would then pass management and operation to owners of the irrigated lands.

Section 7—Eminent domain would be used for obtaining sites and rights-of-way.

Section 8—State water laws would prevail; no vested rights could be affected.

Section 9—The major portion of funds from sale of public lands was to be expended in that state, so that 51 percent of the proceeds generated in that state by public land sales was spent in state. Gates, *History of Public Land*, 655.

9. Gates, *The Jefferson Dream*, 113–114.

10. Gates, *History of Public Land*, 655.

11. Robbins, *Our Landed Heritage*, 332; Elwood Mead, "The Rise of Irrigation in the United States," *The Bureau of Reclamation: Its Functions and Accomplishments*, pamphlet prepared for the Pan-Pacific Conference on Education, Rehabilitation, and Reclamation, Honolulu, Hawaii, 11–16 April 1927 (Washington, D.C.: Government Printing Office, 1927).

12. Frederick Haynes Newell was appointed to the U.S. Geological Survey staff in 1888 by John Wesley Powell. Newell stayed with the USGS through the 1890s and was a leader in forest conservation. He became chief hydrographer in 1895. In that position Newell measured the volume and condition of streams throughout the nation. Newell became the first director of the Reclamation Service in 1902. Pisani, *To Reclaim a Divided West*, 114.

13. The U.S. Geological Survey was the primary scientific organ of the United States. Gates, *History of Public Law*, 657; Clark, *Water in New Mexico*, 82; Smythe, *Conquest of Arid America*, 297; U.S. Department of Interior, Geological Survey, *First Conference of Engineers of the Reclamation Service* (Washington, D.C.: Government Printing Office, 1904), 17.

14. Most of the engineers were young men. Some had experience in measuring streams and studying ecological possibilities of the West under the U.S. Geological Survey. Others became Reclamation engineers through a competitive exam. Most young engineers, or "engineering aids," entered fresh from college and earned $60 per month. Raises came through merit until the aid became an assistant engineer and made between $1,200 and $1,600 per year and demonstrated the ability to conduct independent work and initiate plans. Full engineers made between $1,800 and $3,600 based on age and experience. Smythe, *The Conquest of Arid America*, 298.

15. Reisner, *Cadillac Desert*, 118, quotation on 119.

16. Wisner had been involved in hydraulic construction for forty years and was a consulting engineer. Sanders, a consulting engineer, had designed and built power and irrigation systems in California. Consulting Engineer Hall was from Georgia and had twenty-five years of experience in constructing canals. Davis was supervising engineer and principal assistant to the chief of Reclamation. Quinton and Savage were also consulting engineers. Government Investigating Board, board

members, sheet no. A069567, U.S. Department of the Interior, Bureau of Reclamation, RG 115, box 441, entry 3; Smythe, *Conquest of Arid America*, 297; U.S. Department of Interior, Geological Survey, *First Conference of Engineers of the Reclamation Service*, 17.

17. On all projects, engineers measured streams to learn the amount of water available and surveyed the lands susceptible to irrigation and promising sites to obtain the exact information on cost and efficiency. They studied all factors pertinent to the economy. When the fieldwork was complete (reports, maps, photos), it was sent to the chief engineer in Washington, D.C. Then a consulting board reviewed all the information. If it was promising, the board visited the locale and verified the conclusions of the field engineers. It then transmitted its findings to the chief engineer. The chief either called for more surveys or sent the proposal to the secretary of the interior for his approval. Smythe, *Conquest of the Arid West*, 299; Gates, *History of Public Land*, 658.

18. Mead drafted a model water code for Wyoming and wrote important water legislation in the 1890s. In 1899 Mead was in charge of developing a corps of irrigation and drainage engineers to help irrigation farmers. Mead was unsure about the Reclamation Act of 1902, but did want government aid going to the West. Lake Mead behind Boulder Dam is named for him. Pisani, *To Reclaim a Divided West*, 113; U.S. Department of Agriculture, Office of Experiment Stations, *The Use of Water*, 15; U.S. Department of Agriculture, *Department and Office of Irrigation Investigation, 1899* (Washington, D.C.: Government Printing Office, 1900), 650; Gates, *History of Public Lands*, 650.

19. *Eddy County News*, 26 September 1947; U.S. Department of Agriculture, Office of Experiment Stations, *The Use of Water*, 15.

20. Granville Richardson, the New Mexico irrigation commissioner, also asked Reclamation to investigate the area. Robert Tansill died at his home in La Huerta of heart failure on 27 December 1902, leaving many of the responsibilities with Pecos Irrigation to his wife. Mrs. Tansill took her late husband's place on the board of directors of the company and continued to pursue avenues that might allow the company and assets to be sold to the federal government. Interestingly, Tansill, who purchased Hagerman Heights, planned to open it as a sanatorium and health resort in 1902. *Eddy County News*, 26 September 1947; Gates, *History of Public Land*, 657; *Eddy County News*, 3 October 1947; *History of Eddy County*, 28; *Roswell Record*, 20 December 1901.

21. *Hearing in the Matter of the Protest . . . against the Construction of the Hondo Reservoir*, 2, 6–7, 9, U.S. Department of the Interior, Bureau of Reclamation, RG 115.

22. Ibid., 7–8.

23. Ibid., 7.

24. Ethan Allen Hitchcock was born in Mobile, Alabama, in 1835. In 1860 he went to China and made a fortune in the commission business of Olyphant and Company in Hong Kong, retiring in 1872. Well traveled, Hitchcock settled in St. Louis, where he involved himself in numerous affairs and contributed heavily to

the Republican Party. In 1897 McKinley appointed him first ambassador to Russia and, in 1898, secretary of the interior, which he continued under Theodore Roosevelt until 1907. Hitchcock initiated a sweeping number of investigations to root out public land fraud and corrupt Interior officials. Robert Sobel, *Biographical Dictionary of the United States Executive Branch, 1774–1977* (Westport, Conn.: Greenwood Press, 1977), 167–168.

25. The Hondo Project was twelve miles southwest of Roswell. *Carlsbad Argus*, 24 June 1904; Smythe, *Conquest of Arid America*, 318; *Roswell Record*, 24 June 1904.

26. *Roswell Record*, 24 June 1904; *Carlsbad Argus*, 24 June 1904.

27. *Carlsbad Argus*, 24 June 1904.

28. Ibid.; Tracy to Newell, 6 July 1904, Tracy Papers.

29. Hagerman to Tracy, 13 September 1904, Tracy Papers; *Hearing in the Matter of the Protest . . . against the Construction of the Hondo Reservoir*, 10, 32, U.S. Department of the Interior, Bureau of Reclamation, RG 115; Pecos Water Users Association to Hitchcock, 19 July 1904, Tracy Papers; Tracy to Newell, 6 July 1904, Tracy Papers.

30. Board of Engineers, U.S. Geological Survey, to Newell, 10 September 1904, RG 115, box 1094, entry 3.

31. *Carlsbad Argus*, 16 September 1904; *Hearing in the Matter of the Protest . . . against the Construction of the Hondo Reservoir*, 26, U.S. Department of the Interior, Bureau of Reclamation, RG 115.

32. *Hearing in the Matter of the Protest . . . against the Construction of the Hondo Reservoir*, 7, 26, U.S. Department of the Interior, Bureau of Reclamation, RG 115.

33. Ibid.

34. Ibid., 7, A-9.

35. *Hearing in the Matter of the Protest . . . against the Construction of the Hondo Reservoir*, 26–27, U.S. Department of the Interior, Bureau of Reclamation, RG 115; New Mexico (Territory), Bureau of Immigration, *Eddy County, New Mexico, Home of the Carlsbad Project of the National Reclamation Service and of the Greatest Artesian Wells on Earth . . .* (Santa Fe, N. Mex.: Bureau of Immigration, 1908), 10; U.S. Department of Agriculture, Office of Experiment Stations, *The Use of Water*, 107, 109.

36. *Hearing in the Matter of the Protest . . . against the Construction of the Hondo Reservoir*, 26–27, U.S. Department of the Interior, Bureau of Reclamation, RG 115.

37. Hagerman to Reed, 1 July 1904, U.S. Department of the Interior, Bureau of Reclamation, RG 115.

38. Ibid.

39. *Hearing in the Matter of the Protest . . . against the Construction of the Hondo Reservoir*, 1–2, U.S. Department of the Interior, Bureau of Reclamation, RG 115.

40. Ibid., 2, 3.

41. Ibid.

42. Ibid., 5–6.

43. Ibid., 6–7.

44. Ibid., 7, 9.

45. Ibid., 8, 24.

46. Hagerman claimed he had retained several thousand acres of land after he left Carlsbad, which he had sold for anywhere from one dollar to two dollars per acre. *Hearing in the Matter of the Protest . . . against the Construction of the Hondo Reservoir,* 16–17, U.S. Department of the Interior, Bureau of Reclamation, RG 115.

47. Ibid., 17.

48. Ibid., 21–23.

49. Ibid., 23–24.

50. Ibid., 8, 24.

51. Ibid., 24–25. Gypsum made the soil highly alkaline, limiting the types of plants grown in the region. Furthermore, the "gyp" substrate tended to dissolve easily, making for porous bedrock where water disappeared.

52. Ibid., 25.

53. Ibid., 25–27.

54. Ibid., 29.

55. Ibid., 29.

56. Ibid., 25–27.

57. Ibid., 24.

58. Ibid., 28–29.

59. Ibid., 23–24.

60. Ibid., 2.

61. Ibid., 9.

62. U.S. Department of Agriculture, Office of Experiment Stations, *The Use of Water,* 107, 109; *Hearing in the Matter of the Protest . . . against the Construction of the Hondo Reservoir,* 20a, 33, U.S. Department of the Interior, Bureau of Reclamation, RG 115.

63. *Hearing in the Matter of the Protest . . . against the Construction of the Hondo Reservoir,* 34, U.S. Department of the Interior, Bureau of Reclamation, RG 115.

64. *Carlsbad Argus,* 16 September 1904.

65. Ibid.

66. Board of Engineers, U.S. Geological Survey, to Newell, 10 September 1904, U.S. Department of the Interior, Bureau of Reclamation, RG 115.

67. Ibid.

68. Reclamation spent over $250,000 on the Hondo Project between 1904 and 1907, when the system was completed. Residents anticipating irrigation were disappointed when the reservoir failed to hold water. The gypsum formations in the floor of the reservoir allowed the water to seep through and disappear. By 1908 the government had spent about $350,000 on the Hondo, but in 1909 no lands were

irrigated because of the drought. The fitness of the reservoir was called into question, and the project in general was deemed a failure. Gibbs, "History, the Farmer, and the Pecos Valley Past," 7; U.S. Congress, Senate, *Investigation of Irrigation Projects*, 61st Cong., 3d sess., 3 March 1911, S.R. 1281, 857; A. P. Davis, *Irrigation Works Constructed by the United States Government* (New York: John Wiley and Sons, 1917), 231.

69. Tracy and the Carlsbad contingent withdrew their protest in November 1904 to facilitate cooperation for their own project. Coan, *History of New Mexico,* 466; Tracy to McGowen and Servien, Washington, D.C., 10 November 1904, Tracy Papers; U.S. Congress, Senate, *Investigation of Irrigation Projects*, 3 March 1911; Davis, *Irrigation Works*, 231.

70. *Carlsbad Argus,* 7 October 1904.

71. *Carlsbad Argus,* 7 October 1904; Coan, *History of New Mexico,* 465.

72. Tracy to Stockholders of Pecos Irrigation Company, 12 October 1904, Tracy Papers; *Carlsbad Argus,* 7 October 1904; *U.S. v. Hope,* vol. 1, 3265; *U.S. v. Hope,* vol. 2, 573; *U.S. v. Hope,* Hearings at Carlsbad, 4 January 1926, 53.

73. Sanford and Kelley, New Bedford, to Honorable William S. Greene, Fall Rivers, Mass., 25 October 1904, U.S. Department of the Interior, Bureau of Reclamation, RG 115, box 1, entry 4.

74. Ibid.; *U.S. v. Hope,* vol. 2, 573.

75. *Carlsbad Argus,* 7 October 1904; Reed to Newell, 10 October 1904, U.S. Department of the Interior, Bureau of Reclamation, RG 115, Carlsbad 651–652, box 443, entry 3.

76. Commercial clubs were local booster and civic clubs. The Roswell and Carlsbad Commercial Clubs included many water users.

77. Reed to Newell, 4 October 1904, U.S. Department of the Interior, Bureau of Reclamation, RG 115.

78. Ibid.

79. Ibid.

80. Mary Tansill personally owned the power dam, originally owned by James Hagerman and then sold to her husband, Robert. Within three months of the flood, the dam was again functioning, having been rebuilt of concrete masonry at a cost of $20,000 to Mary Tansill. *Carlsbad Argus,* 7 October 1904; *Carlsbad Current,* 30 March 1906, RG 115, box 441, entry 3.

81. In 1904, perhaps responding to years of controversy, Secretary of the Interior Ethan Hitchcock held that under the Desert Land Act irrigation actually had to be practiced and entrymen had a right to sufficient water to irrigate the land they had entered under the act. Under the law, company ditch systems built to conduct water over the land had to be adequate for this purpose. Land had to be irrigated for a sufficient period of time to demonstrate an adequate water supply. The government designed the law to protect innocent purchasers taking up homes in the desert near canals unable to supply the necessary water. According to Hitchcock, claimants had to provide absolute evidence that ditch capacity and water

supply irrigated the land, that the ditch company actually supplied the necessary water, and that the land itself was actually cultivated. Irrigation companies, hoping to profit through loopholes in the Desert Land laws, now realized that they must make money not through speculation but through providing water to farmers. *Carlsbad Argus,* 11 March 1904; Gates, *History of Public Land,* 655–656.

82. Water users associations began when water users collectivized to negotiate with companies and later the government. Beginning in 1905, farmers formed irrigation districts to build new projects or acquire established ones. The Bureau of Reclamation required water users organizations for repayment purposes. Dunbar, *Forging New Rights,* 34–35; U.S. Department of Agriculture, Bulletin 1177, *Irrigation Districts, Operations and Financing,* by Wells A. Hutchins (Washington, D.C.: Government Printing Office, 1923); U.S. Department of Agriculture, Bulletin 254, *Irrigation Districts and Their Origins, Operation and Financing,* by Wells A. Hutchins (Washington, D.C.: Government Printing Office, 1931; U.S. Department of Agriculture, Bulletin 103, *Summary of Irrigation District Statutes of Western States,* by Wells A. Hutchins (Washington, D.C.: Government Printing Office, 1931).

83. *Carlsbad Argus,* 14 October 1904, 21 October 1904; Tracy to Newell, 12 October 1904, Tracy Papers; Tracy to Pecos Water Users Association, undated, Tracy papers; Pecos Water Users Association to Newell, 11 November 1905, Tracy Papers.

84. *Carlsbad Argus,* 14 October 1904; Pecos Water Users Association to Tracy, 8 October 1904, Tracy Papers; Tracy to Pecos Water Users Association, undated, Tracy Papers.

85. Ibid.

86. Nothing in the Reclamation Act provided for the Reclamation Service to rehabilitate irrigation works owned by private companies. Thus, Reclamation, although considering the purchase of the Pecos Irrigation works, undertook a project outside its scope.

8. MR. WRECKLAMATION MAN

1. A. P. Davis to Hall, 21 December 1904, U.S. Department of the Interior, Bureau of Reclamation, RG 115, box 443, entry 3.

2. "Reports Covering General Plans and Outline of Work, 1906 through 1910," 12–13, U.S. Department of the Interior, Bureau of Reclamation, RG 115, Carlsbad 651–652, box 443, entry 3, loc. 40/16/2.

3. Ibid., 17.

4. Ibid., 11.

5. E. W. Myers to Hall, 24 June 1905, U.S. Department of the Interior, Bureau of Reclamation, RG 115, box 443, entry 3.

6. "Reports Covering General Plans," 15, U.S. Department of the Interior, Bureau of Reclamation, RG 115.

7. Ibid., 15–16.

8. Hall to Newell, 14 March 1905, U.S. Department of the Interior, Bureau of Reclamation, RG 115, Carlsbad 651–652, box 443, entry 3.

9. "Reports Covering General Plans," 15, U.S. Department of the Interior, Bureau of Reclamation, RG 115.

10. Almost one-half of the investors were foreign, and most were small owners. Tracy to Newell, 12 October 1904, Tracy Papers; Tracy to Otero, 22 February 1905, Tracy Papers.

11. Davis to Hall, 21 December 1904, U.S. Department of the Interior, Bureau of Reclamation, RG 115, Carlsbad 651–652, box 443, entry 3; Tracy to Pecos Water Users Association, undated, Tracy Papers.

12. Newell to Hall, 21 February 1905, U.S. Department of the Interior, Bureau of Reclamation, RG 115, Carlsbad 651–652, box 443, entry 3; Davis to Hall, 12 January 1905, RG 115, Carlsbad 651–652, box 443, entry 3.

13. Grunsky, who was from California, was a leading civil engineer for the Army Corps of Engineers. He served as a member of the Isthmian Canal Commission before being appointed "special advisor." Tracy to G. B. Shaw, Chicago, 29 June 1905, Tracy Papers; Hays, *Conservation and the Gospel of Efficiency*, 8; Smythe, *Conquest of Arid America*, 298.

14. Sanders, Wisner, Bien, and Hall to Newell, undated, Tracy Papers.

15. Hays, *Conservation and the Gospel of Efficiency*, 15; quotation from Newell to Hall, 12 October 1905, U.S. Department of the Interior, Bureau of Reclamation, RG 115, Carlsbad 651–652, box 443, entry 3.

16. Tracy had been complaining and laid the blame on project engineer E. W. Myers. Tracy claimed that Myers had no experience and "was stubborn, obtuse, and never planned ahead." Tracy to Shaw, 29 June 1905, Tracy Papers; Wisner to Davis, 5 September 1905, U.S. Department of the Interior, Bureau of Reclamation, RG 115.

17. Newell to Sanders, 27 September 1905, U.S. Department of the Interior, Bureau of Reclamation, RG 115, Carlsbad 651–652, box 443, entry 3.

18. Ibid.; Newell to Hall, 12 October 1905, U.S. Department of the Interior, Bureau of Reclamation, RG 115, Carlsbad 651–652, box 443, entry 3.

19. Tracy contended that there was no sheet piling under the abutment. Sanders to Newell, 13 October 1905, U.S. Department of the Interior, Bureau of Reclamation, RG 115, Carlsbad 651–652, box 443, entry 3; Tracy to Shaw, 19 June 1907, Tracy Papers.

20. Reed later became Hall's boss and supervised activities in the Pecos Valley. Sanders to Newell, 13 October 1905, U.S. Department of the Interior, Bureau of Reclamation, RG 115.

21. Apparently Reclamation faced criticism in Oklahoma for lack of construction projects in that area. Oklahoma had contributed $1,824,881 to the Reclamation Fund but had no projects, while New Mexico, which had contributed a mere $297,365 to the fund, had several. By 1906 Oklahoma still remained the only state

or territory in the West not to have a federal reclamation project. Gates, *History of Public Land*, 658; Newell to Sanders, 18 October 1905, U.S. Department of the Interior, Bureau of Reclamation, RG 115, Carlsbad 650, box 443, entry 3.

22. Newell to Director USGS, 27 October 1905, U.S. Department of the Interior, Bureau of Reclamation, RG 115, Carlsbad 651–652, box 443, entry 3.

23. Davis to Hydrographer Geological Survey, 11 November 1905, U.S. Department of the Interior, Bureau of Reclamation, RG 115, Carlsbad 651–652, box 443, entry 3.

24. The area of proposed Reservoir Number Three fell into township 20, south range 26 east, and township 21, south range 25 east. Hall believed that Reservoir Number Three might be feasible for storage, since his engineers had not struck enough gypsum to condemn the site. Reservoir Number Three was not part of the proposed Carlsbad Project, which would water 20,000 acres, but Hall believed that another reservoir would be needed if Reclamation approved plans for a larger project. Hall to Newell, 22 November 1905, U.S. Department of the Interior, Bureau of Reclamation, RG 115, Carlsbad 651–652, box 443, entry 3.

25. The first oil well in Eddy County was brought in near Dayton in 1913. The well produced only 52 gallons per day but precipitated an oil boom around Carlsbad. Hall to Newell, 22 November 1905, U.S. Department of the Interior, Bureau of Reclamation, RG 115; *History of Eddy County*, 15.

26. Newell to Hall, 21 December 1905; U.S. Department of the Interior, Bureau of Reclamation, RG 115, Carlsbad 651–652, box 443, entry 3.

27. Various letters to U.S. Reclamation Service, U.S. Department of the Interior, Bureau of Reclamation, RG 115, Carlsbad 650, box 443, entry 3.

28. Tracy had lived on Long Island and reportedly knew the president's brother Elliot. Roberts, "Fifty Eventful Years."

29. Samuel D. Myers, ed., "Charles Littlepage Ballard," *Southwestern Studies* 4 (1966): 15–23; Richard Melzer and Phyllis Ann Mingus, "Wild to Fight: The New Mexico Rough Riders in the Spanish American War," *New Mexico Historical Review* 59 (April 1984): 109–136.

30. Ibid.

31. Ibid.

32. Ibid.

33. Ibid.

34. Ibid.

35. *Carlsbad Daily Current-Argus*, 28 June 1936.

36. Ibid.

37. Ibid.; Roberts, "Fifty Eventful Years," 11.

38. Tracy, like most property owners in the West, complained that Reclamation did not pay enough for lands and canals. Hays, *Conservation and the Gospel of Efficiency*, 242; *U.S. v. Hope*, vol. 2, 574.

39. Carlsbad Irrigation District, Stockholders Meeting, 2 December 1905, Carlsbad Irrigation District Number One, Records; Clark, *Water in New Mexico*, 88.

40. Newell to Hall, 29 November 1905, U.S. Department of the Interior, Bureau of Reclamation, RG 115 Carlsbad 651–652, box 443, entry 3; Clark, *Water in New Mexico*, 88; Carlsbad Irrigation District, Stockholders Meeting, 2 December 1905, Carlsbad Irrigation District Number One, Records.

41. In his 1905 report, Means recalled that this was not his first visit to the valley. In 1899, at the request of settlers, the secretary of agriculture sent two men from the Bureau of Soils to investigate and report on the conditions in the valley and to offer advice concerning improvement. Means and Frank D. Gardner carried out this survey in April, May, and June of 1899. Thomas H. Means, "Report on Agricultural Possibilities of Carlsbad Project, New Mexico," 1905, U.S. Department of the Interior, Bureau of Reclamation, RG 115, entry 3, box 438.

42. Hall to Davis, 2 February 1905, U.S. Department of the Interior, Bureau of Reclamation, RG 115, Carlsbad 651–652, box 443, entry 3.

43. Means, "Report on Agricultural Possibilities," U.S. Department of the Interior, Bureau of Reclamation, RG 115.

44. Ibid.

45. Ibid.

46. Ibid.

47. Ibid.

48. Ibid.

49. Across the West, farm tenantry increased steadily and consistently, decade after decade, from 25 percent in 1880 to 28 percent in 1890, to 36 percent in 1900, and to 38 percent in 1910. Means's report that tenantry was decreasing was another unique feature of the Carlsbad Project. Ibid.; Lawrence Goodwyn, *The Populist Moment: A Short History of the Agrarian Revolt in America* (New York: Oxford University Press, 1978), 295.

50. Means, "Report on Agricultural Possibilities," U.S. Department of the Interior, Bureau of Reclamation, RG 115.

51. *Eddy County News,* 1 August 1947.

52. Francis Tracy signed the agreement on July 31, 1906, on behalf of Pecos Irrigation Company. One of the tenets of Progressivism was efficiency in industry. Samuel Hays says, "The crux of the gospel of efficiency lay in a rational and scientific method of basic technological decisions through a single, central authority." Water users associations were designed to help maintain efficiency and central control. The idea did not work well—each association considered its own particular interest far more important than any other. Pecos Water Users Association, Board of Directors Meeting, 17 July 1908, Carlsbad Irrigation District Number One, Records; Hays, *Conservation and the Gospel of Efficiency,* 271–272; Hall to Pecos Water Users Association, 16 January 1906, Tracy Papers; Hall to Pecos Water Users Association, 2 December 1905, Tracy Papers.

53. Pecos Water Users Association, Board of Directors Meeting, 20 February 1906, Carlsbad Irrigation District Number One, Records; Carlsbad Irrigation District Board of Directors to Hitchcock, 20 February 1906, Carlsbad Irrigation District Number One, Records.

54. Bogener, "Carlsbad Project," 15; "Carlsbad Project: Annual Project History," through 1913 [1 and 2] and 1910–1912, vol. 1, U.S. Department of the Interior, Bureau of Reclamation, RG 115, box 3, vol. 1, 5–9; U.S. Department of Agriculture, Office of Experiment Stations, Bulletin 215, *Irrigation in New Mexico*, 29–30.

55. New steel gates replaced the antiquated wooden ones. Davis, *Irrigation Works*, 227–229; "Carlsbad Project: Annual Project History," through 1913 [1 and 2] and 1910–1912, vol. 1, U.S. Department of the Interior, Bureau of Reclamation, RG 115, box 3, vol. 1, 5–9; U.S. Department of the Interior, National Park Service, *HAER*, 91–93.

56. Davis, *Irrigation Works*, 227–228.

57. Many westerners, including Francis Tracy, sought help from the federal government, then resisted the controls that came with that help and complained that they could not influence agency decisions. Hays, *Conservation and the Gospel of Efficiency*, 241; Means, "Report on Agricultural Possibilities," U.S. Department of the Interior, Bureau of Reclamation, RG 115.

58. Reed to Newell, 15 August 1906, U.S. Department of the Interior, Bureau of Reclamation, RG 115, Carlsbad 301, box 451, entry 7.

59. L. E. Foster, tagged for the job of selecting lands for the project, began his career as a member of the U.S. Geological Survey, and following the creation of the Reclamation Service transferred into that service. Soon afterward he was assigned to Carlsbad where he served as superintendent of the project until 1945, when he retired. Lewis, "Early Settlers."

60. In a letter to the association in January 1906, B. M. Hall, supervising engineer for Reclamation indicated that he had final authority to approve lands for the project according to the Newlands Act. Under the work about to begin, Reclamation included 20,000 acres of land, although Hall realized that twice that amount existed under the company canal system. Hall wanted the project, as much as possible, to contain contiguous lands in one unit. He advised the board of directors of the Pecos Water Users Association to select a committee of three to help in the selection approval process. Reed to Newell, 15 August 1906, U.S. Department of the Interior, Bureau of Reclamation, RG 115; Reed to Newell, 5 June 1909, RG 115, box 441, entry 3, file: Land Class, Farm Units; Hall to Pecos Water Users Association, 16 January 1906, Carlsbad Irrigation District Number One, Records.

61. Reclamation's engineers knew full well that much of the soil in the irrigated area contained salts and was unfit for farming, and recommended against using such lands. Thomas Means questioned the use of such lands to recoup construction and maintenance costs. Reed to Newell, 15 August 1906, U.S. Department of the Interior, Bureau of Reclamation, RG 115.

62. Reed to Newell, 5 June 1909, U.S. Department of the Interior, Bureau of Reclamation, RG 115.

63. U.S. Department of the Interior, Bureau of Reclamation, RG 115, box 449, entry 3. Box 449 includes extensive documentation of the process of perfecting title to property.

64. Pecos Water Users Association, Board of Directors Meeting, 18 December 1905, Carlsbad Irrigation District Number One, Records.

65. Shareholders Meeting, Carlsbad Irrigation District, 18 November 1905, Carlsbad Irrigation District Number One, Records.

66. J. O. Cameron to Richard Bolles, 9 December 1905, Exhibit "C," *Bolles v. Pecos Irrigation Co.*, 23 N.M. 32.

67. Ibid.

68. Ibid.; *Colorado Springs Gazette Weekly,* 22 June 1893; Reed to Newell, 15 August 1906, U.S. Department of the Interior, Bureau of Reclamation, RG 115; Pecos Irrigation Company, Ninth Annual Stockholders Report, 4, Tracy Papers.

69. *Bolles v. Pecos Irrigation Co.*, 23 N.M. 32.

70. The Pecos Irrigation Company faced lawsuits over water rights in the new century—one from C. W. Beeman, president of the water users association and one from Bolles. The Beeman case was settled over a $300 water-right note, and the company did not appeal. Hove to Bolles, 20 July 1906, Carlsbad Irrigation District Number One, Records; Pecos Irrigation Company, Ninth Annual Stockholders Report, 4, Tracy Papers.

71. Most private irrigation companies found vigorous spokesmen to attack Reclamation. Hays, *Conservation and the Gospel of Efficiency*, 243; Reed to Newell, 15 August 1906, U.S. Department of the Interior, Bureau of Reclamation, RG 115; Tracy to Newell, 12 October 1904, Tracy Papers.

72. R. Wells Benson, vice president, and A. M. Hove, secretary, Pecos Water Users Association to Secretary of Interior, undated, Carlsbad Irrigation District Number One, Records; Carlsbad Irrigation District, Board of Directors Meeting, 16 January 1906, Carlsbad Irrigation District Number One, Records.

73. In 1907 one-third of New Mexico's population was engaged in agriculture, although farmers cultivated only 400,000 acres along river valleys out of 78,000,000 acres in the Territory. Out of 12,311 farms in New Mexico, 9,128 were irrigated. Eighty-six acres of irrigated land existed for each mile of ditch laid. The average size of all farms in the Territory was 464 acres. The average size for irrigated farms was 360 acres. Each irrigated farm watered on average twenty-six acres. The average value per acre for irrigated lands in 1907 was $29.26, although irrigated alfalfa land sold for $50 to $100 an acre and fruit lands for $400 to $500 an acre. Clark, *Water in New Mexico*, 192; New Mexico (Territory), Bureau of Immigration, *New Mexico: Its Resources*, 990.

74. Many locals disliked conservation practices and the methods used to calculate resource use. They hated the idea that decisions were made far from local control and feared broader programs would minimize their projects. Hays, *Conservation and the Gospel of Efficiency*, 273.

75. Robbins, *Our Landed Heritage*, 378; Pisani, *To Reclaim a Divided West*, 112; Ganoe, "Desert Land Act since 1891," 269–270; Hays, *Conservation and the Gospel of Efficiency*, 11; Dunbar, *Forging New Rights*, 54.

76. James Rudolph Garfield was born in Hiram, Ohio, in 1865, the son of

President James Garfield. Garfield studied law at Columbia University and was admitted to the Ohio bar in 1888. He practiced law with his older brother and was a member of the state senate from 1896 to 1899, and was appointed to the U.S. Civil Service Commission, 1902–1903. In the Department of Commerce, Garfield served as commissioner of corporations before his appointment by Theodore Roosevelt to the post of secretary of the interior in 1907, where he served for two years. Sobel, *Biographical Dictionary of the United States Executive Branch;* Carlsbad Irrigation District, Board of Directors Meeting, 31 December 1907, Carlsbad Irrigation District Number One, Records.

77. Pecos Water Users Association to Secretary of the Interior James R. Garfield, 13 February 1908, Carlsbad Irrigation District Number One, Records.

78. Pecos Water Users Association to James R. Garfield, Secretary of the Interior, 13 February 1908, Carlsbad Irrigation District Number One, Records.

79. Many westerners complained about the provision that no private landowners could receive water rights for more than 160 acres. Reclamation refused to take projects unless it had written agreements that those with more than 160 acres would not sell the excess land at inflated prices. By 1924 many in the Carlsbad area still had not adhered to the rule. An amendment to the Reclamation Act proposed in 1924 stipulated that any acreage above 160 acres must be deeded to the United States before a person could benefit from the Reclamation Act. Hays, *Conservation and the Gospel of Efficiency,* 242–243; Newlands Reclamation Act, 1902, sec. 8; Tracy to Mary Tansill, Wilmette, Illinois, 20 May 1924, Tracy Papers.

80. F. E. Bryant, "Uncle Sam Is a Water Hog," *Field and Farm,* 29 February 1908, U.S. Department of the Interior, Bureau of Reclamation, RG 115, Carlsbad 301, box 451, entry 7.

81. Ibid.

82. Carlsbad Irrigation District, Board of Directors Meeting, 6 January 1908, Carlsbad Irrigation District Number One, Records; Reisner, *Cadillac Desert,* 120.

83. McLenathen and Tracy Real Estate and Insurance to Pecos Water Users Association, 13 February 1908, Carlsbad Irrigation District Number One, Records.

84. Ibid.

85. Pecos Water Users Association to Professor R. S. Mack, Chicago, Illinois, 22 February 1908, Carlsbad Irrigation District Number One, Records.

86. Carlsbad Irrigation District, Board of Directors to Secretary of the Interior, 21 February 1908, Carlsbad Irrigation District Number One, Records.

87. Reclamation backed down on other projects as well.

88. Franklin Pierce to Pecos Water Users Association, 13 March 1908, Carlsbad Irrigation District Number One, Records.

89. Ibid.

90. Ibid.

91. Pecos Water Users Association, Board of Directors Meeting, 29 March 1908, Carlsbad Irrigation District Number One, Records.

9. SUCH OF THE LANDS ECONOMICALLY
PRACTICABLE TO IRRIGATE

1. Brochure and insert in Tracy to Newell, 9 April 1908, U.S. Department of the Interior, Bureau of Reclamation, RG 115, box 441, entry 3.

2. Pecos Water Users Association, Board of Directors Meeting, 18 December 1906, Carlsbad Irrigation District Number One, Records.

3. Ward and Thomas to Pecos Water Users Association, 13 August 1907, Carlsbad Irrigation District Number One, Records.

4. D. L. Meyers, General Freight and Passenger Agent, Eastern Railway of New Mexico System, to A. M. Hove, Secretary, Pecos Water Users Association, 24 December 1907, Carlsbad Irrigation District Number One, Records; Tracy to C. J. Blanchard, U.S. Bureau of Reclamation Statistician, 2 January 1908, Tracy Papers.

5. Pecos Water Users Association to R. W. Cuff, Peanut, California, 18 December 1907, Carlsbad Irrigation District Number One, Records.

6. Ibid.

7. Notable during the Ninth Annual Pecos Irrigation Shareholders Meeting held 8 February 1910, was the fact that only three shareholders were present: Francis G. Tracy (one share), I. S. Osborne (one share), and C. H. McLenathen (one share). Tracy held the proxy of shareholders in Pecos Irrigation Company. Bonbright and Company, G. B. Shaw, Mary Tansill, Sanford and Kelley, and a new interest, Iselin and Company, which seems to have absorbed the Swiss investments and those of other foreigners, were absent, and had given up control of their investments to Tracy. Bonbright owned 70 shares of Pecos Irrigation Company, Shaw owned 189 shares, Tansill owned 379 shares, and Sanford and Kelley controlled 59 shares. Iselin owned 1,246 shares. Francis Tracy held the fate and control of the company in own hands. Iselin and Company was founded by Adrian Iselin, a banker who helped finance the Union army during the Civil War. Iselin's father was a Swiss capitalist who immigrated to the United States in the early 1800s as part of the silk and glove industries. Wise investments in railroad and mining stocks increased his wealth and social standing in New York. His son, yachtsman C. Oliver Iselin, successfully defended the America's Cup in the early 1900s. The patriarch helped found the Museum of Natural History and the Society for the Prevention of Cruelty to Animals. At his death in 1905 at the age of eighty-six, he left his wife of fifty years and his seven children between $20 million and $50 million. *New York Times,* 29 March 1905; Pecos Irrigation Company, Ninth Annual Stockholders Report, 8 February 1910, Tracy Papers; Pecos Water Users Association to the Honorable W. H. Andrews, Washington, D.C., 1 December 1910, Carlsbad Irrigation District Number One, Records.

8. Pecos Water Users Association to Overholt, 23 October 1907, Carlsbad Irrigation District Number One, Records.

9. Pecos Irrigation Company, Ninth Annual Shareholders Report, 8 February 1910, Tracy Papers; Morris Bien to McLenathen, 3 April 1908, U.S. Department of the Interior, Bureau of Reclamation, RG 115, Carlsbad 651–652, box 443, entry 3, loc. 40/16/2.

10. Supervising Engineer Louis Hill to Newell, 20 June 1908, U.S. Department of the Interior, Bureau of Reclamation, RG 115, Carlsbad 301, box 451, entry 7; Bien to McLenathen, 3 April 1908, RG 115; Reed to Newell, 17 November 1908, RG 115, Carlsbad 651–652, box 443, entry 3, loc. 40/16/2; Hill to Newell, 30 June 1908, RG 115, box 441, entry 3.

11. Hill to Newell, 30 June 1908, U.S. Department of the Interior, Bureau of Reclamation, RG 115.

12. Reed to Newell, 31 March 1908, U.S. Department of the Interior, Bureau of Reclamation, RG 115, Carlsbad 301, box 441, entry 7.

13. Reed to Newell, 17 November 1908, U.S. Department of the Interior, Bureau of Reclamation, RG 115.

14. Reed to Newell, 11 November 1908, U.S. Department of the Interior, Bureau of Reclamation, RG 115, Carlsbad 651–652, box 443, entry 3.

15. Reed to Newell, 17 November 1908, U.S. Department of the Interior, Bureau of Reclamation, RG 115.

16. McLenathen to Bien, 9 April 1908, U.S. Department of the Interior, Bureau of Reclamation, RG 115, Carlsbad 651–652, box 443, entry 3, loc. 40/16/2.

17. Hartshorn, Malaga Lands, to Roy Black, Minco, Oklahoma, 10 December 1908, U.S. Department of the Interior, Bureau of Reclamation, RG 115, Carlsbad 651–652, box 443, entry 3, loc. 40/16/2.

18. Morris Bien was a legal expert and was asked to write a model law based on Wyoming's reclamation program. His code included needing a permit to acquire water rights and a hierarchy of authority, with adjudication by courts. Dunbar, *Forging New Rights*, 120–122.

19. McLenathen to Bien, 9 April 1908, U.S. Department of the Interior, Bureau of Reclamation, RG 115.

20. Ibid.

21. Sanford was the senior partner of Sanford and Kelley, a New Bedford, Massachusetts, investment banking firm, whose investors had bankrolled a good portion of the valley's irrigation activities in the nineteenth century. *New Bedford Evening Standard*, 26 December 1911.

22. McLenathen to Bien, 9 April 1908, U.S. Department of the Interior, Bureau of Reclamation, RG 115.

23. Ibid.

24. Ibid.; Hove to Dr. F. W. Montgomery, Totwin, Kansas, 25 March 1908, U.S. Department of the Interior, Bureau of Reclamation, RG 115, Carlsbad 651–652, box 443, entry 3, loc. 40/16/2; Bien to Albert Thurstin, Malaga, 9 September 1909, RG 115, Carlsbad 651–652, box 443, entry 3, loc. 40/16/2.

25. Bien to McLenathen, 3 April 1908, U.S. Department of the Interior, Bureau of Reclamation, RG 115.

26. Hove to Montgomery, 25 March 1908, U.S. Department of the Interior, Bureau of Reclamation, RG 115.

27. Reed to Newell, 24 June 1908, U.S. Department of the Interior, Bureau of Reclamation, RG 115, box 441, entry 3; Hill to Newell, 30 June 1908, RG 115.

28. Hill to Newell, 30 June 1908, U.S. Department of the Interior, Bureau of Reclamation, RG 115.

29. Ibid.

30. Ibid.

31. Davis, *Irrigation Works*, 229; U.S. Department of the Interior, National Park Service, *HAER*, 107–108; "Carlsbad Project: Annual Project History," through 1913 [1 and 2] and 1910–1912, vol. 1, U.S. Department of the Interior, Bureau of Reclamation, RG 115, box 3, vol. 1, 122.

32. Ballinger was born in Iowa in 1858. President William Howard Taft appointed him secretary in 1909. He served until 1911 and died in 1922. *The Secretaries of the Department of the Interior* (Washington, D.C.: Office of Administration Services, 1986), 13.

33. Pecos Irrigation Company, Ninth Annual Stockholders Report, Tracy Papers.

34. The acreage seemed so small the company could not justify spending money to create a demand for it. Ibid.

35. Ibid.

36. The first payment was originally due in 1908, but had been delayed until 1909.

37. Tracy to Blanchard, 2 January 1908, Tracy Papers; Ad Hoc Committee of Water Users to Pecos Water Users Association, 8 October 1908, Tracy Papers.

38. The charge of incompetency stemmed from the Hondo Project, which never worked well. Tracy claimed the main problem was inadequate preliminary investigations. Tracy also wanted the third reservoir, which Davis opposed. Clark, *Water in New Mexico*, 192; Bien to Ballinger, 27 May 1909, U.S. Department of the Interior, Bureau of Reclamation, RG 115, Carlsbad 651–652, box 443, entry 3.

39. Newell to Supervising Engineer in Phoenix (Hall?), 1 July 1910, U.S. Department of the Interior, Bureau of Reclamation, RG 115, Carlsbad 651–652, box 443, entry 3, loc. 40/16/2.

40. Pecos Water Users Association to Honorable W. H. Andrews, 1 December 1910, Carlsbad Irrigation District Number One, Records.

41. The Pecos Water Users Association was most concerned with water delivery on the west side of the Pecos River from just north of Carlsbad to the Delaware River on the Texas line, where at one time Pecos Irrigation and Improvement Company intended to irrigate 100,000 acres of land. Pecos Water Users Association to Honorable W. H. Andrews, 1 December 1910, Carlsbad Irrigation District Number One, Records.

42. "651 Carlsbad—Reports Covering General Plans and Outline of Work, 1911," U.S. Department of the Interior, Bureau of Reclamation, RG 115, Carlsbad 651–652, box 443, entry 3, loc. 40/16/2.

43. Ibid.

44. Later active in many civic and business affairs, Etter was a member of various fraternal organizations, serving as secretary of the Carlsbad Chamber of Commerce, mayor of the city for two years, chairman of the New Mexico Board of

Water Commissioners, and member of the Executive Board of the National Federation of Water Users. Lewis, "Early Settlers."

45. Scott Etter, Secretary and Treasurer, Pecos Water Users Association to Dry Land Holders, 9 May 1911, Carlsbad Irrigation District Number One, Records.

46. Ibid.; Davis to Newell, 4 November 1911, U.S. Department of the Interior, Bureau of Reclamation, RG 115, box 441, entry 3.

47. Carlsbad Irrigation District Secretary to Reed, 18 December 1911, Carlsbad Irrigation District Records.

48. Unknown to Reed, 11 October 1911, U.S. Department of the Interior, Bureau of Reclamation, RG 115, box 441, entry 3.

49. McShane to Reed, 11 October 1911, U.S. Department of the Interior, Bureau of Reclamation, RG 115, box 441, entry 3.

50. Davis to Newell, 4 November 1911, U.S. Department of the Interior, Bureau of Reclamation, RG 115, box 441, entry 3.

51. Ibid.

52. Ibid.; McShane to Davis, 4 November 1911, U.S. Department of the Interior, Bureau of Reclamation, RG 115, box 441, entry 3; Davis to Reed, 4 November 1911, RG 115, box 441, entry 3; Carlsbad Irrigation District Meeting, 18 November 1911, Carlsbad Irrigation District Number One, Records.

53. Reed, corresponding with Davis, noted that McShane's Irish background gave him plenty of fighting blood—more than enough to handle Tracy's group. Reed to Davis, 18 November 1911, U.S. Department of the Interior, Bureau of Reclamation, RG 115, box 441, entry 3.

54. Carlsbad Irrigation District Meeting, 18 November 1911, Carlsbad Irrigation District Number One, Records.

55. Fisher was born in Virginia in 1862. He was appointed secretary by Taft. He served until 1913 and died in 1935. *The Secretaries of the Department of the Interior*, 13.

56. Carlsbad Irrigation District, "Carlsbad, New Mexico" and "This is to Certify," in Minutes, 18 November 1911, Carlsbad Irrigation District Number One, Records.

57. McLenathen proved more successful than Tracy in remaining in the background. Although his name was seldom mentioned in the local newspaper or other documents, McLenathen stayed active in the valley until his death in 1919. Eddy County, Files and Records of Wills, book 1, p. 225; Reed to Davis, 18 November 1911, U.S. Department of the Interior, Bureau of Reclamation, RG 115.

58. Reed to Davis, 18 November 1911, U.S. Department of the Interior, Bureau of Reclamation, RG 115.

59. Ibid.

60. Newell to Adams, 8 February 1912, U.S. Department of the Interior, Bureau of Reclamation, RG 115, Carlsbad 301, box 451, entry 7.

61. Newell to Acting Director, 2 October 1911, U.S. Department of the Interior, Bureau of Reclamation, RG 115, box 441, entry 3; Tracy to Taft, 9 September 1911, Tracy Papers.

62. Tracy to Taft, 9 September 1911, Tracy Papers.

63. Newell to Adams, 8 February 1912, U.S. Department of the Interior, Bureau of Reclamation, RG 115.

64. Ibid. If the Charles Ballard stories are to be believed, he and Theodore Roosevelt had much to do with the project's approval.

65. Reed to Newell, 12 February 1912, U.S. Department of the Interior, Bureau of Reclamation, RG 115, Carlsbad 301, box 451, entry 7.

66. Ibid.

67. Newell to Davis, 2 October 1911, U.S. Department of the Interior, Bureau of Reclamation, RG 115, box 441, entry 3.

68. Ibid. General William L. Marshall was consulting engineer to the secretary of the interior at the time.

69. Davis to Newell, 7 October 1911, U.S. Department of the Interior, Bureau of Reclamation, RG 115, box 441, entry 3.

70. *Carlsbad Argus,* 3 November 1911.

71. Adams to Tracy, 27 May 1912, Tracy Papers.

72. Newell to Department of Interior, 21 June 1912, U.S. Department of the Interior, Bureau of Reclamation, RG 115, Carlsbad 301, box 451, entry 7.

73. Acting Secretary of the Interior A. A. Jones to Tracy, 16 September 1912, U.S. Department of the Interior, Bureau of Reclamation, RG 115, Carlsbad 301, box 451, entry 7.

74. U.S. Department of the Interior, Bureau of Reclamation, RG 115, box 441, entry 3.

75. John W. Armstrong of Attorneys Armstrong and Botts to H. B. Ferguson, 19 February 1914, U.S. Department of the Interior, Bureau of Reclamation, RG 115.

76. Andrius Aristieus Jones was born 16 May 1862, in Albion County, Tennessee. He was a teacher and lawyer before serving as mayor of Las Vegas, New Mexico, from 1893 to 1894. From 1894 until 1898, Jones served as special assistant U.S. attorney. In 1896 he was delegate to the Democratic National Convention and Democratic territorial chairman from 1906 to 1908 and 1911, and national committee man from 1908 until 1922. Jones served as first assistant secretary of the interior from 1913 to 1916, and U.S. senator from 1917 until his death, 20 December 1927, in Washington, D.C. Dan and Inez Morris, *Who Was Who in American Politics, A Biographical Dictionary of Over 4000 Men and Women Who Contributed to the United States Scene from Colonial Days up to and Including the Immediate Past* (New York: Hawthorn Books, 1974), 346.

77. Ferguson to Jones, 23 February 1914, U.S. Department of the Interior, Bureau of Reclamation, RG 115, Carlsbad 301, box 451, entry 7.

78. Carlsbad Commercial Club Letterhead, 18 February 1914, Carlsbad Irrigation District Number One, Records.

79. "The Carlsbad, N.M. Project of U.S. Reclamation Service," brochure by McLenathen and Tracy, Real Estate and Insurance, Carlsbad, NM, U.S. Department of the Interior, Bureau of Reclamation, RG 115, Carlsbad 301, box 451, entry 7.

80. *Carlsbad Current,* 18 May 1914, RG 115, box 441, entry 3; *Carlsbad Argus,* 20 March 1912.

81. *Carlsbad Current,* 18 May 1914, RG 115, box 441, entry 3.

82. Ibid.

83. *Carlsbad Current,* c. 1914, U.S. Department of the Interior, Bureau of Reclamation, RG 115, Carlsbad 301, box 451, entry 7.

84. Carlsbad Commercial Club Letterhead, 18 February 1914, Carlsbad Irrigation District Number One, Records.

85. Ibid.

86. Muggeridge to Lane, 13 April 1914, U.S. Department of the Interior, Bureau of Reclamation, RG 115, box 449, entry 3, file 1158.

87. Jones to Frank H. Richards, 13 April 1916, U.S. Department of the Interior, Bureau of Reclamation, RG 115, Carlsbad 301, box 451, entry 7.

88. Around 1910, unique technology was being designed by Frank Teichman, an engineer for Reclamation who had emigrated from Germany in 1882. Teichman's most important contributions to the Avalon Dam were the "cylinder gate" outlet works, originally used on navigation canals. Their advantage over traditional sliding gate valves was that water pressure was exerted from all directions. The gates replaced the original design, which Francis Tracy blamed for damages and lack of irrigation water during the 1906–1907 growing season. Teichman first employed cylinder gates on the Yuma Project in Arizona, then modified them at Avalon and elsewhere. After their success at Avalon, similar gates were used on other projects, including Elephant Butte Dam and within the intake towers at Hoover Dam. Bogener, "Carlsbad Project," 18–19.

89. Robbins, *Our Landed Heritage,* 379; George Thomas, *The Development of Institutions under Irrigation, with a Special Reference to Early Utah Conditions* (New York: Macmillan, 1920), 254; Reisner, *Cadillac Desert,* 118, 120; see also Pecos Irrigation Company, Ninth Annual Stockholders Report, Tracy Papers.

10. GOD PITY THE WATER USER?

1. Hays, *Conservation and the Gospel of Efficiency,* 246.

2. Foster wanted to postpone payment only until Reclamation could resolve the question of lands affected by seepage. F. W. Hanna, Supervising Engineer, Carlsbad Project, to Director, Subject: Postponement of charges on seeped lands—Carlsbad Project, 24 March 1914, U.S. Department of the Interior, Bureau of Reclamation, RG 115, box 449, entry 3, folder 338-A3; Dunbar, *Forging New Rights,* 54.

3. Hays, *Conservation and the Gospel of Efficiency,* 246.

4. Newell to Adams, 8 February 1912, U.S. Department of the Interior, Bureau of Reclamation, RG 115, Carlsbad 301, box 451, entry 7.

5. Hays, *Conservation and the Gospel of Efficiency,* 246–247.

6. Franklin K. Lane was born near Charlottetown, Prince Edward Island, Canada, in 1864. During 1884–1886 Lane attended the University of California as

a special student, later studying law. Lane worked for western newspapers in the early to mid-1890s and ran for several offices in California. In 1906 Theodore Roosevelt appointed Lane to the Interstate Commerce Commission, a position he held until resigning in 1913. President Woodrow Wilson appointed him secretary of the interior. Lane created the National Park Service and established seven new national parks. Lane resigned on 1 March 1920, and died one year later in Rochester, New York. Sobel, *Biographical Dictionary of the United States Executive Branch*, 214.

7. Pecos Water Users Association Annual Report for 1913, 14 January 1914, Carlsbad Irrigation District Number One, Records.

8. Hays, *Conservation and the Gospel of Efficiency*, 248.

9. Letter to Water Users, Carlsbad, from Director of Reclamation, undated, U.S. Department of the Interior, Bureau of Reclamation, RG 115, box 434, entry 3, folder 338-A2; Reisner, *Cadillac Desert*, 121; Dunbar, *Forging New Rights*, 55; Clark, *Water in New Mexico*, 193.

10. Brigadier General William L. Marshall retired as chief of the Army Corps of Engineers and in July 1910 became consulting engineer to the secretary of the interior. As the president's representative to sessions of the National Irrigation Congress, he also replaced Bureau of Reclamation officials. Hays, *Conservation and the Gospel of Efficiency*, 112, 152.

11. Central Board of Review, Carlsbad Project, New Mexico, Elwood Mead et al. to Lane, 2 March 1916, 1, 2, U.S. Department of the Interior, Bureau of Reclamation, RG 115, box 10, entry 4; Report of Local Board of Review, Carlsbad Project, by Scott Etter and T. U. Taylor, RG 115, box 10, entry 4.

12. "In the Matter of the Re-evaluation of the Carlsbad Project, Hearings and Proceedings of the Board of Review at Carlsbad, New Mexico," 13–17 April 1915, U.S. Department of the Interior, Bureau of Reclamation, RG 115, box 11, entry 4, loc. 42/4/7:6.

13. "In the Matter of the Re-Evaluation," 1, 2, U.S. Department of the Interior, Bureau of Reclamation, RG 115.

14. Foster's engineering background, like that of many men of the day, was limited to a short stint at the University of California—about two years. Most of his experience was on-the-job practical experience with the U.S. Reclamation Service and the Geological Survey before the service was formally organized. Before coming to Carlsbad, Foster worked as a field assistant, familiar with the Uncompahgre Valley Project during investigations of that valley. Foster worked there until early 1905. Beginning in 1907 he worked on the Carlsbad Project as assistant engineer in charge of maintenance and operation. "In the Matter of the Re-evaluation," 22–23, U.S. Department of the Interior, Bureau of Reclamation, RG 115.

15. In 1907 construction charges were $31.00 per acre. Operation and maintenance were $0.75 per acre. By 1909 operation costs had risen to $1.35. By 1912 construction costs and charges stood at $45.00 per acre, with operation charges at $1.00. Graduated payments, longer payment periods, and differing charges for

different types of lands made the system more and more complicated. Clark, *Water for New Mexico,* 193; "In the Matter of the Re-evaluation," U.S. Department of the Interior, Bureau of Reclamation, RG 115.

16. Foster raised hay, forage, grain, and other crops to see if they could be grown in other parts of the irrigation district. Reclamation also maintained a nursery with 2,000 to 3,000 trees, 1,600 of them along the Main Canal between the flume and Avalon Dam.

17. "In the Matter of the Re-evaluation," 58–59, U.S. Department of the Interior, Bureau of Reclamation, RG 115.

18. Ibid., 102–111.

19. Ibid., 112–130.

20. Ibid., 151.

21. U.S. Congress, House, *Report of Chief of Engineers, Army, 1910, 3 vols.,* 61st Cong., 3d sess., 1910–1911, H. Doc. 1010, vols., 17, 18, 19, serials 5961, 5962, 5963, sec. 16, p. 97.

22. "In the Matter of the Re-evaluation," 131, 137, U.S. Department of the Interior, Bureau of Reclamation, RG 115.

23. Ibid., 151–152.

24. Ibid., 152.

25. Ibid., 171–172.

26. Ibid., 171–172, 179.

27. Ibid., 171–172, 179.

28. Central Board of Review to Lane, U.S. Department of the Interior, Bureau of Reclamation, RG 115; Report of Local Board of Review, RG 115.

29. Central Board of Review to Lane, 5, U.S. Department of the Interior, Bureau of Reclamation, RG 115.

30. Ibid.; Taylor to Tracy, 5 October 1916, Tracy Papers.

31. Pecos Water Users Association and Independent Water Users to Lane, 23 June 1915, U.S. Department of the Interior, Bureau of Reclamation, RG 115, box 440, entry 3.

32. Pecos Water Users Association to Lane, 23 June 1915, U.S. Department of the Interior, Bureau of Reclamation, RG 115, box 440, entry 3.

33. Ibid.

34. Report of Local Board, U.S. Department of the Interior, Bureau of Reclamation, RG 115; Central Board of Review to Lane, 2, 3, 6, RG 115.

35. Central Board of Review to Lane, 1–2, U.S. Department of the Interior, Bureau of Reclamation, RG 115; Report of Local Board, RG 115.

36. Central Board of Review to Lane, U.S. Department of the Interior, Bureau of Reclamation, RG 115.

37. Ibid., 6, 7, 9.

38. Ibid.

39. Ibid.

40. U.S. Congress, House, *Report of Chief of Engineers, Army, 1910, 3 vols.,* 61st

Cong., 3rd sess., 1910–1911, H. Doc. 1010, vols., 17, 18, 19, serials 5961, 5962, 5963, sec. 16, p. 97.

41. Central Board of Review to Lane, 10–11, U.S. Department of the Interior, Bureau of Reclamation, RG 115.

42. Ibid., 11, 13.

43. Ibid., 13, 16.

44. Ibid., 16–17.

45. Ibid., 17–18.

46. Ibid., 18–19.

47. Ibid., 19.

48. Ibid., 42; Report of Local Board, U.S. Department of the Interior, Bureau of Reclamation, RG 115.

49. The silt survey showed that the Pecos River deposited layers of silt in the reservoir each year, amounting to some 3,900,000 cubic yards by 1916. In another twenty years, the reservoir would be filled with silt, but Reclamation dismissed the idea of dredging the reservoir's floor as technologically impractical. Periodically raising the height of McMillan Dam would become expensive and would add costs to Carlsbad's repayment plan. Project Manager L. E. Foster to Newell, undated, Subject: Supplemental Construction, Carlsbad Project, U.S. Department of the Interior, Bureau of Reclamation, RG 115, box 440, entry 3, file 338-A6: Carlsbad classification of lands—supplemental construction; Bogener, "Carlsbad Project," 20–21.

50. Board of Engineers to Chief Engineer, 12 April 1906, U.S. Department of the Interior, Bureau of Reclamation, RG 115, box 449, entry 3; Foster to Director/Chief of Construction, Denver, 9 August 1916, RG 115, box 449, entry 3.

51. Ibid.

52. "Carlsbad, New Mexico, The Greatest Natural Sanitarium in the Southwest," Carlsbad Chamber of Commerce, 1923, U.S. Department of the Interior, Bureau of Reclamation, RG 115, box 5, entry 8, folder 301.32.

53. Ibid.

54. Francis Tracy was a pioneer in cotton experimentation. Tracy to Mary Tansill, Miles, Michigan, 28 July 1924, Tracy Papers.

55. "Carlsbad, New Mexico, The Greatest Natural Sanitarium," U.S. Department of the Interior, Bureau of Reclamation, RG 115.

11. AN ELUSIVE RESERVOIR

1. *Carlsbad Current,* 1 June 1914, RG 115, box 441, entry 3.

2. Ibid.

3. Ibid.

4. Ibid.

5. *Carlsbad Current,* 17 July 1914.

6. *Carlsbad Current,* 1 June 1914.

7. William Robert Smith was born near Tyler, Texas, on 18 August 1863. In 1888 he moved to Colorado City, Texas, where he practiced law until he was appointed judge of the Thirty-second Judicial District of Texas. He served in that capacity from 1897 until 1903, when he became the Democratic nominee for representative to the Fifty-eighth Congress from the Sixteenth District. He served as representative from 1903 until he lost the election in 1916. In Congress, he served on several committees, including the Committee on Interstate and Foreign Commerce. He was the chairman of the Committee on Irrigation of Arid Lands in 1911 and held that position until 1916. Smith introduced the Elephant Butte Irrigation Project to provide irrigation water for farms in the Rio Grande Valley. After Smith left Congress, President Woodrow Wilson appointed him U.S. district judge for the Western District of Texas. *The New Handbook of Texas*, vol. 5 (Austin: Texas State Historical Association, 1996), 1111; *Texas Almanac, 1994–1995* (Dallas: Dallas Morning News, 1993), 62.

8. *Carlsbad Current,* 12 June 1914, RG 115, box 441, entry 3.

9. Tracy to Ralph Twitchel, Las Vegas, New Mexico, 18 February 1914, Tracy Papers; *Carlsbad Current,* 1 June 1914.

10. When Texas entered the Union in 1845, she kept control of her public lands, the only state to do so. Texas used her public lands to pay surveyors, reward veterans, and to entice settlers. *Carlsbad Current,* 1 June 1914; Richardson to McDonald, 12 March 1914, U.S. Department of the Interior, Bureau of Reclamation, RG 115.

11. Lingle and Linford, *Pecos River Commission,* 66.

12. *Carlsbad Current,* 12 June 1914.

13. *Report on Investigations on the Pecos River in Texas,* 1914, by P. M. Fogg, 18, U.S. Department of the Interior, Bureau of Reclamation, RG 115, box 443, entry 3.

14. Lingle and Linford, *Pecos River Commission,* 91; Dunbar, *Forging New Rights,* 146.

15. The United States and Mexico, and Texas and New Mexico, had long argued over waters of the Rio Grande. In 1906, Reclamation settled disputes with all parties through construction of a storage dam near Elephant Butte in south-central New Mexico. In 1906 Congress extended the Reclamation Act to apply to that portion of Texas bordering the Rio Grande, which could also be irrigated by the proposed dam at Elephant Butte. Also by 1906, a diminishing volume of public land sales no longer adequately financed reclamation projects. Congressional supporters of Reclamation started looking for other means to pay for water projects. Lingle and Linford, *Pecos River Commission,* 91–92.

16. Eventually, Red Bluff Dam and Reservoir was constructed through the efforts of government-sponsored public works programs in the early 1930s under the auspices of the Public Works Administration, a New Deal agency. Lingle and Linford, *Pecos River Commission,* 94–95.

17. Lingle and Linford, *Pecos River Commission,* 94–95.

18. The reports included Fogg's 1914 report on irrigation in Texas, and various other reports, including those of Vernon L. Sullivan, who from 1907 until 1911 was

New Mexico state engineer, Wills T. Lee, E. E. Teeter, Ferd Bonstedt, E. B. Debbler, O. L. McDermith, and O. S. Harper. Lingle and Linford, *Pecos River Commission,* 131.

19. Initially known as the Alamogordo Dam and Reservoir, the government built the facility near Fort Sumner in the 1930s. Ferd Bonstedt and E. B. Debbler, "Report on Investigations, Pecos River and Carlsbad Extensions in New Mexico," 2 June 1923, U.S. Department of the Interior, Bureau of Reclamation, RG 115, box 451, entry 7, file 301; Clark, *Water in New Mexico,* 192; F. E. Weymouth, Chief Engineer, to Tracy, 25 August 1921, Tracy Papers; "Pecos River Investigations, 1925," by C. C. Elder, RG 115, Carlsbad 301, box 451, entry 7; Ray Palmer Teele, *The Economics of Land Reclamation in the United States* (Chicago: W. W. Shaw, 1927), 7–8.

20. Bonstedt and Debbler, "Report on Investigations," 2, a, U.S. Department of the Interior, Bureau of Reclamation, RG 115.

21. Lingle and Linford, *Pecos River Commission,* 68; Mead to H. L. Kent, President of New Mexico Agricultural and Mechanical College, 20 July 1925, Tracy Papers.

22. By 1929 the Bureau of Reclamation responded, in the words of R. F. Walter, that the Avalon Reservoir idea was "definitely abandoned." Walter to Commissioner, Bureau of Reclamation, Subject: Proposed Enlargement of Avalon Dam—Carlsbad Project, 29 October 1929, U.S. Department of the Interior, Bureau of Reclamation, RG 115, box 452, entry 7, loc. 42/14/3:2; Elder, "Pecos Valley Investigations," RG 115; "Re-examination of Avalon Reservoir with Respect to Water Tightness," by Kirk Bryan, 31 August 1929, U.S. Department of Interior, Bureau of Reclamation, RG 115, box 452, entry 7, loc. 42/14/3:2.

23. Smith to Mead, 18 September 1929, U.S. Department of the Interior, Bureau of Reclamation, RG 115, box 452, entry 7. The original statute for raising Avalon was Act of 29 May 1928, *U.S. Statutes at Large* (45 Stat. 902).

24. Bryan, "Re-Examination of Avalon Reservoir," U.S. Department of the Interior, Bureau of Reclamation, RG 115.

25. Walter to Mead, 13 December 1924, U.S. Department of the Interior, Bureau of Reclamation, RG 115, box 452, entry 7, loc. 42/14/3:2.

26. Ibid.

27. Bratten introduced a bill in February 1926 to adjust water-right charges. U.S. Congress, Senate, *A Bill for the Adjustment of Water Right Charges on the Carlsbad Irrigation Project . . . ,* 69th Cong., 1st sess., 23 February 1926, S. Bill 3261; Mead to Bursum, 27 December 1924, U.S. Department of the Interior, Bureau of Reclamation, RG 115, box 452, entry 7; Bratten to Mead, 3 June 1925, RG 115, box 452, entry 7.

28. Lewis, "Early Settlers"; Tracy and Pecos Water Users Association to Honorable Thomas E. Campbell, Chairman, Board of Survey and Adjustments, undated, U.S. Department of the Interior, Bureau of Reclamation, RG 115, box 5E.8, folder 301.32.

29. Proceedings, Board of Survey and Adjustments (open southern division),

Carlsbad Project, New Mexico, visited 18–21 March 1925, U.S. Department of the Interior, Bureau of Reclamation, RG 115, box 5E.8, folder 301.32; Tracy to Hagerman, 2 April 1925, Tracy Papers.

30. Summary of Preliminary Report, Board of Survey and Adjustments, U.S. Department of the Interior, Bureau of Reclamation, RG 115, box 5E.8, folder 301.32.

31. Miles Cannon, "Government Reclamation in New Mexico, Hondo-Carlsbad Project," 3 January 1924, 173, U.S. Department of the Interior, Bureau of Reclamation, RG 115, box 5E.8, folder 301.32.

CONCLUSION

1. Patricia Limerick, *Desert Passages: Encounters with the American Deserts* (Albuquerque: University of New Mexico, 1985).

EPILOGUE

1. Lingle and Linford, *Pecos River Commission,* 121, 122.

2. Ibid.

3. U.S. Department of the Interior, National Park Service, *HAER,* 147, 157.

4. Lingle and Linford, *Pecos River Commission,* 124.

5. Ibid.

6. Ibid.

7. Dunbar, *Forging New Rights,* 147–148; Clark, *Water in New Mexico,* 192; Lingle and Linford, *Pecos River Commission,* 126–128, 138.

8. The decision culminated a sixteen-year legal dispute, which began in 1974. *Lubbock Avalanche Journal,* 24 February 1990.

9. Richard Benke, "Bitterness Lingers in New Mexico over Water Pact, '88 Ruling," *Dallas Morning News,* 7 May 1995, sec. A, 58.

BIBLIOGRAPHY

ARCHIVAL AND MANUSCRIPT SOURCES

Blauvelt, L. D. Biographical File. Colorado Historical Society, Denver, Colorado.

Carlsbad Irrigation District Number One, Carlsbad, New Mexico. Records. Southwest Collection, Texas Tech University, Lubbock, Texas.

Christian, H. F. Papers. Southwest Collection, Texas Tech University, Lubbock, Texas.

Cloud, H. H. Biographical File. Colorado Historical Society, Denver, Colorado.

Cragin, F. W. Papers. Colorado Springs Pioneers Museum, Colorado Springs, Colorado.

Eddy Brothers Journal, 1879–1880. Western History and Genealogy Collection, Denver Public Library, Denver, Colorado.

"Hagerman Family." Vertical File. Colorado Springs Pioneers Museum, Colorado Springs, Colorado.

Hagerman Family Collection. Papers. Rio Grande Historical Collections/Hobson-Hutsinger University Archives, New Mexico State University Library, Las Cruces, New Mexico.

Hendley, Josephine. Papers. Southwest Collection, Texas Tech University, Lubbock, Texas. Microfilm.

Jones, Billy M. Papers. Southwest Collection, Texas Tech University, Lubbock, Texas.

Lipsey, John J. *Colorado's Almost Forgotten Tycoon.* Vertical File. Colorado Springs Pioneer Museum, Colorado Springs, Colorado.

Missouri Historical Society. Necrology Scrapbooks, IIC and IX. St. Louis Public Library, St. Louis, Missouri.

Nelson, Morgan. Papers. Southwest Collection, Texas Tech University, Lubbock, Texas.

Nettleton, Edwin S. Biographical File. Colorado Historical Society, Denver, Colorado.

Pecos Irrigation and Investment. Articles of Incorporation, 18 July 1888. New
  Mexico State Records Center and Archives, Santa Fe, New Mexico.
Pecos Irrigation Company. Certificate of Incorporation, 19 September 1900. New
  Mexico State Records Center and Archives, Santa Fe, New Mexico.
Prince, Governor L. Bradford. Papers. Record Group: Letters Received. New
  Mexico State Records Center and Archives, Santa Fe, New Mexico.
Tracy, F. G. Papers, 1888–1975. Southwest Collection, Texas Tech University,
Lubbock, Texas.

GOVERNMENT DOCUMENTS

Twelfth U. S. Census of Population. Washington, D.C.: Government Printing
  Office, 1900.
Chaves County. Deeds of Record. Roswell, New Mexico.
Chaves County. Records of Wills. "Inventory Containing the Property of James J.
  Hagerman, Deceased, Commended this 21st Day of January A.D. 1910."
  Roswell, New Mexico.
Chaves County. Records of Wills. "Petition for Letters of Administration in the
  Matter of the Estate of John S. Chisum." Roswell, New Mexico.
Eddy County. Deeds of Record. Carlsbad, New Mexico.
Eddy County. Files and Records of Wills. Carlsbad, New Mexico.
Eddy County. Mortgage Records, Water Rights, 1891–1892. Carlsbad, New Mexico.
Eddy County. Records of Patents. Carlsbad, New Mexico.
Eddy County. Warranty Deeds. Carlsbad, New Mexico.
Lincoln County. Deeds of Record. Carrizozo, New Mexico.
Mead, Elwood. "The Rise of Irrigation in the United States." *The Bureau of
  Reclamation: Its Functions and Accomplishments.* Pamphlet prepared for the
  Pan-Pacific Conference on Education, Rehabilitation, and Reclamation. Hon-
  olulu, Hawaii, 11–16 April 1927. Washington, D.C.: Government Printing
  Office, 1927.
New Mexico (Territory). Bureau of Immigration. *Compilation of Facts Con-
  cerning the Pecos Valley: Resume of the Improvement Being Made in South-
  eastern New Mexico, Embracing the Magnificent Pecos Valley.* Santa Fe, N.
  Mex.: New Mexican Printing, 1891, 1978.
New Mexico (Territory). Bureau of Immigration. *Eddy County, New Mexico,
  Home of the Carlsbad Project of the National Reclamation Service and of the
  Greatest Artesian Wells on Earth. . . .* Santa Fe, N. Mex.: Bureau of Immigra-
  tion, 1908.
New Mexico (Territory). Bureau of Immigration. *Eddy County, New Mexico, the
  Most Southeastern County in the Territory. . . .* Santa Fe, N. Mex.: J. S. Duncan,
  1903.
New Mexico (Territory). Bureau of Immigration. *New Mexico: Its Resources, Cli-
  mate, Geography, Geology, History, Statistics, Present Condition, and Future
  Prospects.* Santa Fe, N. Mex.: New Mexican Printing, 1894.

"Report of the Central Board of Review on the Carlsbad Project, New Mexico."
Reclamation Record, July 1916, 298–308.

U.S. Congress. House. *Report of Exploration and Survey of Route from Fort Smith, Arkansas, to Santa Fe, New Mexico, Made in 1849, by First Lieutenant James H. Simpson, Corps of Topographical Engineers.* 31st Cong., 1st sess., 1850. Ex. Doc. 45.

U.S. Congress. House. *Report of the House Committee on the Unlawful Occupancy of the Public Lands.* 48th Cong., 1st sess., 1884. H.R. 1325. *U.S. House Reports,* vol. v., 2, 4, 8.

U.S. Congress. House. *Aliens Owning Lands in the U.S.* 49th Cong., 1st sess., 1885–1886. *U.S. House Reports,* no. 1951.

U.S. Congress. House. *Adjustment of Private Land Claims in New Mexico.* 49th Cong., 1st sess., 1885–1886. Ex. Doc. 209. Vol. 33. Serial 2401.

U.S. Congress. House. *Annual Report of Governor of New Mexico, 1885.* 49th Cong., 1st sess., 1885. Ex. Doc. 1. Vol. 12. Serial 2379.

U.S. Congress. House. *Report of Chief of Engineers, Army, 1910, 3 vols.* 61st Cong., 3rd sess., 1910–1911. H. Doc. 1010. Vols. 17, 18, 19. Serials 5961, 5962, 5963,.

U.S. Congress. House. *Fund for Reclamation of Arid Lands.* 61st Cong., 3rd sess., 1911. H. Doc. 1262. Serial 6022.

U.S. Congress. House. *Hearings on HR 17223.* 70th Cong., 1st sess., 1928.

U.S. Congress. Senate. *A communication of the Secretary of the Interior to Congress with accompanying papers setting forth the urgent necessity of stringent measures for the repression of the rapidly increasing evasion and violation of the land laws relating to the public land, etc.* 47th Cong., 2d sess., 1883. Ex. Doc. 61. Vol. iii, 29–32.

U.S. Congress. Senate. *Fraudulent Acquisition of Titles to Lands in New Mexico.* 48th Cong., 2d sess., 1884–1885. Ex. Doc. 106. Vol. 2. Serial 2263.

U.S. Congress. Senate. *Reports of Committees of the Senate of the United States, 1889–1890, in 10 vols.* S. Doc. 1466. Vol. 6. No. 298.

U.S. Congress. Senate. *Extracts on Irrigation of Arid Lands.* 57th Cong., 1st sess., 1901–1902. S. Doc. 446. Vol. 30. Serial 4249.

U.S. Congress. Senate. *Investigations of Irrigation Projects.* 61st Cong., 3d sess., 3 March 1911. S.R. 1281.

U.S. Congress. Senate. *A Bill for the Adjustment of Water Rights Charges on the Carlsbad Project.* 69th Cong., 1st sess., 23 February 1926. S. Bill 3231. In U.S. Department of the Interior, Bureau of Reclamation, RG 115, box 446, entry 7, loc. 42/14/2:6.

U.S. Department of Agriculture. Bulletin 103. *Summary of Irrigation District Statutes of Western States.* By Wells A. Hutchins. Washington, D.C.: Government Printing Office, 1931.

U.S. Department of Agriculture. Bulletin 254. *Irrigation Districts and Their Origins, Operation and Financing.* By Wells A. Hutchins. Washington, D.C.: Government Printing Office, 1931.

U.S. Department of Agriculture. Bulletin 1177. *Irrigation Districts, Operations and*

Financing. By Wells A. Hutchins. Washington, D.C.: Government Printing Office, 1923.

U.S. Department of Agriculture. "Department and Office of Irrigation Investigation, 1899." Washington, D.C.: Government Printing Office, 1900.

U.S. Department of Agriculture. Office of Experiment Stations. Bulletin 215. *Irrigation in New Mexico.* By Vernon Sullivan. Washington, D.C.: Government Printing Office, 1909.

U.S. Department of Agriculture. Office of Experiment Stations. Bulletin 86. *The Use of Water in Irrigation: Report of Investigations Made in 1899.* By Elwood Mead and C. T. Johnston. Including Reports by Special Agents and Observers, Reed, Code, Stout, and Berry. Washington, D.C.: Government Printing Office, 1900.

U.S. Department of Agriculture. Report No. 64. *Field Operation of the Division of Soils, 1899.* By Milton Whitney. Washington, D.C.: Government Printing Office, 1900.

U.S. Department of the Interior. Bureau of Reclamation. Board of Review. *Report of Local Board of Review, Carlsbad Project, New Mexico.* By T. U. Taylor and Scott Etter, 1915.

U.S. Department of the Interior. Bureau of Reclamation. Record Group 115, General Administration and Projects Records. National Archives, Rocky Mountain Region, Denver, Colorado.

U.S. Department of the Interior. Geological Survey. *Eleventh Annual Report, 1890.* Washington, D.C.: Government Printing Office, 1890.

U.S. Department of the Interior. Geological Survey. *Twelfth Annual Report, 1891.* Washington, D.C.: Government Printing Office, 1891.

U.S. Department of the Interior. Geological Survey. *Thirteenth Annual Report, 1892.* Washington, D.C.: Government Printing Office, 1892.

U.S. Department of Interior. Geological Survey. *First Conference of Engineers of the Reclamation Service.* Washington, D.C.: Government Printing Office, 1904.

U.S. Department of Interior. National Park Service. *Historic American Engineering Record: Carlsbad Irrigation District, HAER No. NM-4,* vol. 2, *Historical and Descriptive Data.* Washington, D.C.: National Park Service, 1991.

U.S. Department of the Interior. National Resources Planning Board. *Pecos River Joint Investigation Reports of Participating Agencies.* Washington, D.C.: Government Printing Office, June 1942.

U.S. General Land Office. *Annual Report, 1883.* Washington, D.C.: Government Printing Office, 1883.

U.S. General Land Office. *Annual Report, 1890.* Washington, D.C.: Government Printing Office, 1890.

U.S. General Land Office. *Annual Report, 1891.* Washington, D.C.: Government Printing Office, 1891.

U.S. General Land Office. *Reports of Registrar Edmond G. Shields and Receiver James Brown, Las Cruces Land Office.* 50th Cong., 2d sess., 4 October 1888, H. Ex. Doc. 1, serial 2636.

U.S. Statutes at Large. Act of 2 October 1888, Vol. XXV, Sec. 1.

U.S. Statutes at Large. Act of 25 February 1905 (33 Stat. 814).

U.S. Statutes at Large. Act of 2 March 1882, Ch. 47, Sec. 22.

U.S. Statutes at Large. Act of 29 May 1928 (45 Stat. 902).

U.S. Statutes at Large. Vol. XXVI, 391.

COURT CASES

Richard J. Bolles v. Pecos Irrigation Company. 23 N.M. 32 (Supreme Court of New Mexico, 1917).

Bolles v. Pecos Irrigation Company. Docket 992, District Court of Eddy County, New Mexico, 23 December 1908.

United States of America v. Hope Community Ditch, et al. District Court, District of New Mexico, #712, Equity. Vol. 1 and 2, and Hearings at Carlsbad. Carlsbad, New Mexico, 4 January 1926, 13 May 1931, 1 June 1932.

NEWSPAPERS AND MISCELLANEOUS

"Biographical data to accompany portrait of Mr. William C. Bradbury." Denver Public Library, Denver, Colorado.

"Brief Incidents of Special Historical Interest." Clippings File, Denver Public Library, Denver Colorado.

Canterbury, Charles. *History of the Fremont County Cattlemen's Association.* 1989. Pamphlet in the Local History Collection, Cañon City Public Library, Cañon City, Colorado.

Carlsbad Argus, 11 March 1904, 24 June 1904, 16 September 1904, 7 October 1904, 14 October 1904, 21 October 1904, 20 March 1914.

Carlsbad Current, 18 May 1914, 1 June 1914, 12 June 1914, 18 June 1914, 17 July 1914.

Carlsbad Current-Argus, various 1892, 16 May 1947, 23 May 1947, 30 June 1947, 17 July 1966, 20 March 1988.

Carlsbad Daily Current-Argus, 28 June 1936.

Cheyenne Mountain Country Club. "Certificate of Incorporation." Colorado Springs: Out West Printing, 1901. Penrose Public Library, Colorado Springs, Colorado.

Christiansen, Paige W. "The Quest for Water in New Mexico." Technical Completion Report Project No. A-026-NMex, New Mexico Water Resources Research Institution and Department of Humanities. Socorro, N. Mex.: New Mexico Institute of Mining and Technology, August 1973.

Colorado Springs Daily Gazette, 22 January 1885.

Colorado Springs Gazette, 26 March 1917.

Colorado Springs Gazette Telegraph, 10 July 1994.

Colorado Springs Gazette Weekly, 22 June 1893.

Cripple Creek Weekly Journal, 8 April 1894.

Dallas Morning News, 7 May 1995.

Denver Post, 9 November 1931.

Denver Republican, 15 September 1909.

Denver Times, 12 October 1898, 11 September 1901, 26 January 1902, 20 February 1902, 11 July 1902, 26 December 1902.

Eddy Argus, 12 October 1889–19 October 1894, 7 October 1904.

Eddy County News, 6 June 1947, 13 June 1947, 27 June 1947, 4 July 1947, 18 July 1947, 25 July 1947, 1 August 1947, 5 September 1947, 26 September 1947, 3 October 1947.

Eddy Current, 2 February 1895, 27 April 1895, 13 June 1895, 5 July 1895, 11 July 1895, 15 August 1895, 19 September 1895.

Eddy Daily Current, 1 September 1894.

Eddy Weekly Current, 16 December 1892–2 June 1893.

El Paso Daily Times, 14 January 1902.

El Paso Herald, 19 December 1891, 4 December 1915.

"Facts of Colorado Springs." Vol. 3, 19 February 1898. Penrose Public Library, Palmer Wing, Colorado Springs, Colorado.

The Fortunes of a Decade: A Graphic Recital of the Struggles of the Early Days of Cripple Creek, The Greatest Gold Camp on Earth With Stories of Its Mines, Biographies of the Men Who Made Them, With Many New and Hitherto Unpublished Anecdotes and Incidents of Their Lives. Written, compiled, and published under the direction of Sargent and Rohrabacher for the *Evening Telegraph,* Colorado Springs, Colorado, October 1900. Western History and Genealogy Collection, Denver Public Library, Denver, Colorado.

Frank Leslie's Illustrated Newspaper. No. 1797, vol. 70, 22 February 1890. Western History and Genealogy Collection, Denver Public Library, Denver, Colorado.

Goodnight, Charles. "Starting the Longhorn Westward." *Fort Worth Star Telegram and Sunday Record,* 15 September 1929.

Hagerman, Percy. *Cheyenne Mountain Country Club: The First Twenty-five Years.* Unpublished manuscript, c. 1947. Penrose Public Library, Palmer Wing, Colorado Springs, Colorado.

"History of Chemical Banking Corporation." Fact sheet issued by Chemical Banking Corporation, New York, New York, November 1994.

"History of Otsego County, New York, 1840–1878." Milford Free Library, Milford, New York.

"History of Southwestern Public Service Company." Public Information Offices, Southwestern Public Service Company, Amarillo, Texas. Provided by Bill Crenshaw, Amarillo office, SPS.

Las Vegas Gazette, 4 May 1873, 25 November 1875.

Las Vegas Optic, 15 July 1881, 16 July 1881.

Lewis, Charles W., Jr. "Early Settlers in Carlsbad, New Mexico and Vicinity." Compiled for the Southeastern New Mexico Historical Society, 1976.

Lincoln County Leader, 8 December 1888.

Lincoln Golden Era, 1884–1886.

Lincoln Independent, 1886.

Lubbock Avalanche Journal, 24 February 1990.

McPhee, William. "The Eddy Brothers." Typescript in History File (County) #14. New Mexico State Records Center and Archives, Santa Fe, New Mexico.

New Bedford Evening Standard, 26 December 1911.

New York Times, 25 August 1903, 29 March 1905, 20 July 1912, 27 March 1917, 8 December 1937.

Parker, William Thorton, M.D., Munich, "Shall We Have a National Sanitarium for Consumptives?" From microfilm in Billy M. Jones Collection of Misc. Articles, 1838–1906, SWC.

Pecos Valley Argus, 4 February 1898.

Roberts, S. I. "Fifty Eventful Years." In WPA Files #199, State Records Center and Archives, Santa Fe, New Mexico.

Rocky Mountain News, 5 June 1892, 20 October 1880.

Roswell Daily Record, 29 July 1904, 7 October 1937.

Roswell Record, 12 January 1894, 3 February 1894, 6 April 1894, 24 April 1894, 20 May 1892, 28 April 1893, 20 December 1901, 24 June 1904.

Roswell Register, 18 April 1891.

"Township Sections of Mini-Biographies." Available at http://www.rootsweb.com/ ~nyotsego/bios/minimilford.htm

White Oaks Golden Era (newspaper). 1880–1885.

BOOKS

Anderson, George. *The History of New Mexico: Its Resources and People*. Los Angeles: State Publishers, 1907.

Andreas, A. T. *History of Chicago from the Earliest Period to the Present Time in Three Volumes*. Vol. 3, *1871–1885*. Chicago: A. T. Andreas, 1886.

Baxter, John O. *Dividing New Mexico's Water, 1700–1972*. Albuquerque: University of New Mexico, 1997.

Bolton, Herbert Eugene. *Spanish Exploration in the Southwest, 1542–1706*. New York: C. Scribner's Sons, 1916.

Briggs, Charles L., and John R. Van Ness, eds. *Land, Water, and Culture: New Perspectives on Hispanic Land Grants*. New Mexico Land Grant Series. Albuquerque: University of New Mexico Press, 1987.

Bryant, Keith L., Jr. *History of the Atchison, Topeka, and Santa Fe Railway*. Lincoln: University of Nebraska Press, 1974.

Cafky, Morris. *Colorado Midland*. Denver: World Press, 1965.

Carstensen, Vernon, ed. *The Public Lands: Studies in the History of the Public Domain*. Madison: University of Wisconsin Press, 1963.

Chandler, Alfred D., Jr. *The Visible Hand: The Managerial Revolution in American Business*. Cambridge: Harvard University Press, Belknap Press, 1977.

Clark, Ira. *Water in New Mexico: A History of Its Management and Use*. Albuquerque: University of New Mexico Press, 1987.

Clark, Norman. *Deliver Us from Evil: An Interpretation of American Prohibition.* New York: W. W. Norton, 1976.

Coan, Charles D. *A History of New Mexico.* Vol. 1. Chicago: American Historical Society, 1925.

Colorado Mines. Denver: Carson, Hurst, and Harper, Art Printers, undated. Western History and Genealogy Collection, Denver Public Library, Denver, Colorado.

Cramer, T. Dudley. *The Pecos Ranchers in the Lincoln County War.* Oakland: Branding Iron Press, 1996.

Daniels, Betty Marie, and Virginia McConnell. *The Springs of Manitou.* Denver: Sage Books, 1969.

Dary, David. *Cowboy Culture: A Saga of Five Centuries.* New York: Knopf, 1981.

Davis, Arthur Powell. *Irrigation Works Constructed by the United States Government.* New York: John Wiley and Sons, 1917.

Dearen, Patrick. *Castle Gap and the Pecos Frontier.* Fort Worth: Texas Christian University Press, 1988.

DeBuys, William. *Salt Dreams: Land and Water in Low-Down California.* Albuquerque: University of New Mexico Press, 1999.

DeLeón, Arnoldo. *The Mexican American Image in Nineteenth Century Texas.* Texas History Series, gen. ed., Robert J. Rosenbaum. Boston: American Press, 1982.

Dictionary of American Biography. Edited by Dumas Malone. New York: C. Scribner's Sons, 1936.

Dunbar, Robert G. *Forging New Rights in Western Waters.* Lincoln: University of Nebraska Press, 1983.

Ebright, Malcolm. *Land Grants and Lawsuits in Northern New Mexico.* New Mexico Land Grant Series, gen. ed., John R. Van Ness. Albuquerque: University of New Mexico Press, 1994.

Ellis, Leonard Bolles. "History of New Bedford and Its Vicinity, 1602–1892." Syracuse, N.Y.: D. Mason, 1892.

Epperson, Harry A. *Colorado As I Saw It.* [1944?]. Southwest Collection, Texas Tech University, Lubbock, Texas.

Everett, George. *Cattle Cavalcade in Central Colorado.* Denver: Golden Bell Press, 1966.

Faragher, John Mack, Mari Jo Buhle, Daniel Czitrom, and Susan H. Armitage. *Out of Many: A History of the American People.* 2d ed. Upper Saddle River, N.J.: Prentice Hall, 1997.

Fleming, Elvis E., and Minor S. Huffman, eds. *Roundup on the Pecos.* Roswell, N. Mex.: Chaves County Historical Society, 1978.

Flores, Dan. *Caprock Canyonlands: Journeys into the Heart of the Southern Plains.* Austin: University of Texas Press, 1990.

———. *Horizontal Yellow: Nature and History in the Near Southwest.* Albuquerque: University of New Mexico Press, 1999.

Flynn, Norma L. *Early Mining Camps of South Park*. Published by the author, 1952.

Forman, Grant. *Marcy and the Goldseekers*. Norman: University of Oklahoma Press, 1935.

Friedman, Jack P., Jack C. Harris, and J. Bruce Lindeman. *Dictionary of Real Estate Terms*. Woodbury, N.Y.: Barron's, 1984.

Fulton, Maurice Garland. *History of the Lincoln County War*. Edited by Robert N. Mullin. Tucson: University of Arizona Press, 1969.

Garcia, Mario T. *Desert Immigrants: The New Mexicans of El Paso, 1880–1920*. New Haven: Yale University Press, 1981.

Gates, Paul Wallace. *History of Public Land Law Development*. Written for the Public Land Law Review Commission. Washington, D.C.: Government Printing Office, 1968.

Gates, Paul Wallace. *The Jeffersonian Dream: Studies in the History of American Land Policy and Development*. Edited by Allan G. and Margaret Beattie Boague. Historians of the Frontier and American West Series, ed. Richard Etulain. Albuquerque: University of New Mexico Press, 1996.

Gilbert, Beth. *Alamogordo: The Territorial Years, 1898–1912*. Albuquerque: Starline Printing, 1988.

Gillett, James B. *Fugitives from Justice: The Notebook of Texas Ranger Sergeant James B. Gillett*. Austin: State House Press, 1997.

Gilpin, Alan. *Dictionary of Economic Terms*. 2d ed. New York: Philosophical Library, 1970.

Gómez-Quiñones, Juan. *Mexican American Labor, 1790–1990*. Albuquerque: University of New Mexico Press, 1994.

Goodwyn, Laurence. *The Populist Moment: A Short History of the Agrarian Revolt in America*. New York: Oxford University Press, 1978.

Gras, Norman Scott Brien, and Henrietta M. Larson. *Casebook in American Business History*. New York: F. S. Crofts, 1939.

Haley, J. Evetts. *Charles Goodnight: Cowboy and Plainsman*. New York: Houghton Mifflin, 1936.

Hansen, Harry, ed. *Colorado: A Guide to the Highest State*. New rev. ed. (Originally compiled by the Federal Writers' Program of the Works Projects Administration of the State of Colorado.) New York: Hastings House, 1970.

Harkey, Dee R. *Mean As Hell*. Albuquerque: University of New Mexico Press, 1948.

Hays, Samuel P. *Conservation and the Gospel of Efficiency: The Progressive Conservation Movement, 1890–1920*. Cambridge: Harvard University Press, 1959.

Heilprin, Angelo, and Louis Heilprin, eds. *A Complete Pronouncing Gazetteer or Geographical Dictionary of the World*. Philadelphia: J. B. Lippincott, 1931.

Hill, Robert Tudor. *The Public Domain and Democracy: A Study of Social, Economic and Political Problems in the United States in Relation to Western*

Development. Studies in History, Economics, and Public Law, vol. 38, no. 1, whole number 100. New York: Columbia University Press, 1910.

History of Eddy County, New Mexico to 1981. Southeastern New Mexico Historical Society, Carlsbad, New Mexico. Lubbock, Tex.: Craftsman Printers, 1982.

History of the Chemical Bank, 1823–1913. Privately published, 1913.

Huffstetler, Mark, and Lon Johnson. *Watering the Land: The Turbulent History of the Carlsbad Irrigation District*. Denver: National Park Service, Rocky Mountain Region, National Preservation Programs, 1993.

Hundley, Norris, Jr. *The Great Thirst: Californians and Water, 1770s–1990s*. Berkeley and Los Angeles: University of California Press, 1992.

Huntley, Paul L. *Black Mountain Cowboys and Other Stories*. Cañon City, Colo.: Master Printers, 1976.

Jackson, Donald C. *Building the Ultimate Dam: John S. Eastwood and the Control of Water in the West*. Lawrence: University Press of Kansas, 1995.

Jones, Virgil Carrington. *Roosevelt's Rough Riders*. New York: Doubleday, 1971.

Keleher, William A. *The Fabulous Frontier: Twelve New Mexico Items*. Santa Fe, N. Mex.: Rydal Press, 1945.

———. *Violence in Lincoln County, 1869–1881*. Albuquerque: University of New Mexico Press, 1957.

Kraut, Alan M. *The Huddled Masses: The Immigrant in American Society*. The American History Series, gen. ed., Arthur Shink. Arlington Heights, Ill.: Harlan Davidson, 1982.

Larousse Concise Spanish-English Dictionary. London: Larousse, 1993.

Larson, Carole. *Forgotten Frontier: The Story of Southeastern New Mexico*. Albuquerque: University of New Mexico Press, 1993.

Larson, Robert. *New Mexico Populism: A Study of Radical Protest in a Western Territory*. Boulder: Colorado Association University Press, 1974.

———. *New Mexico's Quest for Statehood, 1846–1912*. Albuquerque: University of New Mexico Press, 1968.

Lauck, W. Jeff. *The Causes of the Panic of 1893*. Boston: Houghton Mifflin, 1907.

Limerick, Patricia. *Desert Passages: Encounters with the American Deserts*. Albuquerque: University of New Mexico, 1985.

Lingle, Robert, and Dee Linford. *The Pecos River Commission of New Mexico and Texas*. Santa Fe, N. Mex.: Rydal Press, 1961.

Lipsey, John J. *The Lives of James John Hagerman, Builder of the Colorado Midland Railway*. Denver: Golden Bell Press, 1968.

McClean, Robert N. *That Mexican! As He Really Is North and South of the Rio Grande*. New York: Fleming H. Revell, 1928.

McMullin, Thomas A., and David Walker. *Biographical Directory of American Territorial Governors*. Westport, Conn.: Meckler, 1934.

McWilliams, Carey. *North from Mexico: The Spanish Speaking People of the United States*. Updated by Matt S. Meire. New York: Praeger, 1990.

Mead, Elwood. *Irrigation Institutions: A Discussion of the Economic and Legal*

Questions Created by the Growth of Irrigated Agriculture in the West. New York: Macmillan Press, 1903.

Meeting the Train: Hagerman, New Mexico and Its Pioneers. Hagerman, N. Mex.: Hagerman Historical Society, 1975.

The Merck Manual of Diagnosis and Therapy. 13th ed. Rahway, N.J.: Merck, Sharpe, and Dohme Research Laboratories, 1977.

Metz, Leon C. *Pat Garrett: The Story of a Western Lawman.* Norman: University of Oklahoma Press, 1974.

Meyers, Michael C. *Water in the Hispanic Southwest: A Social and Legal History, 1550–1850.* Tucson: University of Arizona Press, 1984.

Middaugh, John. *Frontier Newspapers: The El Paso Times.* El Paso: Texas Western Press, 1958.

Morris, Dan and Inez. *Who Was Who in American Politics, A Biographical Dictionary of Over 4000 Men and Women Who Contributed to the United States Scene from Colonial Days up to and Including the Immediate Past.* New York: Hawthorn Books, 1974.

Morris, Edmund. *The Rise of Theodore Roosevelt.* Toronto: Longman Canada, 1979.

Morris, John Miller. *El Llano Estacado: Exploration and Imagination on the High Plains of Texas and New Mexico, 1536–1860.* Austin: Texas State Historical Association, 1997.

Morris, Richard B., ed. *Encyclopedia of American History, Updated and Revised.* New York: Harper and Row, 1965.

Myers, Lee. *The Pearl of the Pecos: The Story of the Establishment of Eddy, New Mexico and Irrigation on the Lower Pecos River of New Mexico.* Compiled from Eddy newspapers between October 12, 1889 and October 23, 1897. Privately printed, c. 1974.

National Cyclopedia of American Biography. Vol. 32. New York: James H. White and Company, 1936.

The New Handbook of Texas. Austin: Texas State Historical Association, 1996.

Newcombe, W. W. *The Indians of Texas: From Prehistoric to Modern Times.* Austin: University of Texas Press, 1961.

O'Brien, William, H. A. Burleson, and W. F. Greenwood. *Colorado Mines.* Denver: Carson, Hurst, and Harper, Art Printers, undated. Western History and Genealogy Collection, Denver Public Library, Denver, Colorado.

Ormes, Mauly Dayton, and Eleanor R. Ormes. *The Book of Colorado Springs.* Colorado Springs: Dentan Printing, 1933.

Parker, Morris B. *White Oaks: Life in a New Mexican Gold Camp, 1880–1900.* Edited by C. L. Sonnichsen. Tucson: University of Arizona Press, 1971.

The Pecos Valley, the Fruit Belt of New Mexico. Eddy: Pecos Irrigation and Improvement Company, 1891. Microfilm. Western Americana, reel 415, no. 4200. Southwest Collection, Texas Tech University, Lubbock, Texas.

Peake, Ora Brooks. *The Colorado Range Cattle Industry.* Glendale, Calif.: Arthur H. Clark, 1937.

Pearce, T. M. *New Mexico Place Names, A Geographical Dictionary.* Albuquerque: University of New Mexico Press, 1968.

Pisani, Donald J. *To Reclaim a Divided West: Water, Law, and Public Policy, 1848–1902.* Albuquerque: University of New Mexico Press, 1992.

———. *Water, Land, and Law in the West: The Limits of Public Policy, 1850–1920.* Lawrence: University Press of Kansas, 1996.

Powell, John Wesley. *The Exploration of the Colorado River and Its Canyons.* New York: Penguin Books, 1987.

———. *Report on the Lands of the Arid Regions of the United States with a more detailed account of the Lands of Utah.* Edited by Wallace Stegner. Boston: Harvard Common Press, 1983.

Reisner, Marc. *Cadillac Desert: The American West and Its Disappearing Water.* New York: Viking Press, 1986.

Rice, Ruth Kessler. *Letter from New Mexico, 1899–1904.* Edited by Margaret W. Reid. Albuquerque: Adobe Press, 1981.

Robbins, Roy M. *Our Landed Heritage: The Public Domain, 1776–1970.* 2d ed., rev. Lincoln: University of Nebraska Press, 1942.

Robinson, Michael C. *Water for the West: The Bureau of Reclamation, 1902–1977.* Chicago: Public Works Historical Society, c. 1979.

Rocky Mountain Directory and Colorado Gazetteer of 1871. Denver: S. S. Wallihan, 1871.

Scheiber, Harry N., ed. *U.S. Economic History, Selected Readings.* New York: Knopf, 1964.

Schuyler, James Dix. *Reservoirs for Irrigation, Water-Power, and Domestic Water Supply. With an Account of Various Types of Dams and the Methods and Plans of Their Construction. Together with a Discussion of the Available Water-Supply for Irrigation in Various Sections of Arid America; the Distribution, Application, and Use of Water; the Rainfall and Run-off; the Evaporation from Reservoirs; the Effect of Silt Upon Reservoirs, etc.* New York: John Wiley and Sons, 1901.

The Secretaries of the Department of the Interior. Washington, D.C.: Office of Administration Services, 1986.

Sheridan, Tom. *The Bitter River: A Brief Historical Survey of the Middle Pecos River Basin.* Boulder, Colo.: Western Interstate Commission for Higher Education, 1975.

Sherow, James E. *Watering the Valley: Development along the High Plains Arkansas River, 1870–1950.* Lawrence: University Press of Kansas, 1990.

Shinkle, James D. *"Missouri Plaza," First Settled Community in Chaves County.* Roswell, N. Mex.: Hall-Poorbaugh Press, 1972.

Smith, Karen L. *The Magnificent Experiment: Building the Salt River Reclamation Project, 1890–1917.* Tucson: University of Arizona Press, 1986.

Smith, Toby. *Coal Town: The Life and Times of Dawson, New Mexico.* Santa Fe, N. Mex.: Ancient City Press, 1993.

Smythe, William E. *The Conquest of Arid America.* New York: Macmillan, 1905.

Sobel, Robert, ed. *Biographical Dictionary of the United States Executive Branch, 1774–1977.* Westport, Conn.: Greenwood Press, 1977.

Sonnichsen, C. L. *The Mescalero Apaches.* 2d ed. Norman: University of Oklahoma Press, 1973.

Sprague, Marshall. *Colorado: A Bicentennial History.* The State and Nation Series. New York: W. W. Norton, 1976.

————. *Newport in the Rockies: The Life and Good Times of Colorado Springs.* Chicago: Swallow Press, 1980.

Stanley, F. *The Seven Rivers, New Mexico Story.* Pep, Texas, 1963.

————. *The White Oaks, New Mexico Story.* Pep, Texas, undated.

Stegner, Wallace. *Beyond the Hundredth Meridian: John Wesley Powell and the Second Opening of the West.* Lincoln: University of Nebraska Press, 1983.

Stratton, Porter S. *Territorial Press of New Mexico, 1834–1912.* Albuquerque: University of New Mexico Press, 1969.

Teele, Ray Palmer. *Economics of Land Reclamation in the United States.* Chicago: W. W. Shaw, 1927.

Temin, Peter. *Causal Factors in American Economic Growth in the Nineteenth Century.* Studies in Economic and Social History. London: Macmillan Press, 1975.

Texas Almanac, 1994–1995. Dallas: Dallas Morning News, 1993.

Thomas, George. *The Development of Institutions under Irrigation, with a Special Reference to Early Utah Conditions.* New York: Macmillan, 1920.

Townsend, David. *You Take the Sundials and Give Me the Sun.* Alamogordo, N. Mex.: Alamogordo Daily News, 1984 (Sun Country Printing).

Tyler, Daniel. *The Last Water Hole in the West: The Colorado–Big Thompson Project and the Northern Colorado Water Conservancy District.* Niwot: University Press of Colorado, 1992.

Utley, Robert M. *High Noon in Lincoln: Violence on the Western Frontier.* Albuquerque: University of New Mexico Press, 1987.

Wallace, Ernest, and E. Adamson Hoebel. *The Comanches: Lords of the South Plains.* Norman: University of Oklahoma Press, 1951.

Webb, Walter Prescott. *The Great Plains.* Boston: Ginn, 1931.

Webster's New Geographical Dictionary. Springfield, Mass.: G & C Merriam, 1977.

Wilkins, Thurman. *Clarence King: A Biography.* Albuquerque: University of New Mexico Press, 1988.

Williams, Edwin B. *Holt Spanish and English Dictionary.* New York: Henry Holt, 1955.

Worster, Donald. *Rivers of Empire: Water, Aridity, and the Growth of the American West.* New York: Pantheon, 1986.

Zelany, Carolyn. *Relations between the Spanish-Americans and Anglo-Americans in New Mexico.* New York: Arno Press, 1974.

DISSERTATIONS AND THESES

Gravel, Richard H. "The Pecos Valley Railroad, 1889–1906." Master's thesis, Eastern New Mexico University, Portales, New Mexico, 1924.

PERIODICALS, ARTICLES, AND PARTS OF BOOKS

Ballenger and Richards 25th Annual Denver City Directory. Denver: Ballenger and Richards, 1897.

Bogener, Stephen. "Carlsbad Project." In *Project 2000.* Forthcoming by the Bureau of Reclamation.

Bryant, F. E. "Uncle Sam Is a Water Hog." *Field and Farm,* 29 February 1908.

Bogue, Allan G., and Margaret Beattie Bogue. "'Profits' and the Frontier Land Speculator." In Carstensen, ed., *The Public Lands,* 369–394. First published in *Journal of Economic History* 17 (March 1957): 1–24.

Crylie, M. R. "The IM Ranch, 1890–1899. In Everett, *Cattle Calvalcade.*

Day, James M. "Buggy Days in the Permian Basin." *Permian Historical Annual* 17 (December 1977): 3–20.

Denver Field and Farm. 21 August 1897. Denver Public Library, Denver, Colorado.

Dunham, Harold H. "Some Crucial Years of the General Land Office, 1875–1890." In Carstensen, ed., *The Public Lands,* 181–201. First published in *Agricultural History* 11 (1937): 117–141

Ebright, Malcolm. "New Mexico Land Grants: The Legal Background." In Briggs and Van Ness, eds., *Land, Water, and Culture.*

Fleming, Elvis E. "J. J. Hagerman and the Pecos River Railroad." *Permian Historical Annual* 33 (December 1977): 21–36.

Flores, Dan. "Bison Ecology and Bison Diplomacy: The Southern Plains from 1800 to 1850." *Journal of American History* 78 (September 1991): 464–485.

Ganoe, John. "The Beginnings of Irrigation in the United States." *Mississippi Valley Historical Review* 25 (June 1938): 59–78.

———. "The Desert Land Act in Operation, 1877–1891." *Agricultural History* 11 (1937): 142–157.

———. "The Desert Land Act since 1891." *Agricultural History* 11 (1937): 266–77.

Gates, Paul Wallace. "The Homestead Law in an Incongruous Land System." In Carstensen, ed., *The Public Lands,* 315–348. First published in *American Historical Review* 41 (July 1936): 652–681.

Gibbs, William E. "History , the Farmer, and the Pecos Valley Past." *Southwest Heritage* 11 (fall 1981): 2–7, 21.

Gordon, John Steele. "The Farthest Fall." *American Heritage,* July–August 1997, 14, 18.

Hall, Sharlot M. "The Burden of the Southwest." *Out West Magazine,* 28 January 1908, 3–4.

Hammond, Ruth. "Germans Sit Out War in Desert Drydock." *New Mexico Magazine,* November 1992, 86–91.

Harrington, Earl G. "Cadastral Surveys for the Public Land of the United States." In Carstensen, ed., *The Public Lands*, 35–41. First published in *Surveying and Mapping* 9: 82–86.

"A History of Cleveland, Ohio." *Biographical Illustrated*, vol. 2. Chicago-Cleveland: S. J. Clarke, 1910.

Howell, L. B. "Pecos Valley Irrigation System." *Engineering News and American Railway Journal 36* (17 September 1896): 181–182.

Hundertmark, C. A. "Reclamation in Chaves and Eddy Counties, 1887–1912." *New Mexico Historical Review* 47 (1972): 301–316.

Julian, George W. "Land Stealing in New Mexico." *North American Review* 145 (1887): 2–31.

Lang, Herbert H. "New Mexico Bureau of Immigration, 1880–1912." *New Mexico Historical Review* 51 (July 1976): 193–212.

LeDuc, Thomas. Introduction. In Carstensen, ed., *The Public Lands*, 45–46.

Melzer, Richard, and Phyllis Ann Mingus. "Wild to Fight: The New Mexico Rough Riders in the Spanish American War." *New Mexico Historical Review* 59 (April 1984): 109–136.

Rickards, Colin, ed. "Charles Littlepage Ballard." *Southwestern Studies* 4 (1966): 5–40.

Roark, Denis. "Roswell's History in Newsprint." *Greater Llano Estacado Southwest Heritage* 3 (June 1973): 9–12.

Rosebaum, Robert J., and Robert J. Larson. "Mexicano Resistance to the Expropriation of Grant Lands in New Mexico." In Briggs and Van Ness, eds., *Land, Water, and Culture*.

Shannon, Fred A. "The Homestead Act and the Labor Surplus." In Carstensen, ed., *The Public Lands*, 297–313. First published in *American Historical Review* 41 (1936): 637–651.

St. Louis City Directories for 1880, 1885–1886, 1895–1896, 1889–1890, St. Louis Public Library, St. Louis, Missouri.

Sterling, Everett W. "The Powell Irrigation Survey, 1888–1893." *Mississippi Valley Historical Review* 27 (December 1940): 421–434.

The Southwesterner. 90th anniv. ed., November–December 1994. Provided by Bill Crenshaw, Amarillo office, Southwestern Public Service, Amarillo, Texas.

Spidle, Jake W., Jr. "Coughing, Spitting, and New Mexico History." In *Essays in 20th Century New Mexico History*, edited by Judith Boyce DeMark. Albuquerque: University of New Mexico Press, 1994.

The Trail. Vol. 18, no. 22 (25 November 1935), Denver Public Library, Denver, Colorado.

Tracy, Francis G., Sr. "Pecos Valley Pioneers." *New Mexico Historical Review* 33 (July 1958): 187–204.

Turner, Frederick Jackson. "The Significance of the Frontier in American History." In *The Turner Thesis, Concerning the Role of the Frontier in American History*, 3d ed., edited by George Rogers Taylor. Lexington, Mass.: D. C. Heath, 1977.

Westphall, Victor. "The Public Domain in New Mexico, 1854 to 1891." *New Mexico Historical Review* 33 (January 1958): 24–52; 33 (April 1958): 128–143.

Wilson, Herbert M. "Pecos Valley Canals." *Engineering News and American Railway Journal* 26 (17 October 1891): 350–351.

Wirth, Fremont P. "The Operation of the Land Laws in the Minnesota Iron District." In Carstensen, ed., *The Public Lands*, 93–107. First published in *Mississippi Valley Historical Review* 13 (1927): 438–498.

INDEX

INDEX

Nash, Joe, 262
National Irrigation Association, 177
National Water Users Association, 203
Navajo, 13, 15
Nettleton, Edwin S., 61, 85, 86, 87, 90, 91,
    110, 114, 152, 153, 224, 263
New Bedford, Massachusetts, 5, 32, 111,
    156, 272, 293, 302
New Era Road Machine, 100
New Mexico Bureau of Immigration, 25,
    27, 47, 75, 76, 117, 123, 232, 238,
    240, 248, 255, 260, 264, 274, 278,
    279
New Mexico Day, World's Fair, Chicago, 76
New Mexico State Engineer's Office, 215
Red Bluff Reservoir construction, 227, 310
New York, New York, 3, 5, 26, 27, 31, 32,
    33, 38, 41, 60, 61, 70, 117, 118, 123,
    130, 134, 135, 139, 140, 169, 184,
    223
Newell, Frederick H., 9, 58, 145, 146, 157
Newlands (Reclamation) Act, 7, 97, 142,
    144, 178, 217, 224, 225, 266, 298,
    300
Newlands, Frances G., 144, 288
Northern Canal, 26, 33, 87, 88, 92, 93, 123,
    140, 150, 151, 155, 285, 286
Nymeyer, B. A., 57, 61, 101, 253, 255, 269

Ohls, Edith, 56, 62
Organ Mountains, 13, 239
Otero, Miguel A., 76, 161, 282, 295
Otis, Charles, 5, 38, 72, 111, 118, 122, 124,
    133, 135, 136, 242, 244, 245, 259,
    272, 273, 280, 281, 283, 285
Otis, New Mexico, 71, 76, 131, 182, 259
Otis Steelworks, Cleveland, Ohio, 258, 259,
    272, 280
Otis, William A., 30, 242, 245, 272, 280

Pabst Brewing Company, 125
Palmer, William A., 28, 38, 61, 87, 241, 242
Pecos and Northern Texas Railway Com-
    pany, 135
Pecos Company, 118, 119, 121, 122, 123,
    124, 126, 130, 132, 133, 134, 136,
    140, 162, 276, 278, 285

Pecos Construction and Land Company,
    111
Pecos Irrigated Farm Company, 75
Pecos Irrigation and Improvement Com-
    pany, 26, 29, 54, 56, 58, 76, 77, 79,
    81, 82, 84, 87, 95, 98, 110, 111, 112,
    113, 114, 117, 118, 119, 120, 126,
    127, 139, 140, 141, 142, 148, 150,
    240, 253, 258, 261, 263, 272, 274,
    276, 277, 278, 281, 282, 285, 286,
    303
Pecos Irrigation and Investment Company,
    33, 35, 53, 54, 55–58, 62, 72, 87, 90,
    108, 109, 110, 111, 148, 150, 255,
    263, 270, 274
Pecos Irrigation Company, 142, 145, 146,
    147, 148, 149, 151, 152, 153, 155,
    156, 157, 158, 159, 161, 163, 165,
    166, 169, 173, 174, 175, 176, 177,
    178, 179, 184, 186, 187, 190, 191,
    196, 197, 199, 208, 219, 250, 272,
    286, 287, 297, 299, 301
Pecos Land and Water Company, 113, 114
Pecos Orchard Company, 75
Pecos Railway Construction and Land
    Company, 136, 137
Pecos River Compact, 10
Pecos River Compact Commission, 221,
    226
Pecos River Compact (1948), 227
Pecos River Valley Flood (1904), 156–59,
    163, 176, 223
Pecos River Flume, 85, 86–89, 90, 92, 93,
    98, 100, 119, 120, 121, 122, 140, 142,
    145, 157, 174, 240, 259, 277
Pecos River Valley Flood (18, 93, 108, 119,
    120, 121, 122, 126, 128, 130, 132,
    276, 277
Pecos Valley and Northeastern Railway
    Company, 135–38, 283
Pecos Valley Beet Sugar Company, 125
Pecos Valley Company, 133, 134, 137, 139,
    140
Pecos Valley Land and Ditch Company, 26,
    29, 62, 85, 108, 109
Pecos Valley of Texas Water Users Associa-
    tion, 217, 218
Pecos Valley Orchard Company, 124

White Oaks, New Mexico, 132, 134, 237, 258, 282
Williams, Elmer E., 108, 175, 240
Williams, George, 41, 54
Witt Brothers Construction, 88, 89, 98

Witt, G. W., 85, 88, 158, 161, 253, 264
Worster, Donald, 7, 8, 224, 229

Xcel Energy, 74, 258, 259